Stem Cell Biology and Regenerative Medicine

Series Editor
Kursad Turksen, Ph.D.
kturksen@ohri.ca

For further volumes:
http://www.springer.com/series/7896

Kursad Turksen
Editor

Adult and Embryonic Stem Cells

Humana Press

Editor
Kursad Turksen
Regenerative Medicine Program
Sprott Centre for Stem Cell Research
Ottawa Hospital Research Institute
Ottawa, ON, Canada

ISBN 978-1-61779-629-6 e-ISBN 978-1-61779-630-2
DOI 10.1007/978-1-61779-630-2
Springer New York Dordrecht Heidelberg London

Library of Congress Control Number: 2012930133

Printed on acid-free paper

Humana Press is part of Springer Science+Business Media (www.springer.com)

Preface

Ondokus Mayis University, located in the beautiful historical location of Samsun, Turkey, hosted the 1st International Stem Cell Meeting during September 29 to October 1, 2010. Prof. Dr. Gülsen Ökten chaired the Organizing Committee, which planned an outstanding meeting that brought together many international speakers and Turkish stem cell researchers and trainees. Attendees at the meeting were treated to outstanding Turkish hospitality at the University and at local restaurants.

In Turkey, stem cell research is one of the most rapidly growing areas in the medical arena, and the enthusiasm of Turkish researchers for performing high-quality studies of basic and translational stem cell biology is evident. One of the gratifying aspects of the meeting was the commitment and enthusiasm of the many young trainees who attended, and I look forward to following their future development.

It was difficult to include in this book all of the subjects covered by the numerous speakers who presented their work at the conference. Therefore, I have attempted to cover representative areas that provide a good summary of the scope of the meeting. I am grateful to all the contributors who helped make this volume a success. I am confident that the book's contents will be an invaluable addition to Springer's Stem Cells and Regenerative Medicine series.

It would have been impossible to put together this volume without the help of Ms. Hande Ozturkatalay of Interium Turkey, who worked tirelessly to help me coordinate the chapters for this volume. I thank her for her outstanding contribution. I also thank Dr. Sibel Yildirim for facilitating my contacts with the contributing authors and my editor Aleta Kalkstein (Springer US) for making this volume possible. I am also grateful to Renata Hutter of Springer US for doing an outstanding job of addressing all the details that I missed.

Ottawa, ON, Canada Kursad Turksen

Acknowledgements

Papers from the First International Stem Cell Meeting

Ondokus Mayis University
Autumn 2010

Chair of Organizing Committee
Prof Gulsen Okten

Contents

Contributors

Sajjad Ahmad Institute of Human Genetics, Newcastle University, Newcastle upon Tyne, UK

North East England Stem Cell Institute, Newcastle University, Newcastle upon Tyne, UK

Royal Victoria Infirmary, Newcastle upon Tyne, UK

Department of Ophthalmology, Newcastle University, Newcastle upon Tyne, UK

Institute of Human Genetics, Newcastle University, International Centre for Life, Newcastle upon Tyne, UK

Kamil C. Akcali Laboratory of Stem Cell Research, Department of Molecular Biology and Genetics, Bilkent University, Bilkent, Ankara, Turkey

Haifa Ali Institute of Human Genetics, Newcastle University, Newcastle upon Tyne, UK

North East England Stem Cell Institute, Newcastle University, Newcastle upon Tyne, UK

Ferit Avcu Gulhane Medical Faculty, Department of Hematology and Research Center, Etlik/Ankara, Turkey

Arianna Bianchi Institute of Human Genetics, Newcastle University, Newcastle upon Tyne, UK

North East England Stem Cell Institute, Newcastle University, Newcastle upon Tyne, UK

Alp Can Laboratory for Stem Cell Science and Reproductive Medicine, Department of Histology and Embryology, Ankara University Stem Cell Institute, Ankara University School of Medicine, Sihhiye, Ankara, Turkey

Necati Findikli Department of Bioengineering, Yildiz Technical University, Istanbul, Turkey

Medicana Bahcelievler Hospital IVF Center, Istanbul, Turkey

Istanbul Genetics Group, Istanbul, Turkey

F. Fiori Dipartimento Di.S.C.O. - Sezione Biochimica Biologia e Fisica, Università Politecnica delle Marche, Via Brecce Bianche, Ancona, Italy

A. Giuliani Dipartimento Di.S.C.O. - Sezione Biochimica Biologia e Fisica, Università Politecnica delle Marche, Via Brecce Bianche, Ancona, Italy

Ihsan Gursel Biotherapeutic ODN Research Lab, Department of Molecular Biology and Genetics, Bilkent University, Bilkent, Ankara, Turkey

Ayfer Haydaroğlu Radiation Oncology Department, Ege University Hospital, Bornova, Izmir, Turkey

Chitra Hosing Department of Stem Cell Transplantation and Cellular Therapy, M.D. Anderson Cancer Center, Houston, TX, USA

Erdal Karaöz Department of Stem Cell, Center for Stem Cell and Gene Therapies Research and Practice, Institute of Health Sciences, Kocaeli University, Kocaeli, Turkey

Jasmeet Kaur Life Technologies, Frederick, MD, USA

Hande Kocak Laboratory of Stem Cell Research, Department of Molecular Biology and Genetics, Bilkent University, Bilkent, Ankara, Turkey

Udaykumar Kolkundkar Primary and Stem Cells Systems, Invitrogren BioServices India Pvt. Ltd., Bangalore, India

Research & Development, Primary and Stem Cell Systems, Frederick, MD, USA

Ilknur Kozanoglu Department of Physiology, Baskent University, Ankara, Turkey

Baskent University Adana Teaching and Research Hospital Hematology Research Laboratory, Adana, Turkey

Baskent University Adana Teaching and Medical Research Center, Yuregir, Adana, Turkey

Marcos de Lima Department of Stem Cell Transplantation and Cellular Therapy, M.D. Anderson Cancer Center, Houston, TX, USA

A. Manescu Dipartimento Di.S.C.O. - Sezione Biochimica Biologia e Fisica, Università Politecnica delle Marche, Via Brecce Bianche, Ancona, Italy

Erkan Maytalman Baskent University Adana Teaching and Research Hospital Hematology Research Laboratory, Adana, Turkey

Gülsen Ökten Department of Medical Biology, Medical Genetic Branch, Ondokuz Mayis University, Atakum/Samsun, Turkey

Charles Osei-Bempong Institute of Human Genetics, Newcastle University, Newcastle upon Tyne, UK

North East England Stem Cell Institute, Newcastle University, Newcastle upon Tyne, UK

Rıdvan Seçkin Özen Children's Memorial Research Center, Stem Cell Core, Northwestern University`s Feinberg School of Medicine and Istanbul Genetik Grubu, International Reproductive Genetic Diagnosis Center—InteRepGen, Besiktas, Istanbul, Turkey

Shruthi Pal Primary and Stem Cells Systems, Invitrogren BioServices India Pvt. Ltd, Bangalore India

Research & Development, Primary and Stem Cell Systems, Frederick, MD, USA

Mohit Parekh Institute of Human Genetics, Newcastle University, Newcastle upon Tyne, UK

North East England Stem Cell Institute, Newcastle University, Newcastle upon Tyne, UK

Ani Ray-Chaudhuri Institute of Human Genetics, Newcastle University, Newcastle upon Tyne, UK

North East England Stem Cell Institute, Newcastle University, Newcastle upon Tyne, UK

C. Renghini Dipartimento Di.S.C.O. - Sezione Biochimica Biologia e Fisica, Università Politecnica delle Marche, Via Brecce Bianche, Ancona, Italy

F. Rustichelli Dipartimento Di.S.C.O. - Sezione Biochimica Biologia e Fisica, Università Politecnica delle Marche, Via Brecce Bianche, Ancona, Italy

Bakiah Shaharuddin Advanced Medical and Dental Institute, Universiti Sains Malaysia, Penang, Malaysia

Elizabeth J. Shpall Department of Stem Cell Transplantation and Cellular Therapy, M.D. Anderson Cancer Center, Houston, TX, USA

Geetha M. Swamilingiah Primary and Stem Cells Systems, Invitrogren BioServices India Pvt. Ltd, Bangalore, India

Research & Development, Primary and Stem Cell Systems, Frederick, MD, USA

Zeynep Tokcaer-Keskin Laboratory of Stem Cell Research, Department of Molecular Biology and Genetics, Bilkent University, Bilkent, Ankara, Turkey

Kursad Turksen Regenerative Medicine Program, Sprott Centre for Stem Cell Research, Ottawa Hospital Research Institute, Ottawa, ON, Canada

A. Ugur Ural Gulhane Medical Faculty, Department of Hematology, Medical and Cancer Research Center, Etlik- Ankara, Turkey

Mohan C. Vemuri Life Technologies, Frederick, MD, USA

Sibel Yildirim Faculty of Dentistry, Department of Pediatric Dentistry, Selcuk University, Konya, Turkey

Aysel Yurtsever Ege Üniversity Cancer Research Center (consultant), İzmir, Turkey

Chapter 1
Impressions from the International Stem Cell Symposium

Gülsen Ökten

Abstract The importance of endogenous stem cells in homeostasis and repair of various tissues is well recognized. However, their use as therapeutic tools in most potential regenerative medicine applications is still at an early stage. The first International Stem Cell Conference was organized by Ondokus Mayis University in Samsun, Turkey to bring Turkish and international researchers together to discuss recent developments, ongoing challenges, and potential solutions. The meeting engaged not only established investigators but many trainees in diverse aspects of stem cell biology and regenerative medicine. The enthusiasm expressed and multidisciplinary approaches described bode well for the future of basic science and translational medicine.

1.1 Introduction

That stem cell studies and applications will have great repercussions in the world is recognized. In fact, they comprise a subject that has excited both scientists and patients in recent years. Following the results that we plan to obtain in the near future, treatments with stem cell will perhaps qualify as the "greatest discovery in the medicine field" in this century. In contrast to the news published in the media, however, specialists point out that it is too early to make such a prediction. This is because the stem cell studies that are presented as definitive treatments by the media today are in fact still at the stage of trials or clinical research. While the scientists continue their intensive stem cell research, they also organize meetings at definite intervals to share the knowledge they obtain.

G. Ökten (✉)
Department of Medical Biology, Medical Genetic Branch, Ondokuz Mayis University, Atakum/Samsun, Turkey
e-mail: gulsen@omu.edu.tr

K. Turksen (ed.), *Adult and Embryonic Stem Cells*, Stem Cell Biology and Regenerative Medicine, DOI 10.1007/978-1-61779-630-2_1,

The 1st International Attendant Stem Cell Symposium, which is accepted as one of the most important meetings at the international level, was realized at Ondokuz Mayıs University Atatürk Congress and Culture Center on September 29 to October 1, 2010 in Atatürk's city, Samsun. It was made possible by means of simultaneous translation under the support of Ondokuz Mayıs University and Samsum Stem Cell Union. At the symposium, many national and international scientists found an opportunity to share their experiences and background information about subjects ranging from stem cell biology to ongoing clinical studies. Medical and ethical subjects were also discussed, and there was a wide range of attendees, from international leading researchers to students who sought education on the subject.

1.2 Brief Conference Report

Prof. Dr. Gülsen Ökten, President of the Samsum Stem Cell Union and chairman of the Symposium, gave the opening speech. She said that the conference aimed to gather valuable scientists who perform basic and practical stem cell research in the world in the hope of having an information exchange and lead the way for young researchers. He also pointed out that we must emphasize activities encouraging science in our country and award and support young scientists.

In all, 14 scientists from seven countries (United States, Canada, United Kingdom, Germany, Portugal, Italy, Iran), including 35 scientists from our country who are the leading scientists in stem cell research and regenerative medicine field participated in the program as speakers. The main subjects of the symposium were Mesenchymal Stem Cells, Cellular Treatment and Regenerative Medicine, Embryonic Stem Cells, Pluripotent Stem Cells, Neural Stem Cells, Stem Cells in Surgery, Stem Cells in Oncology, Stem Cells and Clinical Applications, Importance of Basic Scientists, and Clinician Cooperation.

The participation and concern at the Congress were above expectations. The participants had a chance to become acquainted with new techniques and approaches that are being developed on the subject of stem cells and recent studies that are being conducted throughout the world and in Turkey specifically. Very positive feedback was obtained from the participants. The symposium targeted such areas as establishing a bridge between researchers and implementers who work in different disciplines, presenting views to young researchers and our colleagues about research opportunities and issues, and obtaining information about stem cell and gene treatment applications.

In addition, the specialists and researchers who attended the symposium stated that the legal dimension of biological treatments and ethical values must be discussed. Important inferences that will determine the way these areas can be addressed in our country were obtained.

There were nine panels and four conferences conducted with the question and answer method during the symposium. In addition free proclamations and poster presentations and round table meetings went on for the 3 days, where stem cell studies and the future of stem cells were discussed. Participants from different branches of interest, stem cell biologists, biochemical engineers, and clinical implementers discussed stem cells and gene treatments. Activities were followed with great interest.

Prof. Dr. Ökten, chairman of the symposium, stated that with the support of the Ondokuz Mayıs University Research Fund and cooperation of the Medical Biology Department, Medical Genetic Science Branch, and Clinical Science Branches, experimental studies at universities have been possible. The areas being addressed are the effectiveness of mesenchymal stem cells on ischemic cerebral paralysis, peripheral nerve damage, ischemic reperfusion damage, and thermal damage, among others. A positive effect of mesenchymal stem cells for treatment of corneal epithelial damage has proved successful in experimental studies. In addition, preparations to establish the OMU-Cell Center, which will carry out studies on "Stem Cell Clinical Applications" at On Dokuz Mayıs University has begun and will be presented to service in the near future.

The scientific announcement of the symposium was published on the main page of the *Stem Cell Review and Report* (http://www.stemcellgateway.net/default.aspx), an international magazine.

Awards was given to the winners of both best three oral and best two poster presentations, which were evaluated by the scientific committee at the end of the symposium. The plaques were given by the Rector of Ondokuz Mayıs University Prof. Dr. Hüseyin Akan, Dean of Faculty of Medicine Prof. Dr. Haydar Şahinoğlu and Ondokuz Mayıs University, Faculty of Medicine President of the Medical Genetics Department, Chairman of the Symposium, and President of the Samsun Stem Cell Society Prof. Dr. Gülsen Ökten.

The winner of the first prize for best oral presentation was Specialist Dr. Ferda Alpaslan from Ondokuz Mayıs University Faculty of Medicine, Medical Biology Department for "The mesenchymal stem cell in repairing of cornea epithelium, creationist growing factor, and otology serum use." The names of the faculty in this presentation were Prof. Dr. Gülsen Ökten, Assoc. Prof. Dr. Tunç Fışgın, Assoc. Prof. Dr. Ümit Beden, Assis. Prof. Dr. Mehmet Kefeli, Assoc. Prof. Dr. Nurten Kara, Prof. Dr. Feride Duru, and Assoc. Prof. Dr. Leman Tomak.

The winner of the second prize for best oral presentation was Dr. Gökhan Duruksu from Kocaeli University, Stem Cell and Gene Therapy AUM, for "Can telomerase enzyme activity be used as a pre-indicator in perpetual gene transplantation to mesenchymal stem cells?" The names of the faculty in this presentation were Dr. Ayça Aksoy, Dr. Alparslan Okçu, Dr. Gülçin Gacar, and Prof. Dr Erdal Karaöz.

The third prize for best oral presentation was shared by two presentations. The first one was Dr. Osman Kelahmetoğlu from Ondokuz Mayıs University Faculty of Medicine, Plastic Reconstructive and Aesthetic Surgery Department, for "The effect of mesenchymal stem cells and sildenafilin on flap viability in perforator base flaps for ischemia reperfusion damage." The names of the faculty in this presentation were Prof. Dr. Gülsen Ökten, Specialist Dr. Ferda Alpaslan Pınarlı, Assoc. Prof. Dr. Ahmet Demir, Researcher Rukiye Demir, Prof. Dr. Tolga Güvenç and Assoc. Prof. Dr. Emine Duramaz. The second third prize winner was Dr. R. Seda Tığlı from Hacettepe University, Department of Chemical Engineering for "The research of chondrogenesis potentials of stem cells in silk-fibroin tissue frameworks." The names of the faculty in this presentation were Dr. Sourabh Ghosh, Dr. Menemse Gümüşderelioğlu, and David l Kaplan.

The winner of the first prize for best poster presentation was Dr. Özlem Bingöl Akpınar from Marmara University Faculty of Medicine, Department of Biochemistry and Marmara University Faculty of Medicine, Department of Hematology, for "The megakaryocytic differentiation of hematopoietic stem cell in ex vivo." The names of the faculty in this presentation were Dr. Anne Marie Maurer, Dr. Cafer Adıgüzel, Dr. Mahmut Bayık, and Dr. Fikriye Uras.

The winner of the second prize for best poster presentation was Dr. C. Teoman Karahasanoğlu from Ondokuz Mayıs University Faculty of Medicine, Medical Biology Department, Medical Genetic Branch, for "The comparison of mesenchymal stem cell application effects on the rat sciatic nerve damage in different times." The names of the faculty in this presentation were Prof. Dr. Gülsen Ökten, Specialist Dr. Ferda Alpaslan Pınarlı, Assoc. Prof. Dr. Cengiz Çokluk, Assoc. Prof. Dr. Kerameddin Aydın, Prof. Dr. Tolga Güvenç, Assoc. Prof. Sezgin Güneş, and Prof. Dr. Feride Duru.

1.3 Current Status of Stem Cell Studies

Stem cell research and their applications comprise one of the most important and highly discussed subjects of the current science and technology agenda. These studies contain research that is attracting attention from many medical and basic science fields as the studies provide information about formation mechanisms and genetic structures of living beings. Stem cell research is developing rapidly and is providing the opportunity for new cellular treatments in addition to updating and developing basic information on cell biology. These studies are a highly competent model system that lets us examine embryonic development mechanisms. With its superior potential for tissue and organ renewal, it raises scientific and social expectations related to "potential treatments" in the near future for patients with tissue damage or loss for which definitive treatment methods have not yet been found.

Embryo-based stem cell studies are still being discussed on many platforms in terms of their religious, legal, ethical, and hypothetical aspects because of such factors as the need to break an embryo when obtaining cells, the procedures used for growing them and the feed lot preparations, histocompatibility problems, and the risk of tumor formation in experimental animals when they are transplanted experimentally. For this reason, stem cell studies are either prohibited in some countries depending on their point of view or experimental research is permitted under controlled permission. Both the scientific community and governments consider studies on stem cells obtained from persons with their permission more positively.

Stem cell transplantation has been used for medical purposes against many diseases in Turkey since the 1980s. Application of stem cells in regenerative medicine (i.e., in injured tissues for the purpose of reparation) has been discussed in recent years.

In this regard, the results obtained in many countries are promising. Stem cell studies in regenerative medicine will be epochal in the near future.

During the past 5–10 years, developments in this field have offered hope for the many neuromuscular and degenerative diseases, such as amyotrophic lateral sclerosis (ALS), Parkinson's, Alzheimer's, and Huntington's diseases, which today are not curable with traditional treatment methods. The possibility of the use of stem cell applications for treating heart attacks, chronic diseases, and organ transplantation in the near future are exciting. However, before these approaches can be used in the clinic, the following problems must be surmounted: Which stem cells can be used for which diseases? How can stem cells be given? Can we resolve this problem by isolating stem cells from a person and transplanting them into the same person when the disease is a genetic disorder? When transplanted cells start to form other type cells instead of specific required cells, which functions should be edited? What must be done to prevent an immune response to the transplanted stem cells? At this stage, particularly ethical discussions about the cell type are continuing around the world because stem cells can be obtained from embryos or growing humans.

The Turkish Ministry of Health, in two letters published and circulated in 2005 and 2006, prohibited human embryonic stem cell studies until necessary infrastructure and ethical conditions are established by preparing a convenient regulation. It allows non-embryonic-based adult stem cell studies provided that they are in conformity with the regulation determined within the framework of the circulated letter.

As an alternative to the embryonic stem cell, in 2006 as a result of gene transplantations (transcription factors) to fibroblast cells (skin) taken from rats by Japanese scientists, it was proved that these cells could be transformed to embryonic-like stem cells (the induced pluripotent stem cell, or iPS) by being reprogrammed. In 2007, Japanese and American scientists simultaneously reported that they had successfully produced embryonic-like stem cells from human skin cells. In the scientific world it is thought that obstacles confronting cellular gene treatment will be eliminated as a result of these scientific developments, especially for clinical applications. However, in all the studies carried out so far, gene transplantation into the adult stem cell has been realized by means of viral vectors. As it is possible for these vectors to jump inside the cells and establish mutation in the genome, the embryonic-like stem cells that are reprogrammed with these methods cannot be used for clinical purposes. A group of scientists successfully transplanted the genes necessary for reprogramming mesenchymal stem cells obtained from human teeth without using a viral vector. It was determined with laboratory experiments that the gene-transplanted adult stem cells had a more rapid separation and reproductive nature. With these reprogrammed cells, highly successful results were obtained in the cell culture environment, and model animal experiments on wound healing were undertaken. Based on the results, valuable stem cells can be easily obtained from follicle precursor cells of wisdom teeth. This study also revealed that the desired genes could be transplanted to follicles of wisdom teeth by means of non-virus-origin vectors. Thus, reprogramming was possible and the obtained reprogrammed cells could be used for the cell–gene treatment purpose. These results are hopeful, especially for diabetics whose wounds heal late. Adult stem cells reprogrammed by means of nonviral methods are being tested in the other disease models such as those for heart disorders, paralysis, Parkinson's disease, and cancer; and their treatment potential is being researched.

Only embryonic stem cells studies have been executed in countries with different viewpoints under prohibition or controlled permission. Our Ministry of Health has prohibited studies until the necessary scientific background is obtained, and then it provides the opportunity for studies under the conditions that meet the required regulations.

Because of the successful execution of hematopoietic stem cell applications owing to many years of adult-type stem cell research, we have more information about those kind of cells, making it easier to have applications in this area. Clinical applications of hematopoietic stem cells and mesenchymal stem cells are ongoing. Applications in the areas of cardiology and neurology are remarkable.

1.4 In Which Phase Are Stem Cell Studies in Regard to Clinical Treatment?

The stem cell studies that are sometimes presented as a treatment method by the media are in fact at an experimental phase (in areas outside of hematology). As a result of the hematology studies, it has become apparent that stem cells can be used not only to treat hematological disorders but also diseases that develop due to the loss of cells. The specialists who participated in the stem cell symposium emphasized that although the stem cell studies are promising many of these studies are still in the experimental phase. These researchers gave the following replies to the questions about the future of the stem cell.

- *Will treatment of diabetes be possible with stem cell transplantation*? Many research studies are focusing on type 1 diabetes. Insulin production can be provided in rats. However, research on humans has not been completed. It is thought that embryonic stem cells will be used in that field in the near future. Positive results will be obtained in about 5 years at the earliest.
- *Can stem cells be used to treat spasticity*? There is ongoing intensive stem cell research on diseases that cause brain damage, but there are no finalized studies in humans. Spasticity is one of the most-studied subjects.
- *Will diseases such as hypertension and obesity be treated with stem cells*? These diseases are systemic and treatable. For this reason, their treatment with stem cells is difficult. Stem cell treatment will be used for such diseases as diabetes and Parkinson's in the future.
- *Will stem cells be used to treat paraplegia*? Although the animal studies related to paraplegia are promising and have been published, they are still in the experimental phase. Stem cell treatment of paraplegia in humans has not yet been proved scientifically.

- *Is stem cell research promising for treating cancer?* Cancer is a stem cell disease. These cells exist everywhere. When they do not perform their duties or functions correctly, some diseases occur. It is already known that hematopoietic cancers are due to stem cell disease. There are positive data on whether cancers of solid organs (e.g., liver, ovarian, prostate) are stem cell diseases. It is thought that when something goes wrong with the stem cells responsible for continuance of tissues or organs, the stem cells transform to cancer cells.
- *What is the cancer–stem cell relation?* Many options are available for treating cancer, including chemotherapy and radiotherapy. An analogy may be the following: So far, we can kill the bees in the beehive—in other words, the cells—but we cannot kill the queen bee—in other words, the tumor cells. For this reason, even if one cancer-producing cell remains, the cancer can recur many years later. The important thing here is that we find that cancer-producing stem cell and destroy it. Researchers who are studying that matter focus on that issue. Some researchers believe that they will completely change the treatment methods in the world within a decade. At present, according to the pharmaceutical industry, universities, and hospitals, the treatments can be restructured. During that period, cellular treatment has an important place.

1.5 Preimplantation Genetic Diagnosis and Embryonic Stem Cell Applications

When gene mapping of the human was completed, the passwords to many diseases were solved. Many specialists agree that the most striking development of the future will be treatment with stem cells. Many problems—from cancer to heart disorders to problems of the disabled to eye disorders—will no longer exist. Nano-carriers are considered another milestone. Thanks to medicine targeting, nano-carriers will transport the medicine to the problem area and treat it directly, thereby preventing the destruction of useful cells in other areas of body. Biotechnology and genetic and cellular treatments are revolutionary developments of the twenty-first century.

Today, the information obtained after completion of the Human Genome Project plays an important role in diagnosing genetic diseases, studying their types of formation, and determining the proper medical treatment. When it was determined during the embryo studies performed at the end of the 1990s that genetic information could be obtained from an embryo during the implantation phase, a different diagnosis–treatment approach was born.

Preimplantation genetic testing (PGT) is a genetic diagnostic process using samples obtained by oocyst or embryo biopsy before pregnancy. Advanced maternal age, consecutive abortions, and repeated in vitro fertilization (IVF) failures are the usual indications for chromosome testing. Single gene studies can be done indirectly for all the autosomal and X-transmitted dominant and recessive diseases by directly determining if there is a mutation in the family and/or identifying the chromosome that carries the mutation by using the genetic markers. In this group, in addition to postnatal or early-age genetic diseases, there are genetic situations that create maternofetal incompatibility, genes creating a cancer predisposition, genetic diseases that arise at an advanced age, immune insufficiency diseases, and genetic disorders that cause infertility. When parents who have a child with β-thalassemia or Fanconi anemia and needs bone marrow transplantation, giving birth to another child who is free of disease (determined by PGT) and with HLA tissue compatible with the β-thalassemia child can be life-saving for the first child.

Using preembryos that have been determined by PGT to carry a genetic disease, it is possible to obtain diseased human embryonic stem cells. They can then be used to establish a cell series at a bank, where they can be characterized and used in research projects on basic biology. The other potential advantage of this approach is that it can be used in diagnosis and treatment of diseases. By using the obtained research results, information about the basic biological grounds of genetic diseases can be obtained; and new approaches to the diagnosis, treatment, and prevention of these diseases can be developed. The results obtained when wider information about the early embryonic development period is added to the information already known about the basic biology will enable us to recognize new biological mechanisms and to obtain more detailed information about the existing biochemical–genetic–physiological treatments. This will also enable us to develop new pharmacological and regenerative treatment methods and leap forward in many areas we do not recognize at present, thereby opening the door to the future.

During the preimplantation, prenatal, and postnatal periods, different stem cells become active. During these processes, stem cells differentiate in different directions and are transformed to adult cells that have physiological functions on the one hand and form the advanced phase adult stem cells on the other. The adult stem cells, through advanced cell partitions, ensure continuance of stem cells like them and form the advanced differentiating functional cells. Studies made on adult stem cells ensured that treatments were developed that aimed at cellular replacement/support in various tissues under the name of "regenerative medicine."

Obtaining adult stem cells is easier than obtaining prenatal stem cells. Obtaining stem cells during the uterine implantation period is especially difficult and requires advanced techniques. While the pluripotent character of stem cells that belong to that period gives them an ability to differentiate into all of the body's cells, more advanced stem cells have the ability to differentiate into only some cells. These features of the human embryonic stem cells (hESCs) make them unique when compared to other stem cells. Although hESCs can be also obtained by means of backward differentiation of some adult stem cells, we cannot say that these cells completely substitute for natural hESCs. This situation increases the importance of formation of hESCs and their availability for research purposes. Finding new differentiation factors that play a role in establishing different tissues with an hESC series and more detailed determination of the places for existing ones in biological mechanisms will ensure the development of new regenerative medicine applications and pharmacological systems. The preventive and treatment approaches carried out on genetic diseases currently comprise the most important place in health programs of developed countries. Standardized procedures must be established to ensure collaboration between the experiments made in this rapidly developing field. For this purpose, common-use cell banks are needed.

When genetic differentiation and polymorphisms in general of the society are considered, the necessity of establishing many hESC series having different genetic contents is clear. For a better understanding of genetic diseases, studies that subject the gene to mutation by means of genetic engineering or make it inactive biologically is a frequently used method in genetic science at both the cellular and organism level. Thanks to that method, the cause-and-effect relation in a diseased biological system can be better understood. For this reason, production of hESC series containing a genetic disease—in addition to hESC series containing ordinary genetic structure—is important.

To obtain hESC series containing a genetic disease, before the embryos established by means of IVF are implanted in the uterus of the mother, PGT is done and the diseased embryos are used. It was demonstrated that processes applied to embryos with PGT do not have any effect on the development of the other embryos; furthermore, the chance of getting pregnant with selected normal embryos is higher. The number of centers that obtain the diseased hESC series using PGT is diminishing across the globe. Turkey has internationally competent laboratories and personnel in this area, which gives us an advantage worldwide.

Establishing a new hESC costs about USD10,000. Maintenance of the established hESC banks is also costly. The establishment phase of the hESC is important, and it is generally not as well supported financially as it should be. Although some financial support can be obtained for genetic research for the hESC banks because of the newly enforced laws, there is financial distress regarding the establishment phase of a new hESC series, which prevents the establishment of hESC banks in sufficient number or that are rich enough in terms of genetic content. The other problem is the difficulty in establishing joint studies with science groups capable of performing genetic research and producing hESCs. As the methods used by the centers to create hESC banks do not completely conform to the regulations that permit use of these hESCs for the purpose of genetic research, difficulties occur even in using the existing hESC series.

The final declaration of the symposium includes the following points.

- Human embryonic stem cell research, while ensuring the necessary ethical and scientific control mechanisms, must be permitted.
- In parallel with The Scientific and Techonological Research Council of Turkey (Tübitak) vision 2023 report, stem cell and tissue-organ engineering studies must be supported by the Turkish government (Tübitak and DPT). Many preclinical and clinical studies worldwide, established under the scope of regenerative medicine, have demonstrated that in future years stem cell-based treatment and tissue engineering applications will be converted to applicable treatment protocols in many fields of medicine. In this context, many research groups abroad supported by state or private enterprises are obtaining patents in that area. During the application process of stem cell-based treatments, organizations that apply these

treatment protocols will make serious resource transfers to the enterprises having these patents. For this reason, Turkey must establish its physical infrastructure for this newest field of medicine and make planned investments in the active working groups so the process can evolve. These investments will be a model for centers to be established in the future. It is must be noted that because of the existing ethical and legal restrictions the only stem cell source that can be used for cellular treatment and tissue-organ engineering is the mesenchymal stem cell obtained from the bone marrow or fat tissue of patients. For this reason, organizations such as Tübitak, The Turkish Academy of Sciences (Tüba), and the Ministry of Health in Turkey must undertake some strategic planning in that field and cooperate with the existing and to-be-established Research and Development (AR-GE) centers and biotechnology companies.

- Today, there are many disease groups that have expectations for stem cell treatments. Many of these patients hope to be treated by going abroad and paying serious money. It is because of the restrictions resulting from the existing situation in our country. With this regard, our Ministry of Health immediately intervened in the situation and at least assigns and supports the centers that are capable of conducting both preclinical and clinical studies. For instance, the Ministry of Health must financially support a center that focuses on only one disease, completes experimental animal studies, and brings the process to clinical application. It is hoped that with such planning results will be obtained more easily.

Chapter 2
Searching for In Vivo Traces of Mesenchymal Stem Cells and Their Ancestors

Alp Can

Abstract Mesenchymal stem cells (MSCs) have been well identified in cultures obtained from various human tissues. However, they give no clue as to their native identity, frequency, or anatomical location. Based on in vivo and in vitro experimental studies, the most promising candidate for the MSC niche is the vicinity of blood vessels. Capillaries to large-caliber arteries and veins house multipotent progenitor cells that share many morphological, phenotypical, and developmental features with freshly isolated or cultured MSCs. In this mini-review, results from our and other laboratories are summarized and suggest that MSCs originate from tissue sites where pericytes reside, although we do not rule out the possibility that only a small portion of pericytes give rise to MSCs. Understanding the MSC niches can definitely help us to take many parameters into account when designing isolation, expansion, and differentiation protocols for using these cells in future therapeutic applications.

2.1 Introduction

A mesenchymal stem cell (MSC) is defined as a type of adult stem cell (ASC) with an intrinsic potential to give rise to various types of mesenchyme-derived cells such as osteoblasts, chondrocytes, adipocytes, myocytes, and others. This classic definition of an MSC is still found in nearly every article trying to define these heterogeneous

A. Can (✉)
Laboratory for Stem Cell Science and Reproductive Medicine, Department of Histology and Embryology, Ankara University Stem Cell Institute, Ankara University School of Medicine, Sihhiye, Ankara, Turkey
e-mail: alpcan@medicine.ankara.edu.tr

K. Turksen (ed.), *Adult and Embryonic Stem Cells*, Stem Cell Biology and Regenerative Medicine, DOI 10.1007/978-1-61779-630-2_2,

cell populations since the introduction of its concept by Caplan in 1991 (Caplan 1991). Historically, research involving cells currently referred to as MSCs dates back to the 1960s and 1970s. Friedenstein et al. (1974) were the first to report that fibroblast-like cells elaborated from bone marrow (BM) via attachment to tissue culture flasks were inherently osteogenic in rodents and rabbits. Initially, these cells were not called MSCs (they were not even were termed stem cells) but were considered to be fibroblastic precursors derived from an entity with unknown anatomical location in the BM termed the colony-forming unit fibroblast (CFU-F) (Meirelles Lda and Nardi 2009). Later, the fibroblastic colonies derived from BM cells were found to be able to differentiate into cells with characteristics of osteoblasts, chondrocytes, and adipocytes (Phinney and Prockop 2007). Traditionally, MSCs refer to stem cells that are also capable of producing blood cells. However, blood cells were found to derive from a distinct cell population called the hematopoietic stem cells (HSCs) (Dexter et al. 1977). This allowed MSCs to be classified as nonhematopoietic, multipotential stem cells that are capable of differentiating into both mesenchymal and nonmesenchymal cell lineages.

The lack of consensus about the proper nomenclature needed to describe these cells has resulted in an incorrect, but synonymous, use of the terms “marrow stromal cell” and “mesenchymal stem cell” (Horwitz et al. 2005; Dominici et al. 2006). Actually, stromal cells encompass all cells present in the BM that are not part of the hematopoietic system. MSCs, on the other hand, correspond to that rare cell population (MSCs represent only approximately 0.01–0.001% of the total nucleated cells in isolated BM aspirates) that can give rise to mature cells of mesenchymal tissues. A more adequate term for the large number of cell types with the potential to differentiate into mesenchymal tissues would be “mesenchymal progenitor cells” (MPCs), which would include cell types from a hierarchy immediately above the pluripotent MSC but intermediate to that represented by mature mesenchymal cell types. Another point of debate is the fact that the HSC is itself of mesodermic origin and hence a type of MSC. For this reason, some authors prefer the term “nonhematopoietic mesenchymal stem cell.” The fact that these cells may have alternative differentiation pathways that go beyond the normal limits of mesoderm and ectoderm formation renders the term “mesenchymal” inadequate. Probably the best nomenclature to define this cell type would be “adult nonhematopoietic stem cell” followed by “plastic adherent, BM-derived stem cells.” All these concepts, however, are already included when the term “mesenchymal stem cell” is used, and there is a tendency to accept this terminology (Horwitz et al. 2005), even though it is inadequate.

Clonal studies have shown that plastic adherent populations isolated from BM and other sources are functionally heterogeneous and contain undifferentiated stem/progenitors and lineage-restricted precursors with varying capacities to differentiate into connective tissue cell types. Therefore, many unanswered questions remain about the true nature and identity of MSCs, including location, origin, and multipotential capacity. This review particularly aims to draw attention to the hypotheses and concrete findings related to the developmental origin and in vivo correlates of MSCs.

2.2 In Vivo Correlates of Mesenchymal Stem Cells

Because of the difficulty of defining MSCs other than by the operational definition of in vitro self-renewal and differentiation potential, our knowledge of MSCs is based solely on the characterization of cultured cells. Cells bearing MSC characteristics have been derived from different locations of the body including the BM, adipose tissue, tendon, skin, bone, muscle, brain, liver, kidneys, lungs, spleen, pancreas, thymus, synovial membrane, and umbilical cord (reviewed in Salem and Thiemermann [2010]). Those heterogeneous cell populations regarded as MSCs derived from the various organs exhibited many characteristics in common despite some differences regarding differentiation potential.

The confusion regarding the identity of the MSC in vivo and the apparent MSC plasticity observed in vitro prompted researchers to determine the identity and the location of the in vivo correlates of those cells in a living organism. Despite huge number of studies related to MSC biology, they are still defined on an operational basis—i.e., positive or negative selection of MSCs due to their cluster of differentiation (CD) markers, their adherence to culture vessels, their ability to self-renew (expand to a certain extent) and differentiate into at least three mesenchymal cell types, known as golden standards (Dominici et al. 2006)—almost universally in all laboratories. However, there are still serious challenges to identifying a unique population of MSCs from a specific organ because of the lack of definitive markers. Some of the foremost cell surface markers (reviewed in (Meirelles Lda and Nardi 2009; da Silva Meirelles et al. 2008)), which would allow certain cell populations isolated from the others using a cell sorting technology, are specific only in a particular context, or they are redundantly expressed by other stem cell types.

As clearly summarized by da Silva Meirelles et al. (2008), there may be three scenarios to explain the origin of MSCs. The first hypothesis is that the MSCs exist in only one specific tissue or organ (e.g., bone marrow), from which they exit and circulate to other sites to replenish cell populations when they are needed. However, there are consistent results that, under physiologic conditions, no MSC is present in circulating blood (Lazarus et al. 1997; Wexler et al. 2003; da Silva Meirelles et al. 2006), but in the case of hypoxia MSCs may be mobilized to peripheral blood, a fact that argues with the above hypothesis. The second assumption is based on the fact that in addition to BM MSCs can be isolated from many fetal and adult tissues even after the blood washed out from vessels prior to cell isolation (da Silva Meirelles et al. 2006). Therefore, one might think of the fact that tissue-specific stem cells from different sources might phenotypically and biochemically behave as MSCs when characterized in vitro. The third possibility is that all MSCs from different sources originate from or at least have common ancestor with perivascular cells (i.e., the pericytes). This hypothesis has gained substantial support by emerging evidence in recent years (da Silva Meirelles et al. 2006; Crisan et al. 2008, 2009; Diaz-Flores et al. 2009; Zimmerlin et al. 2010), and it can explain why MSCs can be isolated from many tissues, from head to toe. With this hypothesis, in the case of any tissue damage perivascular cells would give rise to MSCs, which would then migrate to the injury site, proliferate if needed, and secrete bioactive compounds to activate the autocrine and paracrine regulatory pathways.

2.3 Perivascular Mesenchymal Stem Cell Niche and Pericytes

Since Schofield first introduced the concept of a stem cell niche in 1978 (Schofield 1978), the niche concept has gained attention in regard to defining the specific anatomical locations that regulate how stem cells participate in tissue generation, maintenance, and repair (Can 2008). The primary characteristic of a stem cell niche is the ability to maintain a compartment of stem cells in an undifferentiated state (Scadden 2006). The niche also contributes a regulatory system, which maintains and governs the location, adhesiveness, retention, homing (recruiting) and mobilization, quiescence or activation, rate of division, orientation of mitotic axes, types of division (symmetrical or asymmetrical), and differentiation of the stem cells. The term "vascular niche" is often used to define the BM microenvironment, where MSCs and HSCs interact with vascular and/or nonvascular cells (i.e., reticular cells) around BM venules; and "endosteal niche" refers to a microenvironment that serves as a milieu for the interactions between osteoblasts and HSCs.

Perivascular cells in close association with capillaries were first noted almost 130 years ago by Eberth and then Rouget (reviewed in Hirschi and D'Amore [1996]). In 1923, Zimmermann introduced the term "pericyte" to describe these cells as adjacent to capillaries in a variety of tissues and continuing with vascular smooth muscle cells of arteries and veins, thus forming a continuous network throughout the entire body. They are embedded in a basement membrane, which surrounds the capillaries. Their long cytoplasmic processes penetrate the basement membrane to go directly to the underlying endothelium. In a reciprocal manner, endothelial processes penetrate the pericytes (Diaz-Flores et al. 2009; Tilton et al. 1979). The number of pericytes varies significantly in different tissues and among different-sized vessels. In general, pericytes are more numerous and have more extensive processes on venous capillaries and postcapillary venules (Simionescu et al. 1976). Specialized pericytes in liver are called Ito cells, hepatic satellite cells, or hepatic lipocytes (Pinzani 1995). Another organ-specific pericyte, the mesangial cell, is found in the kidney glomeruli (Schlondorff 1987). In BM, cells exhibiting pericytic characteristics are referred to as adventitial reticular cells (Funk et al. 1995) or myoid cells, as they express smooth muscle α-actin (Charbord et al. 1996; Andreeva et al. 1998). The differences in distribution and structure among pericytes suggest that they may have vessel- or tissue-specific roles. Hence, pericytes have a variety of proposed functions, including regulation of capillary blood flow (Yemisci et al. 2009), phagocytosis, and regulation of new capillary growth (Hirschi and D'Amore 1996). One of the main and important functions has been raised since 1982, when it was demonstrated that these cells differentiate into mesenchymal cell lineages (i.e., adipocytes, osteocytes, and chondrocytes) (Richardson et al. 1982;

Table 2.1 Cell markers used to trace the pericytes mostly in ex vivo preparations

Marker	Remarks	Pericyte	MSC
α-SMA (smooth muscle α-actin)	Displays expression differences between species	+	+
3G5 antibody (ganglioside)	Particularly specific to microvessel pericytes	+	+
NG2 (nerve/glial antigen-2)	Proteoglycan found particularly in venule pericytes	+	+
Desmin	Intermediate filament protein specific to muscle cells	–	+
Nestin	Type IV intermediate protein expressed mostly in nerve cells	+	+
Vimentin	Intermediate filament proteins especially found in mesenchyme-derived cells	+	+
Stro-1	Antibody that recognizes bone marrow stromal and erythroid cells	+	+
CD73	Also known as ecto-5′-nucleotidase originally found in placenta, peripheral blood lymphocytes, and endothelial cells	+	+
CD90 (Thy-1)	Member of the immunoglobulin supergene family and highly expressed in connective tissue and various fibroblast and stromal cell lines	+	+
CD105 (TGFβ3 receptor)	Also known as endoglin, it serves as the modulator of cellular responses to TGFβ1	+	+
CD146 (MUC18)	Member of the immunoglobulin supergene family and shows subcellular localization at the cell–cell junction	+	+
Angiopoietin-1	Group of growth factors that promote angiogenesis and the formation of blood vessels from preexisting blood vessels	+	–
Annexin A5	Detects cells that have expressed phosphatidylserine on the cell surface, a feature found in apoptosis and other forms of cell death	+	–

Many of the protein markers are shared by MSCs, strongly suggesting that the MSCs are derived from pericytes or are even the same cells depending on the tissue of origin. Note that not all pericyte markers are found in all pericytes

Diaz-Flores et al. 1991, 1992). Therefore, the notion that pericytes are the true in vivo ancestors of various cells types has gained great support among the stem cell field. Interstitial Leydig cells of the testis, which secrete testosterone, were also shown to originate from pericytes after drug-induced Leydig cell death in an animal model (Davidoff et al. 2004).

Recent studies (Shi and Gronthos 2003; Schwab and Gargett 2007; Covas et al. 2008; Zannettino et al. 2008; Robin et al. 2009) have documented the existence of similarities between MSCs and pericytes. A series of cell surface or intracytoplasmic structural proteins—some of which are site-, tissue-, and species-specific—are used to detect pericytes in vivo and in vitro (da Silva Meirelles et al. 2008; Crisan et al. 2008, 2009). Table 2.1 summarizes those pericyte markers, most of which were shared by MSCs.

Human pericytes sorted from diverse sources regenerate muscle, bone, and even skin in vivo and in organ cultures (Crisan et al. 2008; Paquet-Fifield et al. 2009; Sarugaser et al. 2009). da Silva Meirelles et al. (2006) reported that MSC cultures from decapsulated glomeruli were evidence that cultured MSCs are derived from pericytes in vivo, as previously suggested (Brighton et al. 1992; Bianco et al. 2001). Pericytes also behave as stem cells in vivo in periodontal ligament (McCulloch 1985), endometrium (Chan and Gargett 2006), and brain (Yamashima et al. 2004). Crisan et al. (2008) demonstrated the similarities between MSCs and cultured pericytes in terms of developmental potential. As with MSCs, pericytes were successfully differentiated into bone, cartilage, and fat cells when cultured under similar inductive conditions. This evidence in addition to data concerning the behavior of pericytes during tissue repair obtained from the literature (Richardson et al. 1982; Diaz-Flores et al. 1992) and reports showing the broad differentiation capabilities of MSCs, especially when in contact with mature cell types (Kopen et al. 1999; Pittenger et al. 1999; Choi et al. 2005), provided a basis for the proposition of a model in which pericytes are stem cells throughout the vasculature, contributing to the replenishment of lost cells under physiological conditions and possibly assuming a more active role during tissue injury (da Silva Meirelles et al. 2006).

In further support of this concept, intact pericytes in their tissue of origin natively express the MSC markers CD44, CD90, CD73, CD105, and CD146 (Crisan et al. 2008; Schwab and Gargett 2007). This physical interaction between blood vessels and multilineage progenitors may have been acquired early during evolution because a population of vascular mural cells has also been described around the lateral dorsal aorta and anterior mesenteric arteries of the developing zebra fish (Santoro et al. 2009); and these cells share many of the morphological, molecular, and functional characteristics of vascular smooth muscle cells and pericytes found in higher vertebrates.

In a study by McCulloch (1985), slow-cycling cells were observed more frequently within a distance of 10 μm from the blood vessels, whereas proliferating cells were often more distant. The results also indicated that a small fraction of the perivascular cells enter the cell cycle and migrate to a paravascular location, where they undergo proliferation. A likely interpretation is that stem cells reside in a perivascular location; at times, some of them divide perpendicularly in relation to the blood vessel, giving rise to progenitor cells that take up a paravascular site. There, the perivascular-born progenitors proliferate to provide differentiated progeny. Depending on the results of the latter and other studies, it is therefore possible to assume that under various physiological conditions pericytes serve as a reservoir of cells responsible for tissue homeostasis. Under certain conditions, however, they tend to proliferate and leave the niche to migrate to a site (Fig. 2.1) where they undergo differentiation and/or execute many cellular tasks such as immunomodulation, antiapoptosis, or antifibrosis by their physical interactions and/or by secreting soluble factors.

In recent years, stem cells have been found to be severely influenced by local oxygen concentrations in their niches (reviewed in (Mohyeldin et al. 2010)). Comparison of human MSCs cultured in hypoxic versus normoxic conditions (2% and 20% oxygen, respectively) showed that their proliferative capacity was better

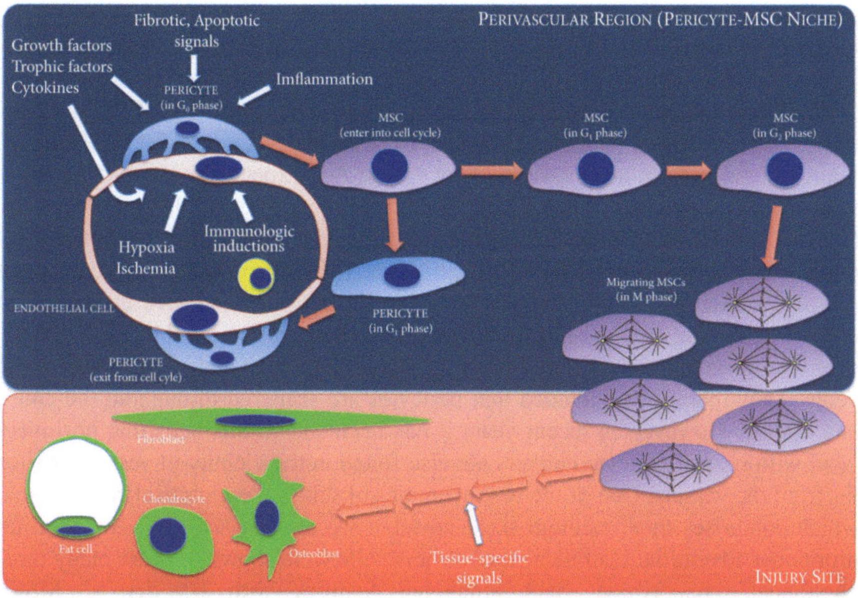

Fig. 2.1 Hypothetical model depicting in vivo pericyte–mesenchymal stem cell (MSC) trafficking. In the perivascular region, considered a pericyte–MSC niche (*dark blue box*), various factors (i.e., growth factors, trophic factors, cytokines) and fibrotic/apoptotic signals, have an effect on pericytes directly (*blue cells*) or through endothelial cells (*pink cells*). Hypoxic or ischemic conditions directly influence pericytes through endothelial cells. Blood-borne immunological cells (*yellow cells*) give secondary responses to ongoing processes. Upon induction, the pericyte exits G_0 phase, enters into the cell cycle, and differentiates into an MSC. Presumably, asymmetrical cell division results in forming two daughter cells: One is a candidate to differentiate into an MSC (*MSC in G_1 phase*), and the other resides in the vicinity of the vessel wall to back up the existing pericytes (*Pericyte in G_1 phase*), which will then give rise to a highly differentiated pericyte around the vessel. Shortly after formation of the MSCs, they remains in the cell cycle and proliferate either on the migration route or upon reaching the injury site (*red box*) where they exhibit therapeutic effects and/or differentiate into tissue-specific cell types (*green cells*) by the aid of various tissue-specific signals, such as hormones and growth factors

maintained in the former (Grayson et al. 2006). In addition, hypoxia at least doubled the number of CFU-Fs present while enhancing the expression of *Oct-4* and *rex-1*, genes expressed by embryonic stem cells and thought to be pivotal in maintaining their stemness. These data suggest that hypoxia enhances not only the proliferative capacity but also the plasticity of MSCs. The mechanism of action of hypoxia on MSCs is currently unknown, although Oct-4 up-regulation by the transcription factor HIF-2α (hypoxia-induced factor 2α) is possible (Mohyeldin et al. 2010; Covello et al. 2006). Recently, we have shown that ischemia induces sustained contraction of pericytes on microvessels in the intact mouse brain (Yemisci et al. 2009). Moreover, pericytes remain contracted despite successful reopening of the cerebral artery after 2 h of ischemia. We also showed that the microvessel wall is the major source of oxygen and nitrogen radicals that cause ischemia and reperfusion–induced microvascular dysfunction. Taken together, oxygen levels in the vessel wall microenvironment may also alter pericyte behavior, which may in turn be associated with MSC metabolism.

The perivascular cell, as a general term, also implies cells apart from the pericytes. The behavior of pericytes as stem cells in the testis (Davidoff et al. 2004) and brain (Yamashima et al. 2004) does not reflect that expected for a mesenchymal stem cell. This leads to a broader perspective, where perivascular stem cells are distributed throughout adult tissues, and these can be viewed as MSCs in mesenchymal tissues. This view does not necessarily imply that perivascular stem cells from different tissues are equivalent despite their similarities. Andreeva et al. (1998) demonstrated that among muscle cells and fibroblast cells that exhibit a 3G5 pericyte marker are found all three layers of large, medium, and small arteries and veins. They therefore concluded that pericytes are scattered throughout the entire vasculature. However, in recent years it has been shown that cells that positively label with several pericyte markers are also found in the vicinity of vessels or sites far from the vessels. One of the best examples of this is the umbilical cord stroma, which comprises three medium-sized vessels (two arteries and one vein) with no prominent adventitia. Umbilical cord stroma cells were shown to share many morphological, phenotypical, and functional features with BM-derived MSCs (Karahuseyinoglu et al. 2007) and were successfully differentiated into many cells types including neural precursors (reviewed in (Can and Karahuseyinoglu 2007; Troyer and Weiss 2008)). Therefore, from a regenerative medicine point of view, they are now considered a good source of cells for allogeneic transplantation. In fact, two clinical trials (www.clinicaltrials.gov) have been started to examine the therapeutic effects of these cells in two sets of patients having myeloblastic syndrome and aplastic anemia. From a physiological point of view, these myofibroblastic cells might have the potential to serve as MSC-like cells during normal tissue turnover in fetal life. Likewise, multipotent progenitors displaying an MSC phenotype and developmental properties have also been described in the bovine artery wall (Tintut et al. 2003) and have recently been isolated from the tunica adventitia of the human pulmonary artery (Hoshino et al. 2008). Corselli et al. (2010) reported, as an unpublished finding, that they isolated cells from the stromal vascular fraction of human adipose tissue that exhibited the same morphology, phenotype, and developmental potential of MSCs, although they did not express a well-known pericyte marker CD146 (Schwab and Gargett 2007). In parallel to this finding, we analyzed perivascular and intervascular (stromal region far from the perivascular compartment) cells using a series of pericyte markers. Interestingly, cells of the perivascular region displayed the whole set of antigens, whereas intervascular cells exhibited the same antigens albeit in lower levels (Fig. 2.2). Given that both cell types successfully displayed many features of MSCs in other tissues, not only pericytes but also nonpericytic cells can behave as multipotent progenitors that could be recruited from tissues showing fetal mesenchyme-derived connective tissue. In other words, as suggested by Caplan (2008), all pericytes are not MSCs. Undoubtedly, a portion of the pericyte population would be highly differentiated cells to execute the given tasks as mentioned above. Taken together, it is possible to conclude that cells of the

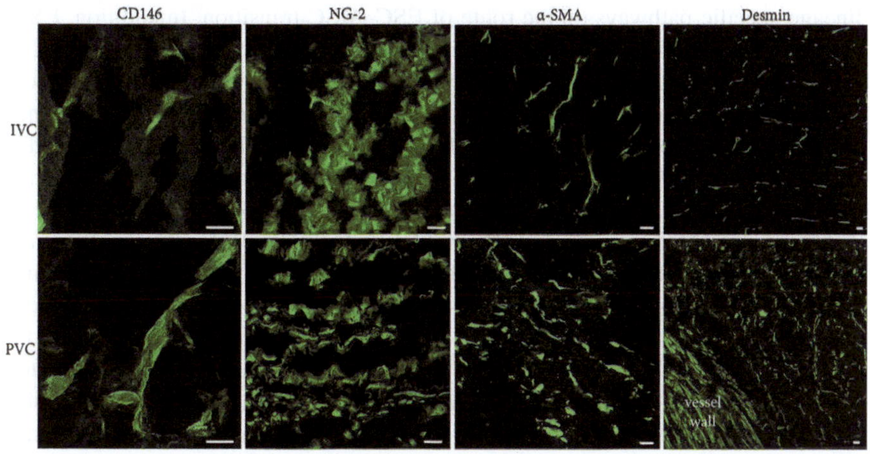

Fig. 2.2 A series of pericyte markers are shown in human umbilical cord stromal cells, which are thought to be fetus-derived MSCs having properties of both adult MSCs and embryonic stem cells. Tissue sections taken from intervascular (*IVC*) and perivascular (*PVC*) stroma exhibit varying degrees of pericyte markers. For instance, CD146 was markedly low in IVC stroma compared to the PVC stroma, whereas NG-2 positivity is dispersed among the entire cell population in both IVC and PVC stromal cells. α-SMA and desmin display more intense staining. *Scale bars* 20 μm

perivascular region serve not only for the sake of the vasculature but also as a niche for the various types of stem cells including MSCs. Coming back to proper nomenclature to describe those cells: MSCs could be termed "perivascular stem cells."

2.4 Lessons from Embryonic Stem Cell–Mesenchymal Stem Cell Differentiation In Vitro

During the development of higher vertebrates, the mesoderm is not the only germ-layer source of mesenchymal cells. For example, in the cranium, the facial bones, jaws, and surrounding connective tissues are derived from the neural crest (NC). NC cells arise from neuroectoderm just after the neural tube closure at days 25–27 in humans. They undergo an epithelial–mesenchymal transition and migrate to diverse regions, where they differentiate into various mature cell types. In the head and neck, NC-derived cells also include nonneural, "mesenchymal" cell types such as chondrocytes, myo/fibroblasts, vascular smooth muscle cells, odontoblasts, and osteoblasts, as shown in mammalian and avian models (Le Douarin et al. 2008; Nagoshi et al. 2009; Dupin et al. 2010).

When embryonic stem cells (ESCs) are cultured, preferably on a feeder layer, they aggregate to form embryoid bodies (EBs), in which the cells are capable of forming ectodermal, mesodermal, and endodermal derivatives. Therefore, successful differentiation and characterization of ESC-derived MSCs could mimic the lineage-specific pathways on the route of ESC–MSC transition. In addition, ESCs are easily genetically modifiable and can be produced in large numbers, thus offering a unique cell culture model to study the earliest steps of mammalian development. Over the last decade, various mature cell types, some of which are functional, have been shown to be derived in vitro from ESCs in mice and human (reviewed in Billon et al. [2008]). Retinoid acid for osteoblast and chondrocyte differentiation (Kawaguchi et al. 2005) and transforming growth factor-β (TGFβ) for myogenic differentiation (Mahmood et al. 2010) were reported to be the key factors for quantitative induction of mesenchymal derivatives. One of the first attempts on human ESCs (hESCs) was reported by Barberi et al. (2005), who successfully induced hESCs to differentiate into CD73+ MSC precursors, which were then further differentiated into adipocytes, chondrocytes, osteoblasts, and skeletal myocytes. Strikingly, genome-wide expression of analysis showed a marked overlap of global gene expression profiles between hESC-derived mesechymal precursors and human bone marrow MSCs.

Although significant progress has been made to derive specific mesenchymal cell types from ESCs, this system is still largely lacking defined culture conditions and good cell-surface markers for the isolation of pure MSCs and mesenchymal precursor populations. The molecular events leading to the formation of such populations remain unclear at present. However, based on recent reports, it is possible to assume that several routes can be used to produce mesenchymal precursors and their derivatives from ESCs, which are likely to involve either mesodermal or neural/NC intermediates. Different routes observed in the ESC system might mimic normal development as mesenchymal precursors can be produced from the mesoderm and the NC. The potential differences that distinguish mesenchymal precursors of mesodermal or neural crest origin remain to be elucidated.

In 2007, an outstanding study by Takashima et al. (2007) revealed that Sox1+ neuroepithelial cells supply the earliest wave of MSC differentiation, which occurs during embryogenesis, but these cells are later replaced by MSCs from other origins during postnatal development. Their results, obtained using mouse ESCs, suggest that in contrast to mesenchymal precursors true MSCs might exclusively originate from a unique source, the neuroepithelium, rather than from the mesoderm, which is generally believed to be the main source of MSCs. To confirm the relevance of these unexpected findings in vivo, Takashima et al. (2007) assessed the origin of MSCs in sox1-gfp embryos. They isolated the trunk of these embryos at E9.5 and purified neuroepithelial cells (Sox1+, GFP+) and mesodermal cells (Sox1−, GFP−, PDGFRα+). They then tested the ability of these cells to give rise to proliferating MSCs and adipocytes in vitro. They demonstrated that although both populations could give rise to adipocytes only neuroepithelial cells could generate PDGFRα+MSCs. These data suggest that in mid-gestation embryos, as seen with mESCs, trunk MSCs originate entirely from neuroepithelium, not from mesoderm. To determine the correlation between Sox1+ neuroepithelial progenitors in E9.5 embryos and MSCs in later life, Takashima et al. (2007) then carried out a persistent labeling of Sox1+ neuroepithelial cells using sox1-cre/rosa26-yfp mice. Considering the possibility that Sox1+ cells give rise to MSCs via the NC, they also permanently labeled NC progeny using P0-cre/yfp mice (P0 being a NC marker). The presence of neuroepithelium/NC-derived YFP+/PDGFRα + cells was assessed in the trunk of E14.5 embryos as was the presence of MSCs in this population. They found that some of the PDGFRα + cells in the E14.5 embryo trunks of both genotypes were also YFP + and thus were derived from neuroepithelium/NC. In addition, some of these PDGFRα+/YFP + cells were MSCs. MSCs could also be established from YFP–/PDGFRα + cells, suggesting that in contrast to E9.5 embryos MSCs in E14.5 embryos derive from both a neural/NC pathway and a nonneural pathway.

Takashima et al. (2007) then analyzed the contribution of these neural/NC-derived MSCs to the postnatal BM. They found that the proportion of YFP+/PDGFRα + cells in the bone marrow of neonates (P0) was far lower than in the embryonic trunk in both sox1-cre/yfp and P0-cre/yfp mice (0.032% and 2.39%, respectively). Furthermore, although the YFP+/PDGFRα + population was able to generate genuine MSCs, this population progressively decreased to negligible levels with aging [0.0 021% and 0.35% for sox1-cre/yfp and P0-cre/yfp mice at P28 (post-natal day 28), respectively]. Finally, the frequency of PDGFRα + MSCs derived from YFP + compared with YFP – populations was nearly the same. Thus, in neonatal and adult bone preparations, MSCs were found within the PDGFRα + population, but most of them derived from a nonneural/NC pathway.

Together, these results suggest that MSCs arise in multiple waves of distinct origins: The first MSCs in the embryo, which are identifiable by the method described by Takashima et al. (2007), are derived from neuroepithelium/NC. Later, this early, but transient, population is eventually replaced by MSCs derived from an as-yet-unidentified pathway. An important question raised by this study is what is the other source of MSCs? Until recently, MSCs have been described to derive from mesoderm, but the results presented by Takashima et al. (2007) strongly argue against this hypothesis because YFP–/PDGFRα + somatic cells in sox1-cre/yfp E9.5 embryos could not generate MSCs. The authors speculate, on the basis of recent reports, that hematopoietic cells might constitute the progenitors of MSCs in adult bone marrow. However, one cannot rule out the possibility that in vitro conditions used in this study (2007) to measure the clonogenicity and the differentiation potential of PDGFRα + cells were not appropriate for revealing the MSC potential of mesoderm-derived cells. Conversely, this study did not test whether neuroepithelium/NC-derived MSCs could form mesenchymal derivatives in vivo.

2.5 Conclusion

The last few years have witnessed a growing optimism by basic scientists and clinicians regarding the clinical application of MSCs for many disease pathologies. Significant advancements have been made in in vitro and in vivo preclinical studies using MSCs. However, unresolved issues such as the lack of conformity with respect to isolation and ex vivo culture-expansion protocols and the heterogeneity by which populations and subpopulations of MSCs are characterized continue to be obstacles.

Therefore, the exact location and in vivo tracing of MSCs are far more important than in past years. Identifying the origin of MSCs in vivo would not only allow us to determine the in vivo state of the existing MSCs in a given organism, it would open many avenues to manipulate these cells for regenerative or self-protective purposes as in diseases such as cancer, Alzheimer's disease, and Parkinson's disease, among others. There is little concern that MSCs are not entirely derived from pericytes. However, the term pericyte has been used in its anatomical, literal sense, without functional connotation. A growing number of studies indicate that nonpericytes in the perivascular region may also play role in MSC biology.

Acknowledgments I thank Drs. Kursad Turksen and Gulsen Ökten for inviting me to write this review. The relevant studies in my laboratory have been supported partly by TUBITAK-SBAG-3314 and the Ankara University Research Fund (09H3330004).

References

Andreeva ER, Pugach IM, Gordon D et al (1998) Continuous subendothelial network formed by pericyte-like cells in human vascular bed. Tissue Cell 30(1):127–135

Barberi T, Willis LM, Socci ND et al (2005) Derivation of multipotent mesenchymal precursors from human embryonic stem cells. PLoS Med 2(6):e161

Bianco P, Riminucci M, Gronthos S et al (2001) Bone marrow stromal stem cells: nature, biology, and potential applications. Stem Cells 19(3):180–192

Billon N, Monteiro MC, Dani C (2008) Developmental origin of adipocytes: new insights into a pending question. Biol Cell 100(10):563–575

Brighton CT, Lorich DG, Kupcha R et al (1992) The pericyte as a possible osteoblast progenitor cell. Clin Orthop Relat Res 275:287–299

Can A (2008) Haematopoietic stem cells niches: interrelations between structure and function. Transfus Apher Sci 38(3):261–268

Can A, Karahuseyinoglu S (2007) Concise review: human umbilical cord stroma with regard to the source of fetus-derived stem cells. Stem Cells 25(11):2886–2895

Caplan AI (1991) Mesenchymal stem cells. J Orthop Res 9(5):641–650

Caplan AI (2008) All MSCs are pericytes? Cell Stem Cell 3(3):229–230

Chan RW, Gargett CE (2006) Identification of label-retaining cells in mouse endometrium. Stem Cells 24(6):1529–1538

Charbord P, Tavian M, Humeau L et al (1996) Early ontogeny of the human marrow from long bones: an immunohistochemical study of hematopoiesis and its microenvironment. Blood 87(10):4109–4119

Choi KS, Shin JS, Lee JJ et al (2005) In vitro trans-differentiation of rat mesenchymal cells into insulin-producing cells by rat pancreatic extract. Biochem Biophys Res Commun 330(4): 1299–1305

Corselli M, Chen CW, Crisan M et al (2010) Perivascular ancestors of adult multipotent stem cells. Arterioscler Thromb Vasc Biol 30(6):1104–1109

Covas DT, Panepucci RA, Fontes AM et al (2008) Multipotent mesenchymal stromal cells obtained from diverse human tissues share functional properties and gene-expression profile with CD146+ perivascular cells and fibroblasts. Exp Hematol 36(5):642–654

Covello KL, Kehler J, Yu H et al (2006) HIF-2alpha regulates Oct-4: effects of hypoxia on stem cell function, embryonic development, and tumor growth. Genes Dev 20(5):557–570

Crisan M, Yap S, Casteilla L et al (2008) A perivascular origin for mesenchymal stem cells in multiple human organs. Cell Stem Cell 3(3):301–313

Crisan M, Chen CW, Corselli M et al (2009) Perivascular multipotent progenitor cells in human organs. Ann N Y Acad Sci 1176:118–123

da Silva Meirelles L, Chagastelles PC, Nardi NB (2006) Mesenchymal stem cells reside in virtually all post-natal organs and tissues. J Cell Sci 119(Pt 11):2204–2213

da Silva Meirelles L, Caplan AI, Nardi NB (2008) In search of the in vivo identity of mesenchymal stem cells. Stem Cells 26(9):2287–2299

Davidoff MS, Middendorff R, Enikolopov G et al (2004) Progenitor cells of the testosterone-producing Leydig cells revealed. J Cell Biol 167(5):935–944

Dexter TM, Allen TD, Lajtha LG (1977) Conditions controlling the proliferation of haemopoietic stem cells in vitro. J Cell Physiol 91(3):335–344

Diaz-Flores L, Gutierrez R, Gonzalez P et al (1991) Inducible perivascular cells contribute to the neochondrogenesis in grafted perichondrium. Anat Rec 229(1):1–8

Diaz-Flores L, Gutierrez R, Lopez-Alonso A et al (1992) Pericytes as a supplementary source of osteoblasts in periosteal osteogenesis. Clin Orthop Relat Res 275:280–286

Diaz-Flores L, Gutierrez R, Madrid JF et al (2009) Pericytes: morphofunction, interactions and pathology in a quiescent and activated mesenchymal cell niche. Histol Histopathol 24(7):909–969

Dominici M, Le Blanc K, Mueller I et al (2006) Minimal criteria for defining multipotent mesenchymal stromal cells: the International Society for Cellular Therapy position statement. Cytotherapy 8(4):315–317

Dupin E, Calloni GW, Le Douarin NM (2010) The cephalic neural crest of amniote vertebrates is composed of a large majority of precursors endowed with neural, melanocytic, chondrogenic and osteogenic potentialities. Cell Cycle 9(2):238–249

Friedenstein AJ, Deriglasova UF, Kulagina NN et al (1974) Precursors for fibroblasts in different populations of hematopoietic cells as detected by the in vitro colony assay method. Exp Hematol 2(2):83–92

Funk PE, Stephan RP, Witte PL (1995) Vascular cell adhesion molecule 1-positive reticular cells express interleukin-7 and stem cell factor in the bone marrow. Blood 86(7):2661–2671

Grayson WL, Zhao F, Izadpanah R et al (2006) Effects of hypoxia on human mesenchymal stem cell expansion and plasticity in 3D constructs. J Cell Physiol 207(2):331–339

Hirschi KK, D'Amore PA (1996) Pericytes in the microvasculature. Cardiovasc Res 32(4):687–698

Horwitz EM, Le Blanc K, Dominici M et al (2005) Clarification of the nomenclature for MSC: the International Society for Cellular Therapy position statement. Cytotherapy 7(5):393–395

Hoshino A, Chiba H, Nagai K et al (2008) Human vascular adventitial fibroblasts contain mesenchymal stem/progenitor cells. Biochem Biophys Res Commun 368(2):305–310

Karahuseyinoglu S, Cinar O, Kilic E et al (2007) Biology of stem cells in human umbilical cord stroma: in situ and in vitro surveys. Stem Cells 25(2):319–331

Kawaguchi J, Mee PJ, Smith AG (2005) Osteogenic and chondrogenic differentiation of embryonic stem cells in response to specific growth factors. Bone 36(5):758–769

Kopen GC, Prockop DJ, Phinney DG (1999) Marrow stromal cells migrate throughout forebrain and cerebellum, and they differentiate into astrocytes after injection into neonatal mouse brains. Proc Natl Acad Sci USA 96(19):10711–10716

Lazarus HM, Haynesworth SE, Gerson SL et al (1997) Human bone marrow-derived mesenchymal (stromal) progenitor cells (MPCs) cannot be recovered from peripheral blood progenitor cell collections. J Hematother 6(5):447–455

Le Douarin NM, Calloni GW, Dupin E (2008) The stem cells of the neural crest. Cell Cycle 7(8):1013–1019

Mahmood A, Harkness L, Schroder HD et al (2010) Enhanced differentiation of human embryonic stem cells to mesenchymal progenitors by inhibition of TGF-beta/activin/nodal signaling using SB-431542. J Bone Miner Res 25(6):1216–1233

McCulloch CA (1985) Progenitor cell populations in the periodontal ligament of mice. Anat Rec 211(3):258–262

Meirelles Lda S, Nardi NB (2009) Methodology, biology and clinical applications of mesenchymal stem cells. Front Biosci 14:4281–4298

Mohyeldin A, Garzon-Muvdi T, Quinones-Hinojosa A (2010) Oxygen in stem cell biology: a critical component of the stem cell niche. Cell Stem Cell 7(2):150–161
Nagoshi N, Shibata S, Nakamura M et al (2009) Neural crest-derived stem cells display a wide variety of characteristics. J Cell Biochem 107(6):1046–1052
Paquet-Fifield S, Schluter H, Li A et al (2009) A role for pericytes as microenvironmental regulators of human skin tissue regeneration. J Clin Invest 119(9):2795–2806
Phinney DG, Prockop DJ (2007) Concise review: mesenchymal stem/multipotent stromal cells: the state of transdifferentiation and modes of tissue repair: current views. Stem Cells 25(11): 2896–2902
Pinzani M (1995) Hepatic stellate (ITO) cells: expanding roles for a liver-specific pericyte. J Hepatol 22(6):700–706
Pittenger MF, Mackay AM, Beck SC et al (1999) Multilineage potential of adult human mesenchymal stem cells. Science 284(5411):143–147
Richardson RL, Hausman GJ, Campion DR (1982) Response of pericytes to thermal lesion in the inguinal fat pad of 10-day-old rats. Acta Anat (Basel) 114(1):41–57
Robin C, Bollerot K, Mendes S et al (2009) Human placenta is a potent hematopoietic niche containing hematopoietic stem and progenitor cells throughout development. Cell Stem Cell 5(4):385–395
Salem HK, Thiemermann C (2010) Mesenchymal stromal cells: current understanding and clinical status. Stem Cells 28(3):585–596
Santoro MM, Pesce G, Stainier DY (2009) Characterization of vascular mural cells during zebrafish development. Mech Dev 126(8–9):638–649
Sarugaser R, Hanoun L, Keating A et al (2009) Human mesenchymal stem cells self-renew and differentiate according to a deterministic hierarchy. PLoS One 4(8):e6498
Scadden DT (2006) The stem-cell niche as an entity of action. Nature 441(7097):1075–1079
Schlondorff D (1987) The glomerular mesangial cell: an expanding role for a specialized pericyte. FASEB J 1(4):272–281
Schofield R (1978) The relationship between the spleen colony-forming cell and the haemopoietic stem cell. Blood Cells 4(1–2):7–25
Schwab KE, Gargett CE (2007) Co-expression of two perivascular cell markers isolates mesenchymal stem-like cells from human endometrium. Hum Reprod 22(11):2903–2911
Shi S, Gronthos S (2003) Perivascular niche of postnatal mesenchymal stem cells in human bone marrow and dental pulp. J Bone Miner Res 18(4):696–704
Simionescu N, Simionescu M, Palade GE (1976) Recent studies on vascular endothelium. Ann N Y Acad Sci 275:64–75
Takashima Y, Era T, Nakao K et al (2007) Neuroepithelial cells supply an initial transient wave of MSC differentiation. Cell 129(7):1377–1388
Tilton RG, Kilo C, Williamson JR (1979) Pericyte-endothelial relationships in cardiac and skeletal muscle capillaries. Microvasc Res 18(3):325–335
Tintut Y, Alfonso Z, Saini T et al (2003) Multilineage potential of cells from the artery wall. Circulation 108(20):2505–2510
Troyer DL, Weiss ML (2008) Wharton's jelly-derived cells are a primitive stromal cell population. Stem Cells 26(3):591–599
Wexler SA, Donaldson C, Denning-Kendall P et al (2003) Adult bone marrow is a rich source of human mesenchymal 'stem' cells but umbilical cord and mobilized adult blood are not. Br J Haematol 121(2):368–374
Yamashima T, Tonchev AB, Vachkov IH et al (2004) Vascular adventitia generates neuronal progenitors in the monkey hippocampus after ischemia. Hippocampus 14(7):861–875
Yemisci M, Gursoy-Ozdemir Y, Vural A et al (2009) Pericyte contraction induced by oxidative-nitrative stress impairs capillary reflow despite successful opening of an occluded cerebral artery. Nat Med 15(9):1031–1037
Zannettino AC, Paton S, Arthur A et al (2008) Multipotential human adipose-derived stromal stem cells exhibit a perivascular phenotype in vitro and in vivo. J Cell Physiol 214(2):413–421
Zimmerlin L, Donnenberg VS, Pfeifer ME et al (2010) Stromal vascular progenitors in adult human adipose tissue. Cytometry A 77(1):22–30

Chapter 3
Isolation and Identification of Mesenchymal Stem Cells

Ilknur Kozanoglu and Erkan Maytalman

Abstract Adult bone marrow is a large organ composed of hematopoietic cells and stromal support cells. Mesenchymal stem cells (MSCs) are stromal stem cells of the bone marrow and the most preferred organ for isolation and proliferation of MSCs is the bone marrow. MSCs grown in culture with appropriate cytokine support are able to differentiate into various tissues such as bone, cartilage and adipose tissue. Clinical use of mesenchymal stem cells has been increasing as they inhibit T lymphocyte proliferation in mixed lymphocyte cultures in vitro and suppress the immune system through numerous soluble factors they release. Identification of these cells and accurate determination of their phenotypic properties will contribute to the understanding of their biological behaviors and increase reliability in clinical and experimental practices.

3.1 Introduction

Mesenchymal stem cells (MSCs) are multi-potent precursor cells with high proliferation and differentiation capacities (Dazzi and Horwood 2007; Deans and Moseley 2000; Pittenger et al. 1999; Abdallah and Kassem 2008). Since they originate from

I. Kozanoglu, M.D. (✉)
Department of Physiology, Baskent University, Ankara, Turkey

Baskent University Adana Teaching and Research Hospital Hematology
Research Laboratory, Adana, Turkey

Baskent University Adana Teaching and Medical Research Center,
Dadaloglu Mahallesi Serin Evler Sokak 39 No: 6, Yuregir, Adana 01250, Turkey
e-mail: ipamuk5@hotmail.com

E. Maytalman
Baskent University Adana Teaching and Research Hospital Hematology
Research Laboratory, Adana, Turkey

K. Turksen (ed.), *Adult and Embryonic Stem Cells*, Stem Cell Biology and Regenerative Medicine, DOI 10.1007/978-1-61779-630-2_3,

the stroma they act as "support cells" and have the potential to be used in many areas of medicine (Fibbe and Noort 2003; Lazarus et al. 1995; Arthur et al. 2008; Horwitz et al. 2002). However, they are scarce in tissues and the necessity of in vitro proliferation in culture media is the major disadvantage of using MSCs in basic science researches and clinical settings. In vitro proliferation in culture media leads to changes in phenotypic, immunological and other biological properties of these cells due to exposure to various stimulators and factors during passage in culture. There is a risk of cell aging, cytogenetic disruption and although low, malignant transformation when cells are proliferated by passaging in culture media. Furthermore, the difficulties in establishing high quality cell processing laboratories complying with the internationally recognized accreditation requirements hinder the widespread use of these cells in clinical settings.

The lack of a consensus on identification of MSCs is another problem for the scientists working on these cells. Currently, there are many studies aiming to accurately identify MSCs and the criteria established by The International Society of Cell Therapy (ISCT) in 2006 are widely recognized and currently used for identifying MSCs (Dominici et al. 2006). These criteria include the ability of adherence to plastic surfaces, expression of stromal cell surface antigens, and the multi-potent differentiation potential (Dominici et al. 2006). Nevertheless, lack of a specific antigen for identification of MSCs is probably the most important issue.

MSCs were shown to have an important role in bone physiology, bone reformation and hematopoiesis (Dazzi et al. 2006; Anselme et al. 2002; Mackay et al. 1998; Caplan 1991). There is growing enthusiasm for using these cells in clinical settings due to their easy proliferation and high differentiation capabilities and immunosuppressive properties (Le Blanc et al. 2003a; Le Blanc and Ringden 2005; Chamberlain et al. 2007; Ohishi and Schipani 2009; Le Blanc et al. 2003b; Le Blanc and Ringden 2007).

3.2 History and Definitions

Mesenchymal stem cells are the main cells of the connective tissue and may differentiate into adipose, bone, cartilage and muscle cells. In addition, they represent the origin of stromal cells which are the supportive cells in all tissues (Chamberlain et al. 2007; Ohishi and Schipani 2009). These cells were first identified by Friedenstein in 1976 (Friedenstein et al. 1976). Friedenstein identified cell colonies which had adhesion ability, were similar to fibroblasts in morphology and were able to differentiate into bone and fat cells in fetal calf serum (FCS) containing bone marrow cultures (Friedenstein et al. 1970, 1976; Friedenstein et al. 1970; Chailakhian et al. 2006). In subsequent studies, these cells were shown to be non-hematopoietic and to have ability to differentiate into cells originating from all three germ layers. Formerly called CFU-F (colony forming unit fibroblast) and "bone marrow stromal fibroblasts," these cells were then referred to as mesenchymal stem cells.

The International Society of Cell Therapy (ISCT) recommended some criteria for identification of human MSCs both for basic researches and pre-clinical studies (Dominici et al. 2006). The primary features commonly used for identifying MSCs are adherence to plastic surfaces, expression of stromal cell surface antigens, and potential of multipotent differentiation. For these cells, ISCT suggested to use the terms “mesenchymal stromal cells” or “multi-potent mesenchymal stromal cells” (Pittenger et al. 1999; Verfaillie et al. 2002, 2003).

3.3 Sources of Mesenchymal Stem Cells

Bone marrow is one of the largest and most active organs in the body and the richest source of stem cells for the organism. Bone marrow is also considered as the main source of MSCs. An average of 2–100 MSCs were shown to be present in bone marrow aspirations for 1×10^6 mononuclear cells (Dazzi and Horwood 2007; Deans and Moseley 2000; Abdallah and Kassem 2008; Caplan 1991; Chamberlain et al. 2007; Ohishi and Schipani 2009; Pittenger 2008).

MSCs having similar morphological and biological properties can be isolated from different tissues. Due to their adhesion properties, they can be isolated from lipoaspiration materials (Vermette et al. 2007), cord blood (Han 2009; Bieback and Klüter 2007), cord stroma (Han 2009; Bieback and Klüter 2007), amniotic fluid (Tsai et al. 2004), placenta, synovial fluids, dental pulp and even from peripheral blood and they can be proliferated (Deans and Moseley 2000; Abdallah and Kassem 2008; Verfaillie et al. 2002; Tsai et al. 2004; Haniffa et al. 2009; Huang et al. 2009). Regardless of the tissues from which MSCs are obtained, they have many common properties such as the ability to adhere to plastic containers for tissue culture, the ability to differentiate, having fibroblastoid morphology, and having surface antigens (Dazzi and Horwood 2007; Deans and Moseley 2000; Pittenger et al. 1999; Abdallah and Kassem 2008; Dominici et al. 2006; Haniffa et al. 2009; Kozanoglu et al. 2008, 2009). However, some studies showed that the differentiation ability and functional properties may vary depending on the tissue of origin. Presence of MSCs in the peripheral blood is controversial.

3.3.1 Isolation of Mesenchymal Stem Cells

Preparation of MSC cultures from different tissues requires removal of erythrocytes or establishment of cell suspensions for solid tissues (Abdallah and Kassem 2008; Caplan 1991; Chamberlain et al. 2007; Meirelles Lda and Nardi 2009; Nolta 2006). For preparation of cell suspensions from tissues other than bone marrow, enzymatic isolation is performed. A mixture of collagenase type-1 enzyme and dispase are frequently used for this purpose. Three methods are commonly used to obtain MSCs from bone marrow: isolation by density gradient method, isolation by simple centrifuge and positive or negative selection (Abdallah and Kassem 2008; Caplan 1991; Chamberlain et al. 2007; Meirelles Lda and Nardi 2009; Nolta 2006).

The cells prepared as suspensions are transferred into culture containers (flasks of T-25 and T-75) with culture medium and incubated in an incubator with a CO_2 level of 5% at 37°C with 95% humidity. DMEM-LG, RPMI-1640 or alpha-MEM containing 10–15% fetal bovine serum (FBS) is frequently used as the culture medium (Meirelles Lda and Nardi 2009; Nolta 2006).

In our laboratory, we proved that cells cultured with RPMI-1640 medium showed better proliferative properties than equal number of cells cultured with DMEM-LG medium (unpublished data).

3.3.2 Physical Properties

MSCs are scarce in tissues, including the bone marrow. In vitro proliferation in culture media is necessary to obtain sufficient amount of cells for use in clinical practices and basic science researches. When examined under light and phase-contrast microscopy, it was noted that MSCs grown in culture media were spindle shaped and formed groups of cells like fibroblasts. Fibroblasts have an asymmetric nuclear localization, whereas MSCs were shown to have a symmetric nuclear localization (Pittenger et al. 1999; Vermette et al. 2007; Bobis et al. 2006).

3.3.3 Immunophenotypic Properties

The immunophenotypic properties of MSCs obtained from the bone marrow and proliferated in culture were investigated in depth by flow cytometry (Huang et al. 2009; Kozanoglu et al. 2008; Haynesworth et al. 1992; Horwitz et al. 2005; Salem and Thiemermann 2010; Simmons and Torok-Strob 1991). The ratio of cells positive for hematopoietic antigens such as CD45, CD34, CD14, CD19, CD11b and HLA class II in the cell population should not exceed 2%. However, the cell population should be positive for stromal associated antigens CD105, CD73, CD90, CD29 and CD44. Moreover, adhesion molecules such as intercellular adhesion molecule-1 (ICAM-1), CD102 (ICAM-2), L-selectin (CD62L) and ICAM-3 (CD50) on the surface of MSCs should be positive in various ratios. Co-stimulatory molecules such as CD80, CD86 and CD40 are not expressed on surface of MSCs (Haynesworth et al. 1992; Horwitz et al. 2005; Salem and Thiemermann 2010; Simmons and Torok-Strob 1991).

According to ISCT criteria, 95% or more of a MSC population should be positive for CD105, CD73 and CD90 antigens (Dominici et al. 2006; Horwitz et al. 2005). Also, these cells must be shown to be negative for antigens such as CD45, CD34, and CD14 which are specific for hematopoietic cells (Dominici et al. 2006; Horwitz et al. 2005). Currently, there is not a specific antigen for identification of

MSCs. Therefore, we believe that showing positive and negative surface markers simultaneously in the same tube by flow cytometry analysis will increase the reliability of the study and enable performing an optimal analysis with lower number of cells (unpublished data).

Recently, new antigens such as CD271 (nerve growth factor receptor), CD140b (platelet derived growth factor receptor), CD340 (HER2/ErbB2), CD349 (Frizzled-9), MSCA-1 (Mesenchymal stem cell antigen-1; W8B2), GD2 (neural ganglioside) and NG2 (chondroitin sulfate proteoglycan neuro-glial Ag 2 clone 7.1) were identified (Kozanoglu et al. 2008; Bühring et al. 2007; Battula et al. 2008, 2009; Martinez et al. 2007; Xu et al. 2009).

3.3.4 Differentiation Properties

Another way to identify MSCs is to show differentiation capacities of the cells in vitro (Dominici et al. 2006). MSCs can transform into adipocytes, chondrocytes, osteocytes and neural cells with appropriate cytokine support (Deans and Moseley 2000; Pittenger et al. 1999; Abdallah and Kassem 2008; Arthur et al. 2008; Horwitz et al. 2002; Anselme et al. 2002; Mackay et al. 1998; Caplan 1991; Salem and Thiemermann 2010; Jackson et al. 2007; Lin et al. 2003). Showing differentiation into at least two of three germ layers is considered as a proof that these cells may be MSCs. In differentiation studies, histochemical, immunohistochemical or immunofluorescent methods are used to determine specific markers, to show whether the targeted cell differentiation is present or not (Dominici et al. 2006; Nolta 2006). In recent years, use of RT-PCR and DNA micro-array techniques for identification of MSCs and their differentiation products has become widespread and therefore quantitation of numerous non-specific proteins in mRNA level is achieved (Case et al. 2010; Augello and De Bari 2010; Hosogane et al. 2010).

The ability of MSCs to differentiate into other cell lines, different from that of their own has raised an interest particularly in fields related to regenerative medicine, and therefore, many studies were conducted and collected under the title of "stem cell plasticity" (Goodell et al. 2001; Mertelsmann 2000; Peterson 2002; Stanworth and Newland 2001; Stocum 2001; Tsai et al. 2002).

3.4 Discussion

The therapeutic potential of MSCs has generated great interest and enthusiasm during the last few years. MSCs can differentiate into different cell lineages and have successfully been used to repair damaged or genetically defective tissues not only in animal models but also in clinical trials. Mesenchymal stem cells have been proposed to have immunosuppressive properties and reduced inflammation. Human MSCs suppress lymphocyte alloreactivity *in vitro* in mixed lymphocyte cultures.

The lacking of a single specific antigen for isolating and defining MSCs, the ambiguity of issues such as the ideal cell dose, the application route and the timing limits use of these cells in clinical and experimental studies. However, MSCs are preferred over other stem cells for tissue repair and cellular based treatments as they are easy to obtain, have strong proliferation and differentiation capacity and may be used for non-relatives.

References

Abdallah BM, Kassem M (2008) Human mesenchymal stem cells: from basic biology to clinical applications. Gene Ther 15(2):109–116

Anselme K, Broux O, Noel B et al (2002) In vitro control of human bone marrow stromal cells for bone tissue engineering. Tissue Eng 8:941–953

Arthur A, Zannettino A, Gronthos S (2008) The therapeutic applications of multipotential mesenchymal/stromal stem cells in skeletal tissue repair. J Cell Physiol 218(2):237–245

Augello A, De Bari C (2010) The regulation of differentiation in mesenchymal stem cells. Hum Gene Ther 21(10):1226–1238

Battula VL, Treml S, Abele H, Bühring HJ (2008) Prospective isolation and characterization of mesenchymal stem cells from human placenta using a frizzled-9-specific monoclonal antibody. Differentiation 76(4):326–336

Battula VL, Treml S, Bareiss PM, Gieseke F, Roelofs H, de Zwart P, Müller I, Schewe B, Skutella T, Fibbe WE, Kanz L, Bühring HJ (2009) Isolation of functionally distinct mesenchymal stem cell subsets using antibodies against CD56, CD271, and mesenchymal stem cell antigen-1. Haematologica 94(2):173–184

Bieback K, Klüter H (2007) Mesenchymal stromal cells from umbilical cord blood. Curr Stem Cell Res Ther 2:310–323

Bobis S, Jarocha D, Majka M (2006) Mesenchymal stem cells: characteristics and clinical applications. Folia Histochem Cytobiol 44:215–230

Bühring HJ, Battula VL, Treml S, Schewe B, Kanz L, Vogel W (2007) Novel markers for the prospective isolation of human MSC. Ann N Y Acad Sci 1106:262–271

Caplan AI (1991) Mesenchymal stem cells. J Orthop Res 9:641–650

Case N, Xie Z, Sen B, Styner M, Zou M, O'Conor C, Horowitz M, Rubin J (2010) Mechanical activation of β-catenin regulates phenotype in adult murine marrow-derived mesenchymal stem cells. J Orthop Res 28(11):1531–1538

Chailakhian RK, Latsinik NV, Gerasimov YuV et al (2006) Life in research. In memory of AJ Friedenstein. The scientist and the teacher. Cell Transplantol Tissue Eng 4(6):9–12

Chamberlain G, Fox J, Ashton B, Middleton J (2007) Concise review: mesenchymal stem cells: their phenotype, differentiation capacity, immunological features, and potential for homing. Stem Cells 25(11):2739–2749

Dazzi F, Horwood NJ (2007) Potential of mesenchymal stem cell therapy. Curr Opin Oncol 19(6):650–655

Dazzi F, Ramasamy R, Glennie S, Jones SP, Roberts I (2006) The role of mesenchymal stem cells in haemopoiesis. Blood Rev 20(3):161–171

Deans RJ, Moseley AB (2000) Mesenchymal stem cells: biology and potential clinical uses. Exp Hematol 28:875–884

Dominici M, Le Blanc K, Mueller I, Slaper-Cortenbach I, Marini F, Krause D et al (2006) Minimal criteria for defining multipotent mesenchymal stromal cells. The International Society for Cellular Therapy position statement. Cytotherapy 8:315–317

Fibbe WE, Noort WA (2003) Mesenchymal stem cells and hematopoietic stem cell transplantation. Ann N Y Acad Sci 996:235–244

Friedenstein AJ, Chailakhjan RK, Lalykina KS (1970) The development of fibroblast colonies in monolayer cultures of guinea-pig BM and spleen cells. Cell Tissue Kinet 3:393–403

Friedenstein AJ, Goskaja UF, Julagina NN (1976) Fibroblast precursors in normal and irradiated mouse hematopoietic organs. Exp Hematol 4:267–274

Goodell MA, Jackson KA, Majka SM, Mi T, Wang H, Pocius J, Hartley CJ, Majesky MW, Entman ML, Michael LH, Hirschi KK (2001) Stem cell plasticity in muscle and bone marrow. Ann N Y Acad Sci 938:208–218

Han ZC (2009) Umbilical cord mesenchymal stem cells (UC-MSC: biology, banking and clinical applications). Bull Acad Natl Med 193(3):545–547

Haniffa MA, Collin MP, Buckley CD, Dazzi F (2009) Mesenchymal stem cells: the fibroblast new clothes? Haematologica 94:258–263

Haynesworth SE, Baber MA, Caplan AI (1992) Cell surface antigens on human marrow-derived mesenchymal cells are detected by monoclonal antibodies. Bone 13:69–80

Horwitz EM, Gordon PL, Koo WK et al (2002) Isolated allogeneic bone marrow-derived mesenchymal cells engraft and stimulate growth in children with osteogenesis imperfecta: implications for cell therapy of bone. Proc Natl Acad Sci USA 99:8932–8937

Horwitz EM, Ml D et al (2005) Clarification of the nomenclature for MSC: the International Society for Cellular Therapy position statement. Cytotherapy 7:393–395

Hosogane N, Huang Z, Rawlins BA, Liu X, Boachie-Adjei O, Boskey AL, Zhu W (2010) Stromal derived factor-1 regulates bone morphogenetic protein 2-induced osteogenic differentiation of primary mesenchymal stem cells. Int J Biochem Cell Biol 42(7):1132–1141

Huang GT, Gronthos S, Shi S (2009) Mesenchymal stem cells derived from dental tissues vs. those from other sources: their biology and role in regenerative medicine. J Dent Res 88(9): 792–806

Jackson L, Jones DR, Scotting P, Sottile V (2007) Adult mesenchymal stem cells: differentiation potential and therapeutic applications. J Postgrad Med 53(2):121–127

Kozanoglu I, Boga C, Ozdogu H, Maytalman E, Ovali E, Sozer O (2008) A detachment technique based on the thermophysiologic responses of cultured mesenchymal cells exposed to cold. Cytotherapy 10(7):686–689

Kozanoglu I, Boga C, Ozdogu H, Sozer O, Maytalman E, Yazıcı AC, Sahin FI (2009) Human bone marrow mesenchymal stem cells express NG2: may increase in discriminative ability of flow cytometry during MSC identification. Cytotherapy 11(5):527–533

Lazarus HM, Haynesworth SE, Gerson SL, Rosenthal NS, Caplan AI (1995) Ex vivo expansion and subsequent infusion of human bone marrow-derived stromal progenitor cells (mesenchymal progenitor cells): implications for therapeutic use. Bone Marrow Transplant 16:557–564

Le Blanc K, Ringden O (2005) Immunobiology of human mesenchymal stem cells and future use in hematopoietic stem cell transplantation. Biol Blood Marrow Transplant 11:321–334

Le Blanc K, Ringden O (2007) Immunomodulation by mesenchymal stem cells and clinical experience. J Intern Med 262:509–525

Le Blanc K, Tammik C, Rosendahl K, Zetterberg E, Ringden O (2003a) HLA expression and immunologic properties of differentiated and undifferentiated mesenchymal stem cells. Exp Hematol 31:890–896

Le Blanc K, Tammik L, Sundberg B, Haynesworth SE, Ringdén O (2003b) Mesenchymal stem cells inhibit and stimulate mixed lymphocyte cultures and mitogenic responses independently of the major histocompatibility complex. Scand J Immunol 57(1):11–20

Lin JR, Guo KY, Li JQ, Yan DA (2003) In vitro culture of human bone marrow mesenchymal stem cell clonies and induced differentiation into neuron-like cells. Di Yi Jun Yi Da Xue Xue Bao 23(3):251–253

Mackay AM, Beck SC, Murphy JM, Barry FP, Chishester CO, Pittenger MF (1998) Chondrogenic differentiation of cultured human mesenchymal stem cells from marrow. Tissue Eng 4: 415–428

Martinez C, Hofmann TJ, Marino R, Dominici M, Horwitz EM (2007) Human bone marrow mesenchymal stromal cells express the neural ganglioside GD2: a novel surface marker for the identification of MSCs. Blood 109(10):4245–4248

Meirelles Lda S, Nardi NB (2009) Methodology, biology and clinical applications of mesenchymal stem cells. Front Biosci 14:4281–4289

Mertelsmann R (2000) Plasticity of bone marrow-derived stem cells. J Hematother Stem Cell Res 9(6):957–960

Nolta JA (2006) Genetic engineering of mesenchymal stem cells. Springer, Dordrecht

Ohishi M, Schipani E (2009) Bone marrow mesenchymal stem cells. J Cell Biochem 109(2):277–282

Peterson DA (2002) Stem cells in brain plasticity and repair. Curr Opin Pharmacol 2(1):34–42

Pittenger MF (2008) Mesenchymal stem cells from adult bone marrow. Methods Mol Biol 449:27–44

Pittenger MF, Mackay AM, Beck SC et al (1999) Multilineage potential of adult human mesenchymal stem cells. Science 284:143–147

Salem HK, Thiemermann C (2010) Mesenchymal stromal cells: current understanding and clinical status. Stem Cells 28(3):585–596

Simmons PJ, Torok-Strob B (1991) Identification of stromal cell precursors in human bone marrow in human bone marrow by a novel monoclonal antibody, STRO-1. Blood 78:55–62

Stanworth SJ, Newland AC (2001) Stem cells: progress in research and edging towards the clinical setting. Clin Med 1(5):378–382

Stocum DL (2001) Stem cells in regenerative biology and medicine. Wound Repair Regen 9(6):429–442

Tsai RY, Kittappa R, McKay RD (2002) Plasticity, niches, and the use of stem cells. Dev Cell 2(6):707–712

Tsai MS, Lee JL, Hwang SM (2004) Isolation of human multipotent mesenchymal stem cells from second- trimester amniotic fluid using a novel two-stage culture protocol. Hum Reprod 19:1450–1456

Verfaillie CM, Pera MF, Lansdorp PM (2002) Stem cells: hype and reality. Hematology 1:369–391

Verfaillie CM, Schwartz R, Reyes M, Jiang Y (2003) Unexpected potential of adult stem cells. Ann N Y Acad Sci 996:231–234

Vermette M, Trottier V, Ménard V, Saint-Pierre L, Roy A, Fradette J (2007) Production of a new tissue-engineered adipose substitute from human adipose-derived stromal cells. Biomaterials 28(18):2850–2860

Xu J, Liao W, Gu D, Liang L, Liu M, Du W, Liu P, Zhang L, Lu S, Dong C, Zhou B, Han Z (2009) Neural ganglioside GD2 identifies a subpopulation of mesenchymal stem cells in umbilical cord. Cell Physiol Biochem 23(4–6):415–424

Chapter 4
Mesenchymal Stromal Cells and Umbilical Cord Blood Transplantation

Chitra Hosing, Marcos de Lima, and Elizabeth J. Shpall

Abstract In recent years umbilical cord blood has become an important source of stem cells for patients undergoing hematopoietic stem cell transplantation. Ease of collection, ready availability, allowance of higher HLA disparity, and the lower incidence of graft-versus-host disease makes cord blood an attractive source of stem cells, especially for minority populations. One of the major limitations to wider use of umbilical cord stem cells for transplantation in adult patients is the relative low number of progenitor cells in the graft. This results in delayed engraftment, delayed immune reconstitution, and increased rates of infectious complications. This can be partly overcome by ex vivo expansion of cord blood stem cells. There are many techniques for cord blood expansion currently being used in clinical trials, but the optimal expansion protocol has yet to be defined. Here we discuss ex vivo expansion using mesenchymal stromal cells.

4.1 Introduction

Hematopoietic stem cell transplantation (SCT) is curative in a wide variety of hematological and nonhematological malignancies. However, only 30% of patients who are candidates for SCT have suitable donors. Umbilical cord blood (UCB) stem cells provide an alternate source of stem cells. The first successful cord blood transplant was reported in 1989 by Gluckman et al. and since then more than 20,000 cord blood transplantations have been performed worldwide in both pediatric and adult patients (Gluckman et al. 1989; Rubinstein et al. 1998; Laughlin et al. 2001; Gluckman et al. 2007; Kurtzberg et al. 1996; Herr et al. 2010).

C. Hosing, M.D. (✉) • M. de Lima • E.J. Shpall
Department of Stem Cell Transplantation and Cellular Therapy, M.D. Anderson Cancer Center, 1515 Holcombe Boulevard, Unit 423, Houston, TX 77030, USA
e-mail: cmhosing@mdanderson.org

K. Turksen (ed.), *Adult and Embryonic Stem Cells*, Stem Cell Biology and Regenerative Medicine, DOI 10.1007/978-1-61779-630-2_4,

One of the major advantages of using UCB for stem cell transplantation is its ease of collection and immediate availability. Transplants using UCB also result in significantly lower rates of acute and chronic graft-versus-host disease (GVHD) (Rubinstein et al. 1998; Gluckman et al. 1997), despite more human leukocyte antigen (HLA) disparity between UCB units and the recipient. This is attributed to the lower number and relatively naive repertoire of the cord blood T cells (Nitsche et al. 2007; Garderet et al. 1998). However, it seems that the graft-versus-leukemia (GVL) effect is preserved, probably due to the unique natural killer (NK) subsets (Beziat et al. 2010). UCB provides a significantly higher chance of finding a donor for minority populations, who are currently underrepresented in donor registries (Barker et al. 2010).

One of the major limitations of using UCB as a source of stem cells, especially in adult patients, is the low cell dose available for transplantation. This may lead to graft failure, delayed engraftment, and/or profound delay in immune reconstitution and therefore significant morbidity and mortality. The total nucleated cell (TNC) dose transplanted per kilogram of body weight of the recipient has been found to correlate with outcomes. Patients weighing>45 kg who receive a single UCB unit tend to have markedly prolonged time to neutrophil and platelet engraftment and higher rates of graft failure. (Rubinstein et al. 1998; Laughlin et al. 2001; Migliaccio et al. 2000; Gluckman et al. 2004; Cohen et al. 2010). As a consequence, UCB transplantation is used more often in children (Gluckman et al. 1997, 2004) Even in children who receive adequate cell doses, the engraftment is delayed when compared to that with bone marrow or peripheral blood grafts (Kurtzberg et al. 2008; Sawczyn et al. 2005; Martin et al. 2006), and there is delay in immune reconstitution (Szabolcs and Niedzwiecki 2007; Thomson et al. 2000; Komanduri et al. 2007), suggesting that even in the optimal patient population the low stem cell dose infused with UCB transplantation has negative effects on outcomes. Another disadvantage of using UCB as the source of stem cells is the lack of cells available for posttransplant therapy such as donor lymphocyte infusions.

Some of the approaches used to improve outcomes and extend the applicability of UCB transplantation to adult recipients include transplantation of two cord blood units (Weinreb et al. 1998; Barker et al. 2001, 2003; De Lima et al. 2002; Fernandez et al. 2001); ex vivo cord blood expansion to increase the number of stem cells transplanted (Shpall et al. 2002; McNiece et al. 2000); and strategies to improve the homing capabilities of cord blood stem cells (Kollet et al. 2007; Ceradini et al. 2004; Taupin 2010).

4.2 Ex Vivo Cord Blood Expansion

4.2.1 Ex Vivo Expansion Using Liquid Culture Media

Several strategies for ex vivo expansion of UCB are currently under investigation. In static liquid expansion systems, UCB cells are cultured with combinations of cytokines, growth factors, and other growth-promoting compounds in various flasks, bags, or containers. Shpall et al. expanded umbilical cord blood hematopoietic stem cells (HSCs) ex vivo using liquid culture media and then used them for transplantation. They isolated CD34+ HSCs from one of the two fractions of the UCB unit and co-cultured them with a cytokine cocktail [stem cell factor (SCF), thrombopoietin (TPO), and granulocyte–colony-stimulating factor (G-CSF)]. A total of 37 patients were given transplants with one expanded and one unmanipulated UCB fraction. The resulting expansion increased the median total nucleated cell (TNC) dose 56-fold (range 1.03–278.0) and the median total number of CD34+ cells fourfold (range 0.1–20.0). However, there was no significant difference in the times to neutrophil and platelet engraftment between the two groups (Shpall et al. 2002).

McNiece et al. (2000) developed a two-step, 14-day expansion protocol that resulted in a TNC increase of 400-fold and CD34+ increase of 20-fold. The efficacy of this technique was tested in a prospective clinical trial at the M. D. Anderson Cancer Center. A total of 71 patients with hematological malignancies were randomized for transplantation with two unmanipulated UCB units or one unmanipulated and one expanded UCB unit using the above two-step strategy (De Lima et al. 2008). The median TNC expansion was 23-fold (0.44- to 275-fold) and the median CD34+ expansion was 2.3-fold (0 to 957-fold). Patients undergoing a reduced-intensity regimen who received an expanded UCB unit engrafted neutrophils in a median of 7 days (range 4–15 days) versus 14 days (range 5–32 days) in those receiving two unmanipulated units ($P=0.05$). At a median follow-up of 11.3 months (range 2–49 months), 34 (48%) of the patients were still alive. Most of the patients on the manipulated arm had some evidence of expanded UCB chimerism after the transplant (7–82%), but by 14 months all patients had predominance of the unmanipulated cord. This suggests that expansion may affect the durability of the engraftment by ex vivo expanded cells.

Modifications to this liquid ex vivo expansion technique have included attempts to further optimize ex vivo culture conditions (Lazzari et al. 2001a, b; Filip et al. 2000; Vavrova et al. 1999; Mohamed et al. 2006; Piacibello et al. 1998; Yao et al. 2004) and the development of serum-free culture systems (McNiece et al. 2000; Lazzari et al. 2001a; Yao et al. 2006). Other agents that may stimulate the proliferation of HSC are also under investigation, including tetraethylenepentamine (TEPA), a copper chelator thought to modulate the proliferation and differentiation of primitive hematopoietic progenitors (Peled et al. 2002, 2004a, b), histone deacetylase inhibitors (e.g., valproic acid), which transiently block the differentiation of UCB stem cells and thus enhance their proliferation (Young et al. 2004; Bug et al. 2005), and glycogen synthase kinase (GSK)-3 inhibitors that are reported to maintain pluripotency of stem cells (Sato et al. 2004). Delaney et al. utilized an immobilized, engineered form of the Notch ligand Delta1 with recombinant cytokines [SCF, FLT-3 ligand interleukin-6 (IL-6), TPO, IL-3] to stimulate ex vivo UCB expansion (Delaney et al. 2010). Ex vivo expansion could also be enhanced by manipulating newly discovered signaling pathways such as Wnt, bone morphogenetic protein 4, and Tie2/angiopoietin-1 and intracellular mediators such as phosphatase and tensin homolog (Hofmeister et al. 2007).

4.3 Ex Vivo Expansion of Umbilical Cord Blood Stem Cells on Mesenchymal Stromal Cells

The hematopoietic microenvironment is composed of both hematopoietic and non-hematopoietic components (Schofield 1983; Lemischka and Moore 2003; Fuchs et al. 2004). The stem cell "niche" provides complex molecular cues that direct hematopoiesis and are, in part, responsible for regulating the differentiation and maturation of HSCs (Allen and Dexter 1984; Allen et al. 1984; Chang et al. 1989; Dexter et al. 1973, 1990, 1977; Moore et al. 1979; Roberts et al. 1987; Yamazaki et al. 1989; Gartner and Kaplan 1980; Hackney et al. 2002; Etheridge et al. 2004; Kadereit et al. 2002; Rattis et al. 2004; Zhang et al. 2003; Majumdar et al. 1998, 2000). Mesenchymal stromal cells (MSCs) are undifferentiated, multipotential cells that give rise to mesodermal tissue types, including bone, cartilage, tendon, muscle, and fat (Deans and Moseley 2000; Toma et al. 2002). In addition, these MSCs regenerate the mesenchymal marrow stroma itself (Banfi et al. 2001; Koc and Lazarus 2001). These stromal cells constitute the connective tissue matrix in bone marrow that interacts with endothelial cells and hematopoietic cells to provide the microenvironment for promoting essential HSC functions including homing to marrow, proliferation, and differentiation (Deans and Moseley 2000). MSCs have been found to secrete cytokines that influence hematopoiesis, including SCF, FLT-3 ligand, IL-6, IL-11, leukemia inhibitory factor (LIF), and thrombopoietin (Majumdar et al. 2000; Haynesworth et al. 1996). Secretion of these cytokines undoubtedly contributes to the UCB expansion seen in the MSC-CB co-cultures described below. In culture, MSCs are characterized by a spindle-shaped, plastic, adherent morphology. They are phenotypically characterized as HLA-I (ABC), CD105, CD73, CD90, and CD166 positive and HLA-DR (II), CD80, CD31, CD34, and CD45 negative. When cells are expanded ex vivo in liquid culture media, they lose the support and regulations provided by the microenvironment and receive only the specific cytokines and growth factors provided in the culture medium. This could potentially drive differentiation at the expense of self-renewal. Third-party allogeneic MSCs have been shown in NOD-SCID mice to promote engraftment of UCB CD34+ when co-administered (in't Anker et al. 2003; Noort et al. 2002) and to possess immunomodulatory activity (Ahrens et al. 2004; Le Blanc et al. 2003a, 2004; Gotherstrom et al. 2004; Le Blanc 2003; Rasmusson et al. 2003; Le Blanc et al. 2003b; Gotherstrom et al. 2003). UCB in itself is a poor source of MSCs (Wexler et al. 2003), but a recent study suggested that MSCs from the Wharton's jelly of umbilical cords demonstrated surface receptors similar to those of other MSCs; thus, they may be able to support UCB expansion (Bakhshi et al. 2008). Preliminary results indicate that co-culture of UCB with MSCs can restore some of the interaction that occurs between the microenvironment of the marrow stroma and the HSCs (Hackney et al. 2002; Etheridge et al. 2004; Kadereit et al. 2002; Rattis et al. 2004; Zhang et al. 2003). Foci of hematopoiesis and cobblestone areas are visible during co-culture (McNiece et al. 2004), demonstrating that direct HSC–MSC interactions are occurring and that the MSCs are not simply acting as a feeder layer.

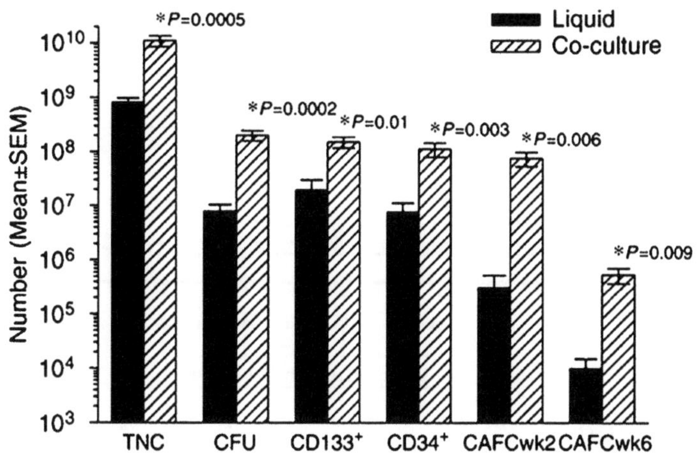

Fig. 4.1 Comparison of total nucleated cell (TNC) and Hematopoietic Progenitor Cell (HPC) output after a 14-day ex vivo liquid culture of CD133+-selected cord blood (CB) cells (*solid bar*) and 14-day ex vivo cord blood mesenchymal stem cell (CB-MSC) co-culture (*shaded bar*). Colony forming unit (CFU), CD133+, CD34+, and Cobblestone area-forming cells (CAFCs) were assayed at weeks 2 and 6 of culture. Data are shown as the mean ± SEM (ex vivo liquid culture, $n=11$; ex vivo CB-MSC co-culture, $n=6$). Ex vivo CB-MSC co-culture generated TNC numbers that were > 13-fold ($P=0.0005$), CFU numbers that were > 25-fold ($P=0.0002$), CD133+ cell numbers that were > sevenfold ($P=0.01$), CD34+ cell numbers that were > 14-fold ($P=0.003$), $CAFC_{wk2}$ numbers that were > 200-fold ($P=0.006$), and $CAFC_{wk6}$ numbers that were > 44-fold ($P=0.009$) those obtained following ex vivo liquid culture. These data demonstrate the superior TNC and HPC expansion obtained following ex vivo CB-MSC co-culture compared with ex vivo liquid culture of CD133+-selected CB cells (Reprinted by permission from Macmillan Publishers Ltd: Bone Marrow Transplantation, Robinson SN, et. al. 37,359–366 ©2006)

For stromal co-culture, mononuclear cells (MNCs) are isolated by density separation and co-cultured with established MSC monolayers in medium containing fetal bovine serum (FBS) and a growth factor cocktail (e.g., SCF, TPO, and G-CSF) (McNiece et al. 2004) The nonadherent cells are removed from the co-culture after 7 days and subjected to secondary expansion on an additional MSC monolayer. The original adherent layer, composed of MSCs and HSCs, is re-fed with fresh medium containing growth factors. Culture is then continued for an additional 7 days (total 14 days). A tenfold increase in total nucleated cells, a 17-fold increase in committed progenitor cells (granulocyte-macrophage colony-forming cells, or GM-CFCs), a 3.5-fold increase in primitive progenitor cells (highly proliferative potential colony-forming cells, HPP-CFC), and a 16- to 37-fold increase in CD34+ cells has been reported using co-culture expansion (McNiece et al. 2004). It may prove clinically beneficial to reinfuse both nonadherent and adherent cells from the expansion process as the MSCs not only can aid engraftment but can provide immunomdulatory effects. When compared to liquid culture, UCB-MSC co-culture requires less cell manipulation, resulting in less initial HSC loss and a markedly improved TNC dose and HSC output (Fig. 4.1) (Robinson et al. 2006).

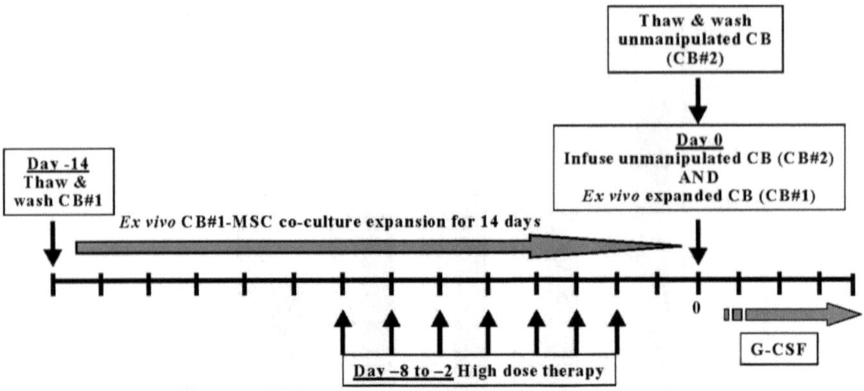

Fig. 4.2 Treatment schema

Based on the above preclinical results, a clinical trial is underway at M. D. Anderson Cancer Center using a UCB unit expanded on MSCs combined with an unmanipulated UCB unit. Initially, bone marrow from a family member (minimum 2/6 HLA match) was used as the source of MSCs. Approximately 100 ml of bone marrow is aspirated, and confluent MSCs are generated over approximately 21 days. The UCB unit with the lower TNC dose is thawed, washed, and divided into ten equal fractions. Each fraction is placed in a flask containing > 70% confluent MSCs and cultured in ex vivo expansion medium. Following incubation for 7 days at 37°C, the nonadherent cells are collected from each flask. The content of a single flask is then placed in a 1-l Teflon-coated culture bag and cultured for an additional 7 days (14 days total). The flasks are then re-fed and incubated as well. Patients receive chemotherapy as shown in Fig. 4.2. On day 0, unmanipulated UCB unit is infused, followed by the expanded UCB cells (from both the bags and the co-culture flasks). A median expansion of 12-fold was seen in both the TNC dose and the CD34+ subsets (de Lima et al. 2009). This resulted in a mean expanded TNC dose of 5.7×10^7/kg and CD34+ cell dose of 3.8×10^5/kg which represents a significant increase when compared to the dose achieved in our previous expansion studies. Thus, the patients received a combined TNC dose of 9.5×10^7/kg and a CD 34+ cell dose of 8.2×10^5 CD34+/kg (manipulated plus unmanipulated unit). Recipients of myeloablative therapy, engrafted neutrophils in a median of 14.5 days (range, 12–23) and platelets in 30 days (range, 25–51). Although promising, the generation of MSCs from a patient's family member is time consuming. It takes an average of 3 weeks to generate the MSC and another 2 weeks to perform CB MNC/MSC co-culture expansion thus delaying the transplant for 5 weeks after enrollment in the study. Moreover, an appropriate family member is not always available to donate marrow. This protocol was later modified to utilize third-party ("off the shelf") mesenchymal progenitor cells (MPCs) from Angioblast Systems, Inc., (New York, NY, USA) ($n = 24$). For ex vivo expansion, one vial of Angioblast™ MPCs was thawed and expanded to confluence in ten flasks within 4 days. The UCB unit with the lowest TNC dose was then thawed and divided into ten fractions. Each fraction was placed in a flask containing the confluent layers of MSCs in expansion medium with SCF, FLT3, G-CSF, and TPO. After 7 days at 37°C, the nonadherent cells were removed from each flask, placed into each of ten 1-l Teflon-coated culture bags (American Fluoroseal, Gaithersburg, MD, USA), and cultured for an additional 7 days (14 days total). Also, 50 ml of medium/growth factors was added to the flasks to culture the remaining adherent layer during that time period. On day 14, the cells from the bags and the flasks were combined, washed, and infused along with a second unmanipulated UCB unit.

Using off the shelf MPCs from Angioblast Inc., a median 14-fold (range, 1–30) for TNC and 40-fold (range, 4–140) for CD34+ was obtained which compares favorably to that seen with MSCs derived from haploidentical family members. A total of 32 patients with refractory hematological malignancies have now been treated on this study. Preliminary results are encouraging (De Lima et al. 2010).

Optimization of stromal co-culture methods continue. It includes defining the ideal growth factor cocktail, the length of MSC and HSC cell co-culture for most effective expansion, and the development of more effective stromal cell lines to support the HSC expansion (De Angeli et al. 2004).

In most studies of transplantation using both the expanded and unmanipulated UCB units, the long-term chimerism analysis shows engraftment with the unmanipulated UCB unit. This has raised concerns that during ex vivo expansion one may be selectively expanding a subset of stem cells that are capable of short-term reconstitution at the expense of long-term reconstituting stem cells. This could result in initial hematopoietic recovery followed by later graft failure (Holyoake et al. 1997). However, it could also be utilized to provide faster short-term engraftment, which can reduce the early deaths associated with UCB transplantation. The unmanipulated UCB would provide sustained long-term hematopoiesis (Pecora et al. 2000).

Some investigators have suggested that ex vivo expansion may be associated with cell cycle abnormalities (Glimm et al. 2000), acquired homing defects (Ramirez et al. 2001; Zhai et al. 2004), and induction of apoptosis (Liu et al. 2003). In contrast, Piacibello et al. observed evidence of self-renewal and amplification of HSCs during ex vivo expansion (Piacibello et al. 1999). Lewis et al. reported that UCB cells are capable of engraftment in primary, secondary, and tertiary xenogeneic recipients and were preserved following ex vivo expansion (Lewis et al. 2001). Guenechea et al. reported a delay in engraftment in a mouse model, suggesting that more primitive, less rapidly engrafting cells may be preserved during ex vivo expansion (Guenechea et al. 1999). Zhai et al. showed that short term ex vivo expansion did not affect the homing of HSCs (Zhai et al. 2004).

4.4 Co-transplantation of Umbilical Cord Blood Stem Cells and Mesenchymal Stromal Cells

Another approach to facilitating engraftment is co-transplantation of HSCs and MSCs. MSCs can be transplanted across the HLA barrier because of their low immunogenicity. Allogeneic MSC co-administration has been shown to promote engraftment of human CD34+ cells in NOD/SCID mice (in't Anker et al. 2003) and fetal sheep (Almeida-Porada et al. 2000). Co-infusion of ex vivo expanded "third party" (nondonor, nonrecipient) bone marrow-derived MSCs and HSCs from peripheral blood and bone marrow has been tested in pilot studies (Ning et al. 2008; Ball et al. 2007). In a randomized clinical trial, patients received HSCs from HLA-identical siblings with or without MSCs. The median number of MSCs infused was 3.4×10^5/kg (range 0.3–15.0×10^5/kg). The MSC infusions were well tolerated. The median time to neutrophil engraftment was 16 days for the MSC group and 15 days for the non-MSC group. The median times to platelet engraftment were 30 and 27 days, respectively. Grade II–IV acute GVHD was observed, respectively, in 11.1% and 53.3% of evaluable patients. Chronic GVHD was found in 14.3% and 28.6% of evaluable patients, respectively. Unfortunately, the relapse rate was higher in the MSC group (60% vs. 20%). The authors concluded that co-transplantation of MSCs and HSCs is feasible and may prevent GVHD, but the higher relapse rate was a concern (Ning et al. 2008). MacMillan et al. co-administered haplo-identical parental MSCs and a single UCB unit to eight children. Three patients received the second dose of MSCs on day 21. All eight patients achieved neutrophil engraftment at a median of 19 days. At a median follow-up of 6.8 years, five patients were alive and disease free (Macmillan et al. 2009). In another study, the "third-party" expanded MSCs and cord blood HSCs were co-administered to nine patients. The outcomes were compared to those of 46 patients who did not receive MSCs. No significant differences in engraftment or incidence of acute GVHD were observed. No adverse events were reported with MSC infusion. Two patients in the MSC group developed steroid-refractory grade II acute GVHD, but therapeutic infusion of MSCs resulted in complete resolution of the GVHD (Gonzalo-Daganzo et al. 2009).

4.5 Conclusions

The ultimate goals of ex vivo expansion of cord blood are to generate sufficient numbers of HSCs to optimize the graft available for transplant and to generate higher numbers of lineage-committed progenitor cells that, although transient, would allow rapid recovery from pancyotpenia, thereby decreasing early morbidity and mortality. At the present time, combining expanded and unmanipulated products may provide the best product for transplantation. Use of MSCs for ex vivo expansion of UCB for SCT is relatively safe, and recent results suggest that outcomes may be better than using unmanipulated units alone. MSCs may also be co-transplanted with UCB-HSC to facilitate engraftment. Ex vivo expansion of UCB may also provide additional cells that may allow adoptive immunotherapy or gene transfer therapy in the setting of UCB transplantation. However, many questions remain, and ongoing clinical trials should answer some of them.

References

Ahrens N, Tormin A, Paulus M, Roosterman D, Salama A, Krenn V, Neumann U, Scheding S (2004) Mesenchymal stem cell content of human vertebral bone marrow. Transplantation 78:925–929

Allen TD, Dexter TM (1984) The essential cells of the hemopoietic microenvironment. Exp Hematol 12:517–521

Allen TD, Simons PJ, Dexter TM (1984) Haemopoietic microenvironments in vitro: which cells are involved? Blood Cells 10:467–471

Almeida-Porada G, Porada CD, Tran N, Zanjani ED (2000) Cotransplantation of human stromal cell progenitors into preimmune fetal sheep results in early appearance of human donor cells in circulation and boosts cell levels in bone marrow at later time points after transplantation. Blood 95:3620–3627

Bakhshi T, Zabriskie RC, Bodie S, Kidd S, Ramin S, Paganessi LA, Gregory SA, Fung HC, Christopherson KW II (2008) Mesenchymal stem cells from the Wharton's jelly of umbilical cord segments provide stromal support for the maintenance of cord blood hematopoietic stem cells during long-term ex vivo culture. Transfusion 48:2638–2644

Ball LM, Bernardo ME, Roelofs H, Lankester A, Cometa A, Egeler RM, Locatelli F, Fibbe WE (2007) Cotransplantation of ex vivo expanded mesenchymal stem cells accelerates lymphocyte recovery and may reduce the risk of graft failure in haploidentical hematopoietic stem-cell transplantation. Blood 110:2764–2767

Banfi A, Bianchi G, Galotto M, Cancedda R, Quarto R (2001) Bone marrow stromal damage after chemo/radiotherapy: occurrence, consequences and possibilities of treatment. Leuk Lymphoma 42:863–870

Barker JN, Weisdorf DJ, Wagner JE (2001) Creation of a double chimera after the transplantation of umbilical-cord blood from two partially matched unrelated donors. N Engl J Med 344:1870–1871

Barker JN, Weisdorf DJ, DeFor TE, Blazar BR, Miller JS, Wagner JE (2003) Rapid and complete donor chimerism in adult recipients of unrelated donor umbilical cord blood transplantation after reduced-intensity conditioning. Blood 102:1915–1919

Barker JN, Byam CE, Kernan NA, Lee SS, Hawke RM, Doshi KA, Wells DS, Heller G, Papadopoulos EB, Scaradavou A, Young JW, van den Brink MR (2010) Availability of cord blood extends allogeneic hematopoietic stem cell transplant access to racial and ethnic minorities. Biol Blood Marrow Transplant 16:1541–1548

Beziat V, Nguyen S, Exley M, Achour A, Simon T, Chevallier P, Sirvent A, Vigouroux S, Debre P, Rio B, Vieillard V (2010) Shaping of iNKT cell repertoire after unrelated cord blood transplantation. Clin Immunol 135:364–373

Bug G, Gul H, Schwarz K, Pfeifer H, Kampfmann M, Zheng X, Beissert T, Boehrer S, Hoelzer D, Ottmann OG, Ruthardt M (2005) Valproic acid stimulates proliferation and self-renewal of hematopoietic stem cells. Cancer Res 65:2537–2541

Ceradini DJ, Kulkarni AR, Callaghan MJ, Tepper OM, Bastidas N, Kleinman ME, Capla JM, Galiano RD, Levine JP, Gurtner GC (2004) Progenitor cell trafficking is regulated by hypoxic gradients through HIF-1 induction of SDF-1. Nat Med 10:858–864

Chang J, Allen TD, Dexter TM (1989) Long-term bone marrow cultures: their use in autologous marrow transplantation. Cancer Cells 1:17–24

Cohen YC, Scaradavou A, Stevens CE, Rubinstein P, Gluckman E, Rocha V, Horowitz MM, Eapen M, Nagler A, Shpall EJ, Laughlin MJ, Daniely Y, Pacheco D, Barishev R, Olmer L, Freedman LS (2010) Factors affecting mortality following myeloablative cord blood transplantation in adults: a pooled analysis of three international registries. Bone Marrow Transplant 46:70–76

De Angeli S, Di Liddo R, Buoro S, Toniolo L, Conconi MT, Belloni AS, Parnigotto PP, Nussdorfer GG (2004) New immortalized human stromal cell lines enhancing in vitro expansion of cord blood hematopoietic stem cells. Int J Mol Med 13:363–371

De Lima M, St John LS, Wieder ED, Lee MS, McMannis J, Karandish S, Giralt S, Beran M, Couriel D, Korbling M, Bibawi S, Champlin R, Komanduri KV (2002) Double-chimaerism after transplantation of two human leucocyte antigen mismatched, unrelated cord blood units. Br J Haematol 119:773–776

De Lima M, McMannis JD, Saliba R, Worth L, Kebriaei P, Popat U, Qazilbash M, Jones R, Giralt S, Silva LDS, Cooper L, Petropoulos D, Lee D, Kelly S, Thall P, Robinson s, Khouri I, Hosing C, Korbling M, Alousi A, Rondon G, Andersson BS, Nieto Y, Ciurea S, Komanduri K, Champlin RE, and Shpall EJ (2008) Double Cord Blood Transplantation (CBT) with and without Ex-Vivo Expansion (EXP): A Randomized, Controlled Study. Blood (ASH Annual Meeting Abstracts), 112:154

De Lima M, McMannis J, Hosing C, Kebraei P, Komanduri K, Worth L, Staba S, Cooper L, Petropolous D, Lee D, Jones R, Nieto Y, Andersson B, Korbling M, Alousi A, Qazilbash M, Popat U, Khouri I, Bollard C, Leen A, Rondon G, Molldrem J, Champlin R, Simmons P, Shpall EJ (2009) Double cord blood transplantation (CBT) with ex-vivo expansion (EXP) of one unit utilizing a mesenchymal stromal cell (MSC) platform. Biol Blood Marrow Transplant 15:47–48

De Lima M, Robinson S, McMannis J, Alousi A, Saliba RM, Munsell M, Kebriaei P, Hosing C, Parmar S, Cooper L, Shah N, Kelly SS, Rondon G, Fernandez-Vina M, Maewall I, Bosque D, Bollard CM, Chen JJ, McNiece I, Komanduri KV, Nieto Y, Jones R, Andersson B, Popat U, Champlin RE, Simmons PJ, Shpall EJ (2010) Mesenchymal Stem Cell (MSC) Based Cord Blood (CB) Expansion (Exp) Leads to Rapid Engraftment of Platelets and Neutrophils. ASH Annual Meeting Abstracts 116:362

Deans RJ, Moseley AB (2000) Mesenchymal stem cells: biology and potential clinical uses. Exp Hematol 28:875–884

Delaney C, Heimfeld S, Brashem-Stein C, Voorhies H, Manger RL, Bernstein ID (2010) Notch-mediated expansion of human cord blood progenitor cells capable of rapid myeloid reconstitution. Nat Med 16:232–236

Dexter TM, Allen TD, Lajtha LG, Schofield R, Lord BI (1973) Stimulation of differentiation and proliferation of haemopoietic cells in vitro. J Cell Physiol 82:461–473

Dexter TM, Allen TD, Lajtha LG (1977) Conditions controlling the proliferation of haemopoietic stem cells in vitro. J Cell Physiol 91:335–344

Dexter TM, Coutinho LH, Spooncer E, Heyworth CM, Daniel CP, Schiro R, Chang J, Allen TD (1990) Stromal cells in haemopoiesis. Ciba Found Symp 148:76–86; discussion 86–95

Etheridge SL, Spencer GJ, Heath DJ, Genever PG (2004) Expression profiling and functional analysis of wnt signaling mechanisms in mesenchymal stem cells. Stem Cells 22:849–860

Fernandez MN, Regidor C, Cabrera R, Garcia-Marco J, Briz M, Fores R, Sanjuan I, McWhinnie A, Querol S, Garcia J, Madrigal A (2001) Cord blood transplants: early recovery of neutrophils from co-transplanted sibling haploidentical progenitor cells and lack of engraftment of cultured cord blood cells, as ascertained by analysis of DNA polymorphisms. Bone Marrow Transplant 28:355–363

Filip S, Vavrova J, Vokurkova D, Blaha M, Vanasek J (2000) Myeloid differentiation and maturation of SCF+IL-3+IL-11 expanded AC133+/CD34+ cells selected from high-risk breast cancer patients. Neoplasma 47:73–80

Fuchs E, Tumbar T, Guasch G (2004) Socializing with the neighbors: stem cells and their niche. Cell 116:769–778

Garderet L, Dulphy N, Douay C, Chalumeau N, Schaeffer V, Zilber MT, Lim A, Even J, Mooney N, Gelin C, Gluckman E, Charron D, Toubert A (1998) The umbilical cord blood alphabeta

T-cell repertoire: characteristics of a polyclonal and naive but completely formed repertoire. Blood 91:340–346

Gartner S, Kaplan HS (1980) Long-term culture of human bone marrow cells. Proc Natl Acad Sci USA 77:4756–4759

Glimm H, Oh IH, Eaves CJ (2000) Human hematopoietic stem cells stimulated to proliferate in vitro lose engraftment potential during their S/G(2)/M transit and do not reenter G(0). Blood 96:4185–4193

Gluckman E, Broxmeyer HA, Auerbach AD, Friedman HS, Douglas GW, Devergie A, Esperou H, Thierry D, Socie G, Lehn P et al (1989) Hematopoietic reconstitution in a patient with Fanconi's anemia by means of umbilical-cord blood from an HLA-identical sibling. N Engl J Med 321:1174–1178

Gluckman E, Rocha V, Boyer-Chammard A, Locatelli F, Arcese W, Pasquini R, Ortega J, Souillet G, Ferreira E, Laporte JP, Fernandez M, Chastang C (1997) Outcome of cord-blood transplantation from related and unrelated donors: Eurocord Transplant Group and the European Blood and Marrow Transplantation Group. N Engl J Med 337:373–381

Gluckman E, Rocha V, Arcese W, Michel G, Sanz G, Chan KW, Takahashi TA, Ortega J, Filipovich A, Locatelli F, Asano S, Fagioli F, Vowels M, Sirvent A, Laporte JP, Tiedemann K, Amadori S, Abecassis M, Bordigoni P, Diez B, Shaw PJ, Vora A, Caniglia M, Garnier F, Ionescu I, Garcia J, Koegler G, Rebulla P, Chevret S (2004) Factors associated with outcomes of unrelated cord blood transplant: guidelines for donor choice. Exp Hematol 32:397–407

Gluckman E, Rocha V, Ionescu I, Bierings M, Harris RE, Wagner J, Kurtzberg J, Champagne MA, Bonfim C, Bittencourt M, Darbyshire P, Fernandez MN, Locatelli F, Pasquini R (2007) Results of unrelated cord blood transplant in Fanconi anemia patients: risk factor analysis for engraftment and survival. Biol Blood Marrow Transplant 13:1073–1082

Gonzalo-Daganzo R, Regidor C, Martin-Donaire T, Rico MA, Bautista G, Krsnik I, Fores R, Ojeda E, Sanjuan I, Garcia-Marco JA, Navarro B, Gil S, Sanchez R, Panadero N, Gutierrez Y, Garcia-Berciano M, Perez N, Millan I, Cabrera R, Fernandez MN (2009) Results of a pilot study on the use of third-party donor mesenchymal stromal cells in cord blood transplantation in adults. Cytotherapy 11:278–288

Gotherstrom C, Ringden O, Westgren M, Tammik C, Le Blanc K (2003) Immunomodulatory effects of human foetal liver-derived mesenchymal stem cells. Bone Marrow Transplant 32:265–272

Gotherstrom C, Ringden O, Tammik C, Zetterberg E, Westgren M, Le Blanc K (2004) Immunologic properties of human fetal mesenchymal stem cells. Am J Obstet Gynecol 190:239–245

Guenechea G, Segovia JC, Albella B, Lamana M, Ramirez M, Regidor C, Fernandez MN, Bueren JA (1999) Delayed engraftment of nonobese diabetic/severe combined immunodeficient mice transplanted with ex vivo-expanded human CD34(+) cord blood cells. Blood 93:1097–1105

Hackney JA, Charbord P, Brunk BP, Stoeckert CJ, Lemischka IR, Moore KA (2002) A molecular profile of a hematopoietic stem cell niche. Proc Natl Acad Sci USA 99:13061–13066

Haynesworth SE, Baber MA, Caplan AI (1996) Cytokine expression by human marrow-derived mesenchymal progenitor cells in vitro: effects of dexamethasone and IL-1 alpha. J Cell Physiol 166:585–592

Herr AL, Kabbara N, Bonfim CM, Teira P, Locatelli F, Tiedemann K, Lankester A, Jouet JP, Messina C, Bertrand Y, Diaz de Heredia C, Peters C, Chaves W, Nabhan SK, Ionescu I, Gluckman E, Rocha V (2010) Long-term follow-up and factors influencing outcomes after related HLA-identical cord blood transplantation for patients with malignancies: an analysis on behalf of Eurocord-EBMT. Blood 116:1849–1856

Hofmeister CC, Zhang J, Knight KL, Le P, Stiff PJ (2007) Ex vivo expansion of umbilical cord blood stem cells for transplantation: growing knowledge from the hematopoietic niche. Bone Marrow Transplant 39:11–23

Holyoake TL, Alcorn MJ, Richmond L, Farrell E, Pearson C, Green R, Dunlop DJ, Fitzsimons E, Pragnell IB, Franklin IM (1997) CD34 positive PBPC expanded ex vivo may not provide durable engraftment following myeloablative chemoradiotherapy regimens. Bone Marrow Transplant 19:1095–1101 in't Anker PS, Noort WA, Kruisselbrink AB, Scherjon SA, Beekhuizen W, Willemze R, Kanhai HH, Fibbe WE (2003) Nonexpanded primary lung and bone marrow-derived mesenchymal cells promote the engraftment of umbilical cord blood-derived CD34(+) cells in NOD/SCID mice. Exp Hematol 31:881–889

Kadereit S, Deeds LS, Haynesworth SE, Koc ON, Kozik MM, Szekely E, Daum-Woods K, Goetchius GW, Fu P, Welniak LA, Murphy WJ, Laughlin MJ (2002) Expansion of LTC-ICs and maintenance of p21 and BCL-2 expression in cord blood CD34(+)/CD38(−) early progenitors cultured over human MSCs as a feeder layer. Stem Cells 20:573–582

Koc ON, Lazarus HM (2001) Mesenchymal stem cells: heading into the clinic. Bone Marrow Transplant 27:235–239

Kollet O, Dar A, Lapidot T (2007) The multiple roles of osteoclasts in host defense: bone remodeling and hematopoietic stem cell mobilization. Annu Rev Immunol 25:51–69

Komanduri KV, St John LS, de Lima M, McMannis J, Rosinski S, McNiece I, Bryan SG, Kaur I, Martin S, Wieder ED, Worth L, Cooper LJ, Petropoulos D, Molldrem JJ, Champlin RE, Shpall EJ (2007) Delayed immune reconstitution after cord blood transplantation is characterized by impaired thymopoiesis and late memory T-cell skewing. Blood 110:4543–4551

Kurtzberg J, Laughlin M, Graham ML, Smith C, Olson JF, Halperin EC, Ciocci G, Carrier C, Stevens CE, Rubinstein P (1996) Placental blood as a source of hematopoietic stem cells for transplantation into unrelated recipients. N Engl J Med 335:157–166

Kurtzberg J, Prasad VK, Carter SL, Wagner JE, Baxter-Lowe LA, Wall D, Kapoor N, Guinan EC, Feig SA, Wagner EL, Kernan NA (2008) Results of the Cord Blood Transplantation Study (COBLT): clinical outcomes of unrelated donor umbilical cord blood transplantation in pediatric patients with hematologic malignancies. Blood 112:4318–4327

Laughlin MJ, Barker J, Bambach B, Koc ON, Rizzieri DA, Wagner JE, Gerson SL, Lazarus HM, Cairo M, Stevens CE, Rubinstein P, Kurtzberg J (2001) Hematopoietic engraftment and survival in adult recipients of umbilical-cord blood from unrelated donors. N Engl J Med 344:1815–1822

Lazzari L, Lucchi S, Porretti L, Montemurro T, Giordano R, Lopa R, Rebulla P, Sirchia G (2001a) Comparison of different serum-free media for ex vivo expansion of HPCs from cord blood using thrombopoietin, Flt-3 ligand, IL-6, and IL-11. Transfusion 41:718–719

Lazzari L, Lucchi S, Rebulla P, Porretti L, Puglisi G, Lecchi L, Sirchia G (2001b) Long-term expansion and maintenance of cord blood haematopoietic stem cells using thrombopoietin, Flt3-ligand, interleukin (IL)-6 and IL-11 in a serum-free and stroma-free culture system. Br J Haematol 112:397–404

Le Blanc K (2003) Immunomodulatory effects of fetal and adult mesenchymal stem cells. Cytotherapy 5:485–489

Le Blanc K, Tammik L, Sundberg B, Haynesworth SE, Ringden O (2003a) Mesenchymal stem cells inhibit and stimulate mixed lymphocyte cultures and mitogenic responses independently of the major histocompatibility complex. Scand J Immunol 57:11–20

Le Blanc K, Tammik C, Rosendahl K, Zetterberg E, Ringden O (2003b) HLA expression and immunologic properties of differentiated and undifferentiated mesenchymal stem cells. Exp Hematol 31:890–896

Le Blanc K, Rasmusson I, Sundberg B, Gotherstrom C, Hassan M, Uzunel M, Ringden O (2004) Treatment of severe acute graft-versus-host disease with third party haploidentical mesenchymal stem cells. Lancet 363:1439–1441

Lemischka IR, Moore KA (2003) Stem cells: interactive niches. Nature 425:778–779

Lewis ID, Almeida-Porada G, Du J, Lemischka IR, Moore KA, Zanjani ED, Verfaillie CM (2001) Umbilical cord blood cells capable of engrafting in primary, secondary, and tertiary xenogeneic hosts are preserved after ex vivo culture in a noncontact system. Blood 97:3441–3449

Liu B, Buckley SM, Lewis ID, Goldman AI, Wagner JE, van der Loo JC (2003) Homing defect of cultured human hematopoietic cells in the NOD/SCID mouse is mediated by Fas/CD95. Exp Hematol 31:824–832

Macmillan ML, Blazar BR, DeFor TE, Wagner JE (2009) Transplantation of ex-vivo culture-expanded parental haploidentical mesenchymal stem cells to promote engraftment in pediatric recipients of unrelated donor umbilical cord blood: results of a Phase I-II clinical trial. Bone Marrow Transplant 43:447–454

Majumdar MK, Thiede MA, Mosca JD, Moorman M, Gerson SL (1998) Phenotypic and functional comparison of cultures of marrow-derived mesenchymal stem cells (MSCs) and stromal cells. J Cell Physiol 176:57–66

Majumdar MK, Thiede MA, Haynesworth SE, Bruder SP, Gerson SL (2000) Human marrow-derived mesenchymal stem cells (MSCs) express hematopoietic cytokines and support long-term hematopoiesis when differentiated toward stromal and osteogenic lineages. J Hematother Stem Cell Res 9:841–848

Martin PL, Carter SL, Kernan NA, Sahdev I, Wall D, Pietryga D, Wagner JE, Kurtzberg J (2006) Results of the Cord Blood Transplantation Study (COBLT): outcomes of unrelated donor umbilical cord blood transplantation in pediatric patients with lysosomal and peroxisomal storage diseases. Biol Blood Marrow Transplant 12:184–194

McNiece I, Kubegov D, Kerzic P, Shpall EJ, Gross S (2000) Increased expansion and differentiation of cord blood products using a two-step expansion culture. Exp Hematol 28: 1181–1186

McNiece I, Harrington J, Turney J, Kellner J, Shpall EJ (2004) Ex vivo expansion of cord blood mononuclear cells on mesenchymal stem cells. Cytotherapy 6:311–317

Migliaccio AR, Adamson JW, Stevens CE, Dobrila NL, Carrier CM, Rubinstein P (2000) Cell dose and speed of engraftment in placental/umbilical cord blood transplantation: graft progenitor cell content is a better predictor than nucleated cell quantity. Blood 96:2717–2722

Mohamed AA, Ibrahim AM, El-Masry MW, Mansour IM, Khroshied MA, Gouda HM, Riad RM (2006) Ex vivo expansion of stem cells: defining optimum conditions using various cytokines. Lab Hematol 12:86–93

Moore MA, Sheridan AP, Allen TD, Dexter TM (1979) Prolonged hematopoiesis in a primate bone marrow culture system: characteristics of stem cell production and the hematopoietic microenvironment. Blood 54:775–793

Ning H, Yang F, Jiang M, Hu L, Feng K, Zhang J, Yu Z, Li B, Xu C, Li Y, Wang J, Hu J, Lou X, Chen H (2008) The correlation between cotransplantation of mesenchymal stem cells and higher recurrence rate in hematologic malignancy patients: outcome of a pilot clinical study. Leukemia 22:593–599

Nitsche A, Zhang M, Clauss T, Siegert W, Brune K, Pahl A (2007) Cytokine profiles of cord and adult blood leukocytes: differences in expression are due to differences in expression and activation of transcription factors. BMC Immunol 8:18

Noort WA, Kruisselbrink AB, in't Anker PS, Kruger M, van Bezooijen RL, de Paus RA, Heemskerk MH, Lowik CW, Falkenburg JH, Willemze R, Fibbe WE (2002) Mesenchymal stem cells promote engraftment of human umbilical cord blood-derived CD34(+) cells in NOD/SCID mice. Exp Hematol 30:870–878

Pecora AL, Stiff P, Jennis A, Goldberg S, Rosenbluth R, Price P, Goltry KL, Douville J, Armstrong RD, Smith AK, Preti RA (2000) Prompt and durable engraftment in two older adult patients with high risk chronic myelogenous leukemia (CML) using ex vivo expanded and unmanipulated unrelated umbilical cord blood. Bone Marrow Transplant 25:797–799

Peled T, Landau E, Prus E, Treves AJ, Nagler A, Fibach E (2002) Cellular copper content modulates differentiation and self-renewal in cultures of cord blood-derived CD34+ cells. Br J Haematol 116:655–661

Peled T, Landau E, Mandel J, Glukhman E, Goudsmid NR, Nagler A, Fibach E (2004a) Linear polyamine copper chelator tetraethylenepentamine augments long-term ex vivo expansion of cord blood-derived CD34+ cells and increases their engraftment potential in NOD/SCID mice. Exp Hematol 32:547–555

Peled T, Mandel J, Goudsmid RN, Landor C, Hasson N, Harati D, Austin M, Hasson A, Fibach E, Shpall EJ, Nagler A (2004b) Pre-clinical development of cord blood-derived progenitor cell graft expanded ex vivo with cytokines and the polyamine copper chelator tetraethylenepentamine. Cytotherapy 6:344–355

Piacibello W, Sanavio F, Garetto L, Severino A, Dane A, Gammaitoni L, Aglietta M (1998) Differential growth factor requirement of primitive cord blood hematopoietic stem cell for self-renewal and amplification vs proliferation and differentiation. Leukemia 12:718–727

Piacibello W, Sanavio F, Severino A, Dane A, Gammaitoni L, Fagioli F, Perissinotto E, Cavalloni G, Kollet O, Lapidot T, Aglietta M (1999) Engraftment in nonobese diabetic severe combined immunodeficient mice of human CD34(+) cord blood cells after ex vivo expansion: evidence for the amplification and self-renewal of repopulating stem cells. Blood 93:3736–3749

Ramirez M, Segovia JC, Benet I, Arbona C, Guenechea G, Blaya C, Garcia-Conde J, Bueren JA, Prosper F (2001) Ex vivo expansion of umbilical cord blood (UCB) CD34(+) cells alters the expression and function of alpha 4 beta 1 and alpha 5 beta 1 integrins. Br J Haematol 115:213–221

Rasmusson I, Ringden O, Sundberg B, Le Blanc K (2003) Mesenchymal stem cells inhibit the formation of cytotoxic T lymphocytes, but not activated cytotoxic T lymphocytes or natural killer cells. Transplantation 76:1208–1213

Rattis FM, Voermans C, Reya T (2004) Wnt signaling in the stem cell niche. Curr Opin Hematol 11:88–94

Roberts RA, Spooncer E, Parkinson EK, Lord BI, Allen TD, Dexter TM (1987) Metabolically inactive 3T3 cells can substitute for marrow stromal cells to promote the proliferation and development of multipotent haemopoietic stem cells. J Cell Physiol 132:203–214

Robinson SN, Ng J, Niu T, Yang H, McMannis JD, Karandish S, Kaur I, Fu P, Del Angel M, Messinger R, Flagge F, de Lima M, Decker W, Xing D, Champlin R, Shpall EJ (2006) Superior ex vivo cord blood expansion following co-culture with bone marrow-derived mesenchymal stem cells. Bone Marrow Transplant 37:359–366

Rubinstein P, Carrier C, Scaradavou A, Kurtzberg J, Adamson J, Migliaccio AR, Berkowitz RL, Cabbad M, Dobrila NL, Taylor PE, Rosenfield RE, Stevens CE (1998) Outcomes among 562 recipients of placental-blood transplants from unrelated donors. N Engl J Med 339:1565–1577

Sato N, Meijer L, Skaltsounis L, Greengard P, Brivanlou AH (2004) Maintenance of pluripotency in human and mouse embryonic stem cells through activation of Wnt signaling by a pharmacological GSK-3-specific inhibitor. Nat Med 10:55–63

Sawczyn KK, Quinones R, Malcolm J, Foreman N, Garrington T, Gore L, Gao D, Giller R (2005) Cord blood transplant in childhood ALL. Pediatr Blood Cancer 45:964–970

Schofield R (1983) The stem cell system. Biomed Pharmacother 37:375–380

Shpall EJ, Quinones R, Giller R, Zeng C, Baron AE, Jones RB, Bearman SI, Nieto Y, Freed B, Madinger N, Hogan CJ, Slat-Vasquez V, Russell P, Blunk B, Schissel D, Hild E, Malcolm J, Ward W, McNiece IK (2002) Transplantation of ex vivo expanded cord blood. Biol Blood Marrow Transplant 8:368–376

Szabolcs P, Niedzwiecki D (2007) Immune reconstitution after unrelated cord blood transplantation. Cytotherapy 9:111–122

Taupin P (2010) Ex vivo fucosylation of stem cells to improve engraftment: WO2004094619. Expert Opin Ther Pat 20:1265–1269

Thomson BG, Robertson KA, Gowan D, Heilman D, Broxmeyer HE, Emanuel D, Kotylo P, Brahmi Z, Smith FO (2000) Analysis of engraftment, graft-versus-host disease, and immune recovery following unrelated donor cord blood transplantation. Blood 96:2703–2711

Toma C, Pittenger MF, Cahill KS, Byrne BJ, Kessler PD (2002) Human mesenchymal stem cells differentiate to a cardiomyocyte phenotype in the adult murine heart. Circulation 105:93–98

Vavrova J, Filip S, Vokurkova D, Blaha M, Vanasek J, Jebavy L (1999) Ex vivo expansion CD34+/AC133+-selected autologous peripheral blood progenitor cells (PBPC) in high-risk breast cancer patients receiving intensive chemotherapy. Hematol Cell Ther 41:105–112

Weinreb S, Delgado JC, Clavijo OP, Yunis EJ, Bayer-Zwirello L, Polansky L, Deluhery L, Cohn G, Yao JT, Stec TC, Higby D, Anderzejewski C (1998) Transplantation of unrelated cord blood cells. Bone Marrow Transplant 22:193–196

Wexler SA, Donaldson C, Denning-Kendall P, Rice C, Bradley B, Hows JM (2003) Adult bone marrow is a rich source of human mesenchymal 'stem' cells but umbilical cord and mobilized adult blood are not. Br J Haematol 121:368–374

Yamazaki K, Roberts RA, Spooncer E, Dexter TM, Allen TD (1989) Cellular interactions between 3 T3 cells and interleukin-3-dependent multipotent haemopoietic cells: a model system for stromal-cell-mediated haemopoiesis. J Cell Physiol 139:301–312

Yao CL, Chu IM, Hsieh TB, Hwang SM (2004) A systematic strategy to optimize ex vivo expansion medium for human hematopoietic stem cells derived from umbilical cord blood mononuclear cells. Exp Hematol 32:720–727

Yao CL, Feng YH, Lin XZ, Chu IM, Hsieh TB, Hwang SM (2006) Characterization of serum-free ex vivo-expanded hematopoietic stem cells derived from human umbilical cord blood CD133(+) cells. Stem Cells Dev 15:70–78

Young JC, Wu S, Hansteen G, Du C, Sambucetti L, Remiszewski S, O'Farrell AM, Hill B, Lavau C, Murray LJ (2004) Inhibitors of histone deacetylases promote hematopoietic stem cell self-renewal. Cytotherapy 6:328–336

Zhai QL, Qiu LG, Li Q, Meng HX, Han JL, Herzig RH, Han ZC (2004) Short-term ex vivo expansion sustains the homing-related properties of umbilical cord blood hematopoietic stem and progenitor cells. Haematologica 89:265–273

Zhang J, Niu C, Ye L, Huang H, He X, Tong WG, Ross J, Haug J, Johnson T, Feng JQ, Harris S, Wiedemann LM, Mishina Y, Li L (2003) Identification of the haematopoietic stem cell niche and control of the niche size. Nature 425:836–841

Chapter 5
Immunoregulatory Functions of Mesenchymal Stromal Cells

Ferit Avcu

Abstract In vitro and in vivo studies have demonstrated that mesenchymal stem cells (MSCs) are not immunogenic and do not have immunomodulatory effects. MSCs do not express class II antigens, FAS ligand, or co-stimulator molecules (CD80, CD86, CD40, CD40L). They have low expression of primary human leukocyte antigens (HLA-I). In vitro, MSCs in bone, cartilage, muscle, or fat differentiate in the case of HLA-I expression but do not express HLA-II antigens. However, the immunogenic effects of in vivo differentiated MSCs are controversial. MSCs suppress naive T lymphocytes, memory T lymphocytes, effector T lymphocytes, B lymphocytes, and natural killer (NK) cell functions. They are ineffective against the natural immune response of T lymphocytes. In the presence of antigen-presenting cells or antibodies, the addition of MSCs to a culture medium that includes HLA mismatch lymphocytes prevents T-lymphocyte stimulation. MSCs increase with B-lymphocyte viability in vitro and suppress interleukin-2 (IL-2)-related NK cell proliferation and the effector functions of NK cells. MSCs partially reduce the expression of major histocompatibility complex II and CD40 and CD86 co-stimulatory molecules from mature dendritic cells (DCs) by secreting a high rate of IL-6 and vascular endothelial growth factor and thus suppress DC-mediated T-lymphocyte proliferation.

F. Avcu, M.D. (✉)
Gulhane Medical Faculty, Department of Hematology and Research Center,
Tevfik Saglam Street, 06018 Etlik/Ankara, Turkiye
e-mail: favcu@gata.edu.tr

K. Turksen (ed.), *Adult and Embryonic Stem Cells*, Stem Cell Biology and Regenerative Medicine, DOI 10.1007/978-1-61779-630-2_5,

5.1 Introduction

Mesenchymal stromal cells (MSCs) are multipotent cells that were first recognized more than four decades ago by Friedenstein et al. (1968). These cells were described as a population of adherent cells in several human tissues that were nonphagocytic, exhibited a fibroblast-like appearance, and could differentiate in vitro into bone, cartilage, adipose tissue, tendon, and muscle (Friedenstein et al. 1968). On culture at low density either as whole bone marrow or following separation over a density gradient, the cells formed characteristic colonies derived from a single precursor, referred to as colony-forming unit fibroblast (CFU-F). MSCs are identified by their adherent properties, immune phenotype, and differentiation potential. MSCs have emerged as a promising therapeutic modality for tissue regeneration and repair because of their ability to migrate to sites of tissue injury (Aggarwal and Pittenger 2005). MSCs are known to secrete a number of cytokines and regulatory molecules implicated in various aspects of hematopoiesis. In addition to providing critical growth factors, MSCs display immunosuppressive properties that might facilitate engraftment (Fibbe et al. 2007). Several in vitro studies have demonstrated their immunosuppressive capacities, and animal studies have indicated that MSCs also have immunosuppressive capacities in vivo. The mechanisms underlying these effects of MSCs have not been clearly identified.

5.2 Mesenchymal Stem Cells and the Hematopoietic Stem Cells Niche

Multipotent MSCs are part of the highly specialized “microenvironment” participating in the regulation of hematopoietic stem cell (HSC) survival, quiescence, and, upon specific triggers, differentiation into mature elements. A review article discussed some areas of growing interest in MSC biology, including their contribution to the HSC niche (Valtieri and Sorrentino 2008). The supportive interaction of MSCs with hematopoietic progenitors in the bone marrow is well exemplified by the fundamental influence of the former cells on early B-cell lymphopoiesis (Uccelli et al. 2006). The close interaction between early B cells and bone marrow stromal cells is crucial for the normal development of progenitor B cells and is supported by cytokines including interleukin-7 (IL-7), stem cell actor, Flt3 ligand, thymic stromal lymphopoietin, and CXCL12 (Bertrand et al. 2000). Several molecules involved in the constitution of the HSC niche synapse (e.g., Gal-1, angiopoietin-1, osteopontin, thrombospondin-1, thrombospondin-2) are highly expressed by MSCs displaying immunomodulating capacity. MSC also express various adhesion molecules, including several integrins (Majumdar et al. 2003). This feature is consistent with their ability to establish firm adhesive interactions with hematopoietic progenitors inside the HSC niche. In addition, molecular cross-talk between HSCs and the cellular constituents of the niches is thought to control the balance between HSC self-renewal and differentiation, indicating that future successful expansion of HSCs for therapeutic use will require three-dimensional reconstruction of a stem cell–niche unit (Wilson and Trumpp 2006). These data support that the hypothesis that MSCs play a direct role in the hematopoietic cells inside the bone marrow stem cell niche.

5.3 Immunological Phenotype and Functions of Mesenchymal Stem Cells

Mesenchymal stem cells can be easily isolated and expanded through passages in plastic culture flasks, where they grow as adherent cells in an appropriately enriched medium, reaching confluence at time intervals related to density (Delorme and Charbord 2007). However, these progenitor cells are capable of differentiating toward different lineages and do not appear to represent a homogeneous population of stem cells (Horwitz et al. 2005). The isolation of MSCs from primary tissue is hampered by the limited selectivity of available markers. Despite the lack of MSC-specific markers, human MSCs do not express the hematopoietic markers CD34, CD14, or CD45, whereas they are positive for CD44, CD71, CD73, CD90, and CD105 (Delorme and Charbord 2007). MSCs also produce a variety of growth factors, cytokines, chemokines, and proteases that are likely to play a role in their immunomodulatory or their migratory function (Kim et al. 2005; Son et al. 2006; Le Blanc and Ringden 2005).

The MSCs are profoundly influenced by microenvironmental factors and respond to some inflammatory cytokines such as IL-1b, IL-17, and interferon γ (IFNγ) and is capable of significantly affecting their function. In this context, it is worth stressing that although under some circumstances IFNγ appears to enhance the immunosuppressive activity of human MSCs (Huang et al. 2006; Krampera et al. 2006), overall, as suggested by *in vitro* and *in vivo* experiments, the functional behavior of MSCs is the result of the combined effect of soluble factors and of mechanisms mediated by cell-to-cell contact.

5.4 Mesenchymal Stem Cells and T Lymphocytes

The MSCs play a crucial role in the development and differentiation of the lympho-hematopoietic system by secreting a number of growth factors and regulatory cytokines and by promoting cell-to-cell interactions (Locatelli et al. 2007). T lymphocytes (T cells) are a major executor of the adaptive immune response, and numerous studies have demonstrated that MSCs modulate the function of T cells (Salem and Thiemermann 2010). MSCs suppress proliferation of activated lymphocytes in vitro in a dose-dependent, non-HLA-restricted, manner (Klyushnenkova et al. 2005). T cells were differently affected by allogeneic MSCs depending on the extent of proinflammatory conditions by adding high concanavalin A (ConA) concentrations or proinflammatory cytokines such as IFNγ, IL-2, and tumor necrosis factor α (TNFα) (Renner et al. 2009).

Most MSC-mediated immune suppression on activated T cells has been attributed to the secretion of antiproliferative soluble factors, such as hepatocyte growth factor, prostaglandin E_2, transforming growth factor-β1 (TGFβ1), indoleamine 2,3-dioxygenase (which causes depletion of tryptophan, an essential factor for lymphocyte proliferation), nitric oxide, and IL-10 (Le Blanc and Ringden 2006; Uccelli et al. 2007; Krampera et al. 2003; Di Nicola et al. 2002; Le Blanc et al. 2003a, b, c; Maccario et al. 2005a, b). MSCs inhibit naive and memory T-cell responses to their cognate antigens. However, the expression of major histocompatibility complex (MHC) molecules and the presence in culture of antigen-presenting cells (APCs) or of CD4+/CD25+ regulatory T cells were not required for MSCs to exert inhibition (Krampera et al. 2003). MSC-mediated inhibition of alloantigen-induced DC1 differentiation and preferential activation of CD4+ CD25+ T-cell subsets with presumed regulatory activity represent important mechanisms contributing to the immunosuppressive activity of MSCs. These findings provide immunological support for the use of MSCs to prevent immune complications related to both HSC and solid organ transplantation (Maccario et al. 2005a). Di Nicola et al. (2002) reported that autologous or allogeneic bone marrow stem cells (BMSCs) strongly suppress T-lymphocyte proliferation. This phenomenon, which is triggered by cellular and nonspecific mitogenic stimuli, has no immunological restriction, and T-cell inhibition is not due to induction of apoptosis (Bartholomew et al. 2002). It is likely due to the production of soluble factors.

5.5 Mesenchymal Stem Cells and B Lymphocytes

In murine studies, MSCs have been reported to inhibit the proliferation of B cells, stimulated with anti-CD40L and IL-4 (Glennie et al. 2005). It has been recently reported that MSCs co-cultured with purified CD19+ B cells in the presence of a cocktail of stimuli significantly inhibited B-cell proliferation and that this effect was mainly due to soluble factors. CXCR4, CXCR5, and CCR7 B-cell expression and chemotaxis to CXCL12 (the CXCR4 ligand) and CXCL13 (the CXCR5 ligand) were significantly down-regulated by MSCs, suggesting that these cells affect chemotactic properties of B cells. However, B-cell co-stimulatory molecule expression and cytokine production were unaffected by human MSCs (Corcione et al. 2006). Allogeneic MSCs have been shown to inhibit the proliferation, activation, and immunoglobulin G (IgG) secretion of B cells from BXSB mice, which are used as an experimental model for human systemic lupus erythematosus (Deng et al. 2005). Krampera et al. (2006) showed that MSCs only reduced the proliferation of B cells in the presence of IFN-γ. The suppressive effect of IFN-γ was possibly related to its ability to stimulate the production of indoleamine 2,3-dioxygenase (IDO) by MSCs, which in turn suppresses the proliferative response of effector cells through the tryptophan pathway (Meisel et al. 2004).

5.6 Mesenchymal Stem Cells and Natural Killer Cells

The MSCs and natural killer (NK) cells have been shown to interact in vitro (Spaggiari et al. 2006; Sotiropoulou et al. 2006; Poggi et al. 2005). The levels of surface expression of activating NK receptors is positively correlated with NK cell function, which can transduce inhibitory or activating signals (Moretta et al. 2001; Moretta and Moretta 2004). One study reported that MSCs can exert a profound inhibitory effect on NK cell function because they can suppress not only IL-2-induced cell proliferation but also the generation of cytolytic activity and production of cytokines. The authors reported a more accurate analysis of MSC-induced inhibition of NK cell function and defined the molecular basis of such inhibitory effect. First, they showed inhibition of the surface expression of NKp30and NKG2D-activating NK receptors that are involved in NK cell activation and target cell killing. In addition, no surface expression of the NKp44 activating receptor (absent in resting NK cells and expressed upon cell activation) occurred in NK cells cultured with MSCs. MSCs sharply inhibited the NK-mediated cytotoxic activity (Spaggiari et al. 2008).

5.7 Mesnchymal Stem Cells and Dendritic Cells

Dendritic cells (DCs) play a key role in the induction of immunity and tolerance, depending on the activation and maturation stage and, as recently proposed, the cytokine milieu at sites of inflammation (Rutella et al. 2006). MSCs have been demonstrated to interfere with DC differentiation, maturation, and function. Addition of MSCs results in inhibition of the differentiation of both monocytes and CD34+ progenitors into CD1a+ DCs, skewing their differentiation toward cells with features of macrophages. Human MSCs display an inhibitory effect on alloantigen-induced DC differentiation and on APC maturation (Beyth et al. 2005). This could be related to their capacity to produce antiinflammatory cytokines, such as TGFβ, known to inhibit *in vitro* activation and maturation of DCs (Strobl and Knapp 1999). DCs generated in the presence of MSCs had an impaired response to maturation signals and exhibited no expression of CD83 or up-regulation of HLA-DR and co-stimulatory molecules (Jiang et al. 2005; Nauta et al. 2006a; Zhang et al. 2004).

The MSCs can suppress monocyte differentiation into DCs, the most potent APCs, indicating the versatile regulation by MSCs of the ultimate specific immune response. Furthermore, mature DCs treated with MSCs had significantly reduced expression of CD83, suggesting their skew to immature status (Jiang et al. 2005). Consistent with these findings, immature DCs generated in the presence of MSCs were strongly hampered in their ability to induce activation of T cells (Aggarwal and Pittenger 2005; Jiang et al. 2005; Beyth et al. 2005) In addition, MSCs suppress the migratory function of DCs, and so they may serve immunoregulatory activities through the modulation of the antigen-presenting function of DCs (Jung et al. 2007).

Taken together, these results suggest that MSCs have a profound inhibitory effect on the generation and function of both CD34+ and monocyte-derived DCs (Nauta et al. 2006b) Also, MSCs impair monocyte-derived DCs differentiation and function by interfering with the cell cycle (Ramasamy et al. 2007). The increased production of IL-10 by DCs upon co-culture with MSCs may also contribute to the suppressive effects of MSCs. Neutralizing antibodies to IL-10, indeed, restored T-cell proliferation, although not completely (Beyth et al. 2005).

5.8 Mesenchymal Stem Cells and HLA

The expression of MHC molecules on all cells on the body allows the immune system to distinguish self from nonself. Adult human MSCs constitutively express low surface densities of MHC class I molecules and are negative for MHC class II as well as for co-stimulatory molecules such as CD80, CD86, and CD40. Western blotting on cell lysates shows that the cells contain intracellular deposits of class II alloantigens (Le Blanc et al. 2003c; Klyushnenkova et al. 2005; Potian et al. 2003). IFNγ stimulation increased both class I and class II molecules (Klyushnenkova et al. 2005). MSCs that have differentiated into adipose, bone, and cartilage cells express HLA class I, but the expression of class II can no longer be induced (Le Blanc et al. 2003c). As a result of the low expression of MHC class I molecules, MSCs can escape detection by T cells. For years, MSCs have been considered cells that potentially can be ignored by the immune system (Le Blanc and Ringden 2006). However, recent *in vivo* data in mice challenge the concept of the immunoprivilege of MSCs because allogeneic MSCs infused into MHC-mismatched mice were sometimes rejected (Nauta et al. 2006; Eliopoulos et al. 2005). Furthermore, both autologous and allogeneic IL-2-activated NK cells can kill MSCs efficiently owing to the low levels of human MHC class I antigen on the latter cells and to the surface expression of ligands recognized by activating NK receptors (Spaggiari et al. 2006).

HLA-G, a nonclassic human leukocyte antigen class I molecule that was initially found on trophoblasts, is a crucial immunosuppressive molecule secreted by adult human mesenchymal stem cells. Trophoblasts are able to express indoleamine IDO and prostaglandin E_2. MSC immunomodulatory properties are similar to those of trophoblasts. This mechanism may relate to tolerance of alloantigens for the prevention of graft rejection after transplantation (Selmani et al. 2009).

5.9 Clinical Use of Mesenchymal Stem Cell-Induced Immunosuppression

The MSCs are considered to be hypoimmunogenic, displaying low expression levels of HLA and MHC class I and, importantly, no expression of co-stimulatory molecules (Tse et al. 2003). At present, little is known about the mechanisms of suppression of graft-versus-host disease (GVHD) by MSCs. In vivo studies demonstrated that MSCs avoid normal alloresponses (Koc et al. 2002). These characteristics support the possibility of exploiting universal donor MSCs for therapeutic applications. However, recent evidence indicates that MSCs can function as APCs and activate immune responses under appropriate conditions (Chan et al. 2006). Furthermore, it has been demonstrated that subcutaneously implanted allogeneic MSCs were rejected in nonimmunosuppressed recipient mice. Splenocytes isolated from mice that had been implanted with allogeneic MSCs displayed a significant IFNα response against allogeneic MSCs in vitro (Eliopoulos et al. 2005).

Clinical use of MSC-induced immunosuppression is allogeneic or autologous in several diseases, such as GVHD and autoimmune disease (Siegel et al. 2009). Indeed, MSCs have been used to treat experimental animal models of multiple sclerosis, diabetes, systemic lupus erythematosus, and with less success rheumatoid arthritis (Uccelli et al. 2007). The multiple immunosuppressive properties of MSCs provide the biological explanation of their efficacy in the treatment of patients with acute GVHD and even those refractory to conventional treatment (Locatelli et al. 2007). MSCs may also have a role in the prevention of GVHD, given that MSCs achieved such prevention in an animal model. We investigated the ability of MSCs to prevent or treat GVHD in a rat bone marrow transplantation (BMT) model (Nevruz et al. 2007). The GVHD model was established by transplantation of Spraque Dawley (SD) rats' bone marrow and spleen cells into lethally irradiated SD × Wistar rat recipients. MSC and GVHD prophylactic regimens were administered using various timing protocols designed to either prevent or treat GVHD. After transplantation, a clinical GVHD scoring system and the survival were monitored. MSCs inhibited lethal GVHD at least with the GVHD prophylactic regimen after allo-BMT. The gross and histopathological findings of GVHD and the ratio of CD4/CD8 expression decreased at the same time the proportion of CD25+ T cells and plasma IL-2 levels increased *in vivo* after allo-BMT with MSC administration compared with conventional allo-BMT. Our results strongly suggested that clinical use of MSCs both prophylactically and after GVHD developed was as effective as the GVHD prophylactic regimen in preventing GVHD (Nevruz et al. 2007).

Immunological studies specifically addressing this issue are needed to improve our understanding of the treatment of acute GVHD. Further studies of the immunogenicity of MSCs are needed, and rejection of MSCs and the clinical consequences should be carefully considered in clinical trials.

5.10 Conclusions

Mesenchymal stem cells are multipotent progenitor cells. Interest in MSC therapy has been raised by the observation that MSCs are able to modulate immune responses *in vitro* and *in vivo*. MSCs preferentially home to damaged tissue and may have therapeutic potential. *In vitro* data suggest that MSCs exert powerful immunomodulatory effects, which include inhibiting the proliferation and function of T cells,

B cells, and NK cells. These unique properties make MSCs of great interest for clinical applications in tissue engineering and immunosuppression. Underlying the MSC-mediated immunomodulatory mechanisms is a nonspecific antiproliferative effect. Of special interest are the molecular mechanisms by which MSCs influence their target cells. Understanding these mechanisms is crucial for future use of MSCs in research and clinical applications. Possible clinical applications include therapy-resistant severe acute GVHD, tissue repair, rejection of organ allografts, and autoimmune disorders.

References

Aggarwal S, Pittenger MF (2005) Human mesenchymal stem cells modulate allogeneic immune cell responses. Blood 105:1815–1822

Bartholomew A, Sturgeon C, Siatskas M et al (2002) Mesenchymal stem cells suppress lymphocyte proliferation in vitro and prolong skin graft survival in vivo. Exp Hematol 30:42–48

Bertrand FE, Eckfeldt CE, Fink JR, Lysholm AS, Pribyl JA, Shah N, LeBien TW (2000) Microenvironmental influences on human B cell development. Immunol Rev 175:175–186

Beyth S, Borovsky Z, Mevorach D et al (2005) Human mesenchymal stem cells alter antigen-presenting cell maturation and induce T cell unresponsiveness. Blood 105:2214–2219

Chan JL, Tang KC, Patel AP et al (2006) Antigen-presenting property of mesenchymal stem cells occurs during a narrow window at low levels of interferon-gamma. Blood 107:4817–4824

Corcione A, Benvenuto F, Ferretti E et al (2006) Human mesenchymal stem cells modulate B cell functions. Blood 107:367–372

Delorme B, Charbord P (2007) Culture and characterization of human bone marrow mesenchymal stem cells. Methods Mol Med 140:67–81

Deng W, Han Q, Liao L et al (2005) Effects of allogeneic bone marrow-derived mesenchymal stem cells on T and B lymphocytes from BXSB mice. DNA Cell Biol 24:458–463

Di Nicola M, Carlo-Stella C, Magni M et al (2002) Human bone marrow stromal cells suppress T-lymphocyte proliferation induced by cellular or nonspecific mitogenic stimuli. Blood 99:3838–3843

Eliopoulos N, Stagg J, Lejeune L, Pommey S, Galipeau J (2005) Allogeneic marrow stromal cells are immune rejected by MHC class I- and class II-mismatched recipient mice. Blood 106: 4057–4065

Fibbe WE, Nauta AJ, Roelofs H (2007) Modulation of immune responses by mesenchymal stem cells. Ann N Y Acad Sci 1106:272–278

Friedenstein AJ, Petrakova KV, Kurolesova AI, Frolova GP (1968) Heterotopic of bone marrow: analysis of precursor cells for osteogenic and hematopoietic tissues. Transplantation 6: 230–247

Glennie S, Soeiro I, Dyson PJ, Lam EW, Dazzi F (2005) Bone marrow mesenchymal stem cells induce division arrest anergy of activated T cells. Blood 105:2821–2827

Horwitz E, Le Blanc K, Dominici M et al (2005) Clarification of the nomenclature for MSC: the International Society for Cellular Therapy position statement. Cytotherapy 7:393–395

Huang W, La Russa V, Alzoubi A, Schwarzenberger P (2006) Interleukin-17A: a T-cell derived growth factor for murine and human mesenchymal stem cells. Stem Cells 24:1512–1518

Jiang XX, Zhang Y, Liu B et al (2005) Human mesenchymal stem cells inhibit differentiation and function of monocyte-derived dendritic cells. Blood 105:4120–4126

Jung YJ, Ju SY, Yoo ES et al (2007) MSC-DC interactions: MSC inhibit maturation and migration of BM-derived DC. Cytotherapy 9(5):451–458

Kim DH, Yoo KH, Choi KS et al (2005) Gene expression profile of cytokine and growth factor during differentiation of bone marrow-derived mesenchymal stem cell. Cytokine 31:119–126

Klyushnenkova E, Mosca JD, Zernetkina V et al (2005) T cell responses to allogeneic human mesenchymal stem cells: immunogenicity, tolerance, and suppression. J Biomed Sci 12:47–57
Koc ON, Day J, Nieder M et al (2002) Allogeneic mesenchymal stem cell infusion for treatment of metachromatic leukodystrophy (MLD) and Hurler syndrome (MPS-IH). Bone Marrow Transplant 30:215–222
Krampera M, Glennie S, Dyson J et al (2003) Bone marrow mesenchymal stem cells inhibit the response of naïve and memory antigen-specific T cells to their cognate peptide. Blood 101:3722–3729
Krampera M, Cosmi L, Angeli R et al (2006) Role for interferon-gamma in the immunomodulatory activity of human bone marrow mesenchymal stem cells. Stem Cells 24:386–398
Le Blanc K, Ringden O (2005) Immunobiology of human mesenchymal stem cells and future use in hematopoietic stem cell transplantation. Biol Blood Marrow Transplant 11:321–334
Le Blanc K, Ringden O (2006) Mesenchymal stem cells: properties and role in clinical bone marrow transplantation. Curr Opin Immunol 18:586–591
Le Blanc K, Tammik L, Sundberg B, Haynesworth SE, Ringden O (2003a) Mesenchymal stem cells inhibit and stimulate mixed lymphocyte cultures and mitogenic responses independently of the major histocompatibility complex. Scand J Immunol 57:11–20
Le Blanc K, Tammik C, Rosendahl K, Zetterberg E, Ringden O (2003b) HLA expression and immunologic properties of differentiated and undifferentiated mesenchymal stem cells. Exp Hematol 10:890–896
Le Blanc K, Tammik L, Zetterberg E, Ringden O (2003c) HLA expression and immunologic properties of differentiated and undifferentiated mesenchymal stem cells. Exp Hematol 31:890–896
Locatelli F, Maccario R, Frassoni F (2007) Mesenchymal stromal cells, from indifferent spectators to principal actors: are we going to witness a revolution in the scenario of allograft and immune-mediated disorders? Haematologica 92:872–877
Maccario R, Podestà M, Moretta A et al (2005a) Interaction of human mesenchymal stem cells with cells involved in alloantigen-specific immune response favors the differentiation of CD4+ T-cell subsets expressing a regulatory/suppressive phenotype. Haematologica 90:516–525
Maccario R, Moretta A, Cometa A et al (2005b) Human mesenchymal stem cells and cyclosporin-A exert a synergistic suppressive effect on in vitro activation of alloantigen-specific cytotoxic lymphocytes. Biol Blood Marrow Transplant 11:1031–1032
Majumdar MK, Keane-Moore M, Buyaner D et al (2003) Characterization and functionality of cell surface molecules on human mesenchymal stem cells. J Biomed Sci 10:228–241
Meisel R, Zibert A, Laryea M, Gobel U, Daubener W, Dilloo D (2004) Human bone marrow stromal cells inhibit allogeneic T-cell responses by indoleamine 2,3 dioxygenase-mediated tryptophan degradation. Blood 103:4619–4621
Moretta L, Moretta A (2004) Unravelling natural killer cell function: triggering and inhibitory human NK receptors. EMBO J 23:255–259
Moretta A, Bottino C, Vitale M et al (2001) Activating receptors and coreceptors involved in human natural killer cell-mediated cytolysis. Annu Rev Immunol 19:197–223
Nauta AJ, Kruisselbrink AB, Lurvink E, Willemze R, Fibbe WE (2006a) Mesenchymal stem cells inhibit generation and function of both $CD34^+$-derived and monocyte-derived dendritic cells. J Immunol 177:2080–2087
Nauta AJ, Westerhuis G, Kruisselbrink AB, Lurvink EG, Willemze R, Fibbe WE (2006b) Donor-derived mesenchymal stem cells are immunogenic in an allogeneic host and stimulate donor graft rejection in a non-myeloablative setting. Blood 108:2114–2120
Nevruz O, Avcu F, Ural AU et al (2007) Immunosuppressive effects of multipotent mesenchymal stromal cells on graft-versus-host disease in rats following allogeneic bone marrow transplantation. Blood 110(11): Abstr. 3248
Poggi A, Prevosto C, Massaro AM et al (2005) Interaction between human NK cells and bone marrow stromal cells induces NK cell triggering: role of NKp30 and NKG2D receptors. J Immunol 175:6352–6360
Potian JA, Aviv H, Ponzio NM, Harrison JS, Rameshwar P (2003) Veto-like activity of mesenchymal stem cells: functional discrimination between cellular responses to alloantigens and recall antigens. J Immunol 171:3426–3434

Ramasamy R, Fazekasova H, Lam EW, Soeiro I, Lombardi G, Dazzi F (2007) Mesenchymal stem cells inhibit dendritic cell differentiation and function by preventing entry into the cell cycle. Transplantation 83(1):71–76

Renner P, Eggenhofer E, Rosenauer A et al (2009) Mesenchymal stem cells require a sufficient, ongoing immune response to exert their immunosuppressive function. Transplant Proc 41(6):2607–2611

Rutella S, Danese S, Leone G (2006) Tolerogenic dendritic cells: cytokine modulation comes of age. Blood 108:1435–1440

Salem HK, Thıemermann C (2010) Mesenchymal stromal cells: current understanding and clinical status. Stem Cells 28:585–596

Selmani Z, Naji A, Gaiffe E et al (2009) HLA-G is a crucial immunosuppressive molecule secreted by adult human mesenchymal stem cells. Transplantation 87(9 Suppl):S62–S66

Siegel G, Schäfer R, Dazzi F (2009) The immunosuppressive properties of mesenchymal stem cells. Transplantation 87(9 Suppl):S45–S49

Son BR, Marquez-Curtis LA, Kucia M et al (2006) Migration of bone marrow and cord blood mesenchymal stem cells in vitro is regulated by stromal-derived factor-1-CXCR4 and hepatocyte growth factor-c-met axes and involves matrix metalloproteinases. Stem Cells 24:1254–1264

Sotiropoulou PA, Perez SA, Gritzapis AP et al (2006) Interactions between human mesenchymal stem cells and natural killer cells. Stem Cells 24:74–85

Spaggiari GM, Capobianco A, Becchetti S, Mingari MC, Moretta L (2006) Mesenchymal stem cell natural killer cell interactions: evidence that activated NK cells are capable of killing MSCs, whereas MSCs can inhibit IL-2-induced NK-cell proliferation. Blood 107:1484–1490

Spaggiari GM, Capobianco A, Abdelrazik H, Becchetti F, Mingari MC, Moretta L (2008) Mesenchymal stem cells inhibit natural killer-cell proliferation, cytotoxicity, and cytokine production: role of indoleamine 2,3-dioxygenase and prostaglandin E_2. Blood 111(3):1327–1333

Strobl H, Knapp W (1999) TGF-β1 regulation of dendritic cells. Microbes Infect 1:1283–1291

Tse WT, Pendleton JD, Beyer WM, Egalka MC, Guinan EC (2003) Suppression of allogeneic T-cell proliferation by human marrow stromal cells: implications in transplantation. Transplantation 75:389–397

Uccelli A, Moretta L, Pistoia V (2006) Immunoregulatory function of mesenchymal stem cells. Eur J Immunol 36(10):2566–2573

Uccelli A, Pistoia V, Moretta L (2007) Mesenchymal stem cells: a new strategy for immunosuppression? Trends Immunol 28:219–226

Valtieri M, Sorrentino A (2008) The mesenchymal stromal cell contribution to homeostasis. J Cell Physiol 217(2):296–300

Wilson A, Trumpp A (2006) Bone-marrow haematopoietic-stem-cell niches. Nat Rev Immunol 6(2):93–106

Zhang W, Ge W, Li C et al (2004) Effects of mesenchymal stem cells on differentiation, maturation, and function of human monocyte-derived dendritic cells. Stem Cells Dev 13:263–271

Chapter 6
Mesenchymal Stem Cells: Possibilities of New Treatment Options

Zeynep Tokcaer-Keskin, Hande Kocak, Ihsan Gursel, and Kamil C. Akcali

Abstract Stem cell research evolved as a new hope and has gained tremendous interest during the last two decades in developing potential strategies for many debilitating diseases. Mesenchymal stem cells (MSCs) are bone marrow-derived multipotent stem cells capable of self-renewal and of differentiating into multiple lineages, such as osteocytes, adipocytes, chondrocytes, myoblasts, cardiomyocytes, and hepatocytes. MSCs are an important source for cellular therapies. They can easily be obtained and expanded in vitro in large numbers without significantly altering their properties. MSCs not only migrate to the injured site in vivo but also have immunomodulatory effects that make their use attractive for allogeneic grafting. MSCs can also be frozen for preservation; and when thawed, they retain their normal physiological function, allowing future "off-the-shelf" therapy approaches. Because of these features, MSCs have high therapeutic value in tissue engineering and regenerative medicine. In this chapter, the contribution of the MSCs to cardiovascular repair and liver regeneration are summarized.

Z. Tokcaer-Keskin • H. Kocak • K.C. Akcali (✉)
Laboratory of Stem Cell Research, Department of Molecular Biology and Genetics, Faculty of Science B-243, Bilkent University, Bilkent, Ankara, 06800 Turkey
e-mail: akcali@fen.bilkent.edu.tr

I. Gursel
Biotherapeutic ODN Research Lab, Department of Molecular Biology and Genetics, Bilkent University, Bilkent, Ankara, 06800 Turkey

K. Turksen (ed.), *Adult and Embryonic Stem Cells*, Stem Cell Biology and Regenerative Medicine, DOI 10.1007/978-1-61779-630-2_6,

6.1 Introduction

Stem cell research has evolved as a new hope and has attracted tremendous interest during the last two decades in developing new strategies for many debilitating diseases. Stem cells are important for living organisms because they have a potential for differentiating into a wide variety of cells under certain circumstances. The incidence of chronic and noninfectious diseases such as cancer, chronic heart disease, and diabetes are the main health concerns of today's world. Thus, treatment of these chronic diseases is important for a healthy future. Applications involving the use of stem cells in humans that might have been considered "science fiction" only less than two decades ago are now being utilized with great success (Akar et al. 2006).

6.2 Stem Cells

Stem cells are unspecialized cells that are capable of self-renewal and give rise to differentiated cells (Till and McCulloch 1961; Morrison et al. 1997; Weissman 2000). Although self-renewal and differentiation to other cells are their common features, stem cells vary in their potential to differentiate, the durations and pathways of self-renewal, the niche in which they exist, and their division properties (Morrison et al. 1997). Stem cells can be categorized based on their time of onset and differentiation potential. Embryonic stem cells (ESCs) are derived from the cells of the inner cell mass (ICM) during early embryonic development. Germline and adult stem cells (ASCs), on the other hand, are groups of cells that are seen, respectively, later in development and postnatally. ESCs are pluripotent, meaning that they can give rise to all of the cell types from three germ layers, where as ASCs are multipotent and able to produce multiple, but not all, lineages. ESCs are important tools for studying embryogenesis and understanding the mechanisms of genetic diseases. Ethical considerations and the risk of tumor formation and immune rejection are main hurdles to the ESCs being used for cellular therapies in humans. The first clinical trial of a human embryonic stem cell (hESC)-based therapy in humans was studied in regard to spinal cord injuries.

Also known as somatic stem cells, ASCs are found in organs and tissues and maintain homeostasis. Compared to ESCs, they are more differentiated, and their potential is more limited (multipotent vs. pluripotent). Among ASCs, mesenchymal stem cells (MSCs) have unique features that make them good candidates in the areas of tissue engineering, regenerative medicine, and cellular therapies.

Also known as bone marrow stromal cells or mesenchymal stromal cells, MSCs were discovered by the pioneering studies of Friedenstein in 1968. Initially, MSCs were shown to attach to culture plates and were able to form colonies; these cells were later named colony forming unit fibroblasts (CFU-Fs) (Lanotte et al. 1981). MSCs are multipotent cells capable of self-renewal and of differentiating into multiple lineages, such as osteocytes, adipocytes, chondrocytes, myoblasts, hepatocytes, and cardiomyocytes (Friedenstein et al. 1968; da Silva Meirelles et al. 2008;

Tokcaer-Keskin et al. 2009). Pittenger et al. (1999) conducted the first characterization study of MSCs. They obtained the MSCs from human bone marrow aspirates. The bone marrow aspirate was first separated according to the density gradient and plated afterward. Attached cells were counted according to their colony formation capacities. When proportioned to the total cell number, only 0.001–0.01% of the nucleated cells formed colonies. The authors also shown that MSCs were positive for CD29, CD90, CD71, and CD106; and they were negative for CD45, CD14, and CD34. The cells were able to undergo 40 population doublings (PDs) in vitro within 10 weeks. In addition to bone marrow, MSCs were isolated from many other sources such as adipose tissue, umbilical cord blood, placenta, and dental pulp. However, studies revealed that MSCs from these sources had different CD marker expression, phenotypes, and PDs. These results forced the scientific community to define certain criteria for identifying these cells. The International Society for Cellular Therapy (ISCT) published a position paper by Dominici et al. (2006) and set forth the following cardinal features for MSCs. According to that seminal report, MSCs should be positive for CD73, CD90, and CD105 and negative for CD19, CD34, CD45, CD11a, and HLA-DR.

An important feature of MSCs is their homing capacity. Homing is defined as the migration of these cells to the site of injury. In the homing process of MSCs, chemokines, cytokines, and the receptors on their surface were shown to be pivotal (da Silva Meirelles et al. 2008). When inflammation occurs at the site of injury, the gradient of cytokines and chemokines increase; and the expression of chemokine receptors mediate the migration of MSCs to the injured tissue (Salem and Thiemermann 2010). Especially CD44 was found to be important in homing of both mouse and human MSCs (Herrera et al. 2004; Sackstein et al. 2008). In addition, CXCR4 and VCAM-1 (CD106) played a critical role in the migration of the MSCs (Segers et al. 2006; Shi et al. 2007; Hung et al. 2007). It was also demonstrated that MSCs can activate matrix metalloproteases so as to enter the tissue from blood to localize the niche at the site of injury (De Becker et al. 2007).

The MSCs also have important antiapoptotic and immunomodulatory effects, which makes them nonimmunogenic. When MSCs were injected into scar tissue in animal models, they were able to reduce the apoptotic rate of the surrounding cells, which was mediated by the secretion of several growth factors including vascular endothelial growth factor (VEGF), fibroblast growth factor 2 (FGF2), and transforming growth factor β (TGFβ), especially in hypoxic conditions (Tögel et al. 2007; Parekkadan et al. 2007; Block et al. 2009). When immunomodulatory effects were considered, it was shown that T-cell proliferation was inhibited during in vitro co-culturing (Di Nicola et al. 2002; Krampera et al. 2003; Le Blanc et al. 2003). Furthermore, not only cytotoxic and helper T cells but also natural killer (NK) cells, B cells, and immature dendritic cells were found to be affected by MSCs. The immunomodulatory effects of MSCs on T cells and NK cells were associated with secretion of such molecules as TGFβ, prostaglandin E_2 (PGE_2), and interleukin-10 (IL-10), whereas their effect on B cells was found to be indirect, via modulation of plasma cells by causing the inhibition of immunoglobulin secretion (Sotiropoulou et al. 2006; Nasef et al. 2007; Rafei et al. 2008; Németh et al. 2009, da Silva Meirelles et al. 2008).

The MSCs have already been used to address osteogenesis imperfecta, hematopoietic recovery, bone tissue regeneration, cardiovascular repair, treatment of lung fibrosis, spinal cord injury, coronary artery disease, local repair, and regeneration of bone cartilage and tendon (Minguell et al. 2001; Salem and Thiemermann 2010). They can form intervertebral disc cartilage, bone, cardiomyocytes, and articular cartilage at knee joints in addition to neurons, skin epithelia, lung, liver, kidney, intestine, and spleen.

In addition to these advantages of MSC use in therapies, however, there are some unknowns, such as their long-term effect and safety, which require more toxicology studies. Also, the efficiency is not clear. There is some evidence on homing and differentiation, but it is insufficient. Large-scale culture, storage, and distribution are important factors in the therapeutic application of MSCs (Minguell et al. 2001).

6.3 Mesenchymal Stem Cells in Cellular Therapies

The literature contains many studies investigating the regenerative capacities of MSCs in various disease models generated by employing nonhuman animal species. Cardiac regeneration, liver regeneration, kidney regeneration, autoimmune diseases and graft-versus-host disease (GVHD), neurological diseases, pulmonary diseases, osteogenic diseases, and cartilage repair are the most widely studied conditions (Salem and Thiemermann 2010). MSCs are also being investigated extensively in clinical trials, mostly in the United States, Europe, and East Asia. Some of the clinical trials are studying MSC use in neurological, liver, bone, heart diseases, GVHD, and some autoimmune diseases such as diabetes and Crohn's disease. In the following section, the application of MSCs in cell-based therapies are discussed, with particular attention given to their roles in heart and liver diseases.

6.3.1 Cardiac Repair

In vivo studies with different animal models were performed to reveal the effect of MSCs in cardiovascular diseases. Shake et al. (2002) demonstrated the differentiation of MSCs expanded from swine bone marrow into functional cardiomyocytes when injected into the infarcted swine myocardium. In a canine model, it was observed that the intracardially injected MSCs differentiated into smooth muscle cells and endothelial cells rather than cardiomyocytes, although they had better functionality at the infarcted area (Silva et al. 2005). It was demonstrated that when the rats were treated with MSCs, the infarcted area became significantly smaller 4 weeks after the treatment, and the MSCs that were labeled prior to the treatment were expressing cardiac markers such as cardiac troponin and smooth muscle actin (Tang et al. 2006). In a similar mouse model, the in vivo effects of human MSCs obtained from a patient with ischemic heart disease were tested regarding the infarct size and heart functions using MRI. The results of the study indicated that MSCs were able to home to the scar tissue and improve the function of the left ventricle by differentiating into smooth muscle and endothelial cells as well as cardiomyocytes (Grauss et al. 2007). In another study, the mode of administration and transplantation times were investigated. Introducing MSCs by transendocardial electromechanically guided delivery was more efficient than intracoronary delivery (Perin et al. 2008), and it was demonstrated that if the MSCs were delivered to the heart 1 week after infarction better results were obtained regarding cardiac function and formation of blood vessels when compared to delivery at 1 h or 2 weeks (Jiang et al. 2008). There are also studies that have attributed the regenerative effect of MSCs to their paracrine effects. MSCs that were genetically modified for overexpression of Akt exerted a better curing effect on the left ventricles of animal myocardial infarction models (Lim et al. 2006; Mangi et al. 2003). According to another study, secretion of hepatocyte growth fact (HGF) from MSCs triggered the migration of cardiac stem cells to the site of infarction and regeneration (Urbanek et al. 2005).

Transplantation of MSCs has also been performed in patients with myocardial infarction, who showed improved myocardial activity after the transplant (Katritsis et al. 2005). The results of one clinical trial revealed that left ventricular function in myocardial infarction patients treated with autologous MSCs was improved (Wollert et al. 2004; Chen et al. 2004). There are ongoing clinical trials that employing MSCs in the treatment of myocardial infarction. A recent trial to be carried out in France is still recruiting patients to investigate the administration of the MSCs intracardially (Phase 1/2 study), and yet another trial is investigating intravenous administration (Phase 2 study) (http://clinicaltrials.gov). Today there are 21 clinical trials either completed, ongoing, or recruiting patients to investigate the treatment capacities of MSCs in heart disease. The theurapeutic potential of MSCs in cardiac repair will be better understood with the help of these in vivo and in vitro studies and clinical trials.

6.3.2 Liver Injuries

Although liver is able to regenerate after injury, end-stage liver injuries may require transplantation, which can take a long time because of having to find a suitable donor. Cellular therapy is considered an important alternative in such cases. As isolation and culture of hepatocytes in efficient amounts is not possible in vitro (Serralta et al. 2003, 2005), cellular therapy involving stem cells has an important role to play in curing liver disease. Among the stem cells, MSCs receive special attention because their differentiation into hepatocyte-like cells has been reported (Chamberlain et al. 2007; Sato et al. 2005; Lee et al. 2004). In addition to their hepatic potential, MSCs can migrate to the injured site, which results in targeting the cells at the site of liver failure to a certain extent. As already mentioned, MSCs have immunoregulatory properties. They are not only nonimmunogenic but are also immunosuppressive (Rasmusson 2006). Because of these features, MSCs shine as a very promising tool for cell-based therapy for liver disease.

Hepatocytes differentiated from MSCs were found to be positive for hepatocyte markers and to be functional as evidenced by their secretion of albumin and glycogen storage (Chamberlain et al. 2007; Lee et al. 2004). Immunomodulatory effects of MSCs were also found to be important during liver regeneration (Bartholomew et al. 2002; Inoue et al. 2006; Aggrawal and Pitteger 2005). Although the mechanism is still unclear, there seem to be several factors playing a role in immunoregulation. Pervsner-Fischer et al. (2007) showed that murine MSCs express several Toll-like receptors (TLRs), and in particular TLR2 was found to be important in the differentiation potential of MSCs. Similar to MSCs in mice, human MSCs express several TLRs (Tomchuck et al. 2008). Our unpublished data suggest that MSC homing to the injury site in liver may be due to TLR expression ability of the injured liver. In this context, we found that the expression levels of TLRs 2, 3, and 9 are increased in hepatectomized rats administered MSCs (Kocak et al. unpublished).

It was shown that engraftment of the MSCs into the liver, but not differentiation, stimulated proliferative and regenerative properties of the liver (Banas et al. 2008; Parekkadan et al. 2007; Caplan and Dennis 2006). The effect of MSCs on hepatic stellate cells have also been demonstrated and shown to reduce the formation of fibrosis (Carvalho et al. 2008; Abdel Aziz et al. 2007; Zhao et al. 2005; Sakaida et al. 2004). Chamberlain et al. (2007) also demonstrated that when the MSCs were administered by intrahepatic injection they formed hepatocytes more efficiently than when given intraperitoneally.

The cytokines responsible in the differentiation, homing, and antifibrotic effects of MSCs were also investigated. FGF-4 and HGF were found to be important in the differentiation of MSCs into hepatocytes (Dong et al. 2010; Parekkadan et al. 2007) and tumor necrosis factor α (TNFα) and IL-6 were critical in antifibrotic effects (Pulavendran et al. 2010; Parekkadan et al. 2007). The expression of matrix metalloproteinases by MSCs was reported to be important in the reduction of fibrosis (Fang et al. 2004; Sakaida et al. 2004; Oyagi et al. 2006). CXCR4 and CCR9 have also been shown to play a role in the homing of MSCs to injured liver (Chen et al. 2010).

6.4 Conclusion and Future Directions

Data strongly point in the direction that MSCs are highly potent multipurpose progenitor cells suitable for several therapeutic applications. It is of great importance to establish the safety and bioactivity in the near future as well as the long-term effects of MSC administration. Another major issue is to better characterize and even subclassify the MSCs following in vitro culturing/expansion. Recent data strongly indicate that MSCs may differentiate into immunostimulatory or immunosuppressive agents depending on the tissue location and/or ongoing inflammation at the sites (Waterman et al. 2010). Differential classification of MSCs will ensure their more successful applications in addition to better utility for treating health problems.

Acknowledgments This work was supported by the Scientific and Technological Research Council of Turkey (TUBITAK): grants SBAG105S393 to K.C.A. and SBAG106S102 and SBAG108S316 to I.G.

References

Abdel Aziz MT, Atta HM, Mahfouz S et al (2007) Therapeutic potential of bone marrow-derived mesenchymal stem cells on experimental liver fibrosis. Clin Biochem 40:893–899

Aggrawal S, Pitteger MF (2005) Human mesenchymal stem cells modulate allogeneic immune cell responses. Blood 105:1815–1822

Akar AR, Durdu S, Corapcioglu T et al (2006) Regenerative medicine for cardiovascular disorders: new milestones—adult stem cells. Artif Organs 30:213–232

Banas A, Teratani T, Yamamoto Y et al (2008) IFATS collection: in vivo therapeutic potential of human adipose tissue mesenchymal stem cells after transplantation into mice with liver injury. Stem Cells 26:2705–2712

Bartholomew A, Sturgeon C, Siatskas M et al (2002) Mesenchymal stem cells suppress lymphocyte proliferation in vitro and prolong skin graft survival in vivo. Exp Hematol 30:42–48

Block GJ, Ohkouchi S, Fung F et al (2009) Multipotent stromal cells are activated to reduce apoptosis in part by upregulation and secretion of stanniocalcin-1. Stem Cells 27:670–681

Caplan AI, Dennis JE (2006) Mesenchymal stem cells as trophic mediators. J Cell Biochem 98:1076–1084

Carvalho AB, Quintanilha LF, Dias JV et al (2008) Bone marrow multipotent mesenchymal stromal cells do not reduce fibrosis or improve function in a rat model of severe chronic liver injury. Stem Cells 26:1307–1314

Chamberlain J, Yamagami T, Colletti E et al (2007) Efficient generation of human hepatocytes by the intrahepatic delivery of clonal human mesenchymal stem cells in fetal sheep. Hepatology 46:1935–1945

Chen SL, Fang WW, Ye F et al (2004) Effect on left ventricular function of intracoronary transplantation of autologous bone marrow mesenchymal stem cell in patients with acute myocardial infarction. Am J Cardiol 94:92–95

Chen Y, Xiang LX, Shao JZ et al (2010) Recruitment of endogenous bone marrow mesenchymal stem cells towards injured liver. J Cell Mol Med 14:1494–1508

da Silva Meirelles L, Caplan AI, Nardi NB (2008) In search of the in vivo identity of mesenchymal stem cells. Stem Cells 26:2287–2299

De Becker A, Van Hummelen P, Bakkus M et al (2007) Migration of culture-expanded human mesenchymal stem cells through bone marrow endothelium is regulated by matrix metalloproteinase-2 and tissue inhibitor of metalloproteinase-3. Haematologica 92:440–449

Di Nicola M, Carlo-Stella C, Magni M et al (2002) Human bone marrow stromal cells suppress T-lymphocyte proliferation induced by cellular or nonspecific mitogenic stimuli. Blood 99:3838–3843

Dominici M, Le Blanc K, Mueller I et al (2006) Minimal criteria for defining multipotent mesenchymal stromal cells: the International Society for Cellular Therapy position statement. Cytotherapy 8:315–317

Dong XJ, Zhang H, Pan RL et al (2010) Identification of cytokines involved in hepatic differentiation of mBM-MSCs under liver-injury conditions. World J Gastroenterol 16:3267–3278

Fang B, Shi M, Liao L et al (2004) Systemic infusion of FLK1(+) mesenchymal stem cells ameliorate carbon tetrachloride-induced liver fibrosis in mice. Transplantation 78:83–88

Friedenstein AJ, Petrakova KV, Kurolesova AI et al (1968) Heterotopic of bone marrow: analysis of precursor cells for osteogenic and hematopoietic tissues. Transplantation 6:230–247

Grauss RW, Winter EM, van Tuyn J et al (2007) Mesenchymal stem cells from ischemic heart disease patients improve left ventricular function after acute myocardial infarction. Am J Physiol Heart Circ Physiol 293:H2438–H2447
Herrera MB, Bussolati B, Bruno S et al (2004) Mesenchymal stem cells contribute to the renal repair of acute tubular epithelial injury. Int J Mol Med 14:1035–1041
Hung SC, Pochampally RR, Hsu SC et al (2007) Short-term exposure of multipotent stromal cells to low oxygen increases their expression of CX3CR1 and CXCR4 and their engraftment in vivo. PLoS One 2:e416
Inoue S, Popp FC, Koehl GE et al (2006) Immunomodulatory effects of mesenchymal stem cells in a rat organ transplant model. Transplantation 81:1589–1595
Jiang CY, Gui C, He AN et al (2008) Optimal time for mesenchymal stem cell transplantation in rats with myocardial infarction. J Zhejiang Univ Sci B 9:630–637
Katritsis DG, Sotiropoulou PA, Karvouni E et al (2005) Transcoronary transplantation of autologous mesenchymal stem cells and endothelial progenitors into infarcted human myocardium. Catheter Cardiovasc Interv 65:321–329
Krampera M, Glennie S, Dyson J et al (2003) Bone marrow mesenchymal stem cells inhibit the response of naive and memory antigen-specific T cells to their cognate peptide. Blood 101:3722–3729
Lanotte M, Allen TD, Dexter TM (1981) Histochemical and ultrastructural characteristics of a cell line from human bone-marrow stroma. J Cell Sci 50:281–297
Le Blanc K, Tammik L, Sundberg B et al (2003) Mesenchymal stem cells inhibit and stimulate mixed lymphocyte cultures and mitogenic responses independently of the major histocompatibility complex. Scand J Immunol 57:11–20
Lee KD, Kuo TK, Whang-Peng J et al (2004) In vitro hepatic differentiation of human mesenchymal stem cells. Hepatology 40:1275–1284
Lim SY, Kim YS, Ahn Y (2006) The effects of mesenchymal stem cells transduced with Akt in a porcine myocardial infarction model. Cardiovasc Res 70:530–542
Mangi AA, Noiseux N, Kong D et al (2003) Mesenchymal stem cells modified with Akt prevent remodeling and restore performance of infarcted hearts. Nat Med 9:1195–1201
Minguell JJ, Erices A, Conget P (2001) Mesenchymal stem cells. Exp Biol Med 226:507–520
Morrison SJ, Shah NM, Anderson DJ (1997) Regulatory mechanisms in stem cell biology. Cell 88:287–298
Nasef A, Chapel A, Mazurier C et al (2007) Identification of IL-10 and TGF-beta transcripts involved in the inhibition of T-lymphocyte proliferation during cell contact with human mesenchymal stem cells. Gene Expr 13:217–226
Németh K, Leelahavanichkul A, Yuen PS et al (2009) Bone marrow stromal cells attenuate sepsis via prostaglandin E(2)-dependent reprogramming of host macrophages to increase their interleukin-10 production. Nat Med 15:42–49
Oyagi S, Hirose M, Kojima M et al (2006) Therapeutic effect of transplanting HGF-treated bone marrow mesenchymal cells into CCl4-injured rats. J Hepatol 44:742–748
Parekkadan B, van Poll D, Megeed Z et al (2007) Immunomodulation of activated hepatic stellate cells by mesenchymal stem cells. Biochem Biophys Res Commun 363:247–252
Perin EC, Silva GV, Assad JA et al (2008) Comparison of intracoronary and transendocardial delivery of allogeneic mesenchymal cells in a canine model of acute myocardial infarction. J Mol Cell Cardiol 44:486–495
Pervsner-Fischer M, Morad V, Cohen-Sfady M et al (2007) Toll-like receptors and their ligands control mesenchymal stem cell functions. Blood 109:1422–1432
Pittenger MF, Mackay AM, Beck SC et al (1999) Multilineage potential of adult human mesenchymal stem cells. Science 284:143–147
Pulavendran S, Vignesh J, Rose C (2010) Differential anti-inflammatory and anti-fibrotic activity of transplanted mesenchymal vs. hematopoietic stem cells in carbon tetrachloride-induced liver injury in mice. Int Immunopharmacol 10:513–519
Rafei M, Hsieh J, Fortier S et al (2008) Mesenchymal stromal cell-derived CCL2 suppresses plasma cell immunoglobulin production via STAT3 inactivation and PAX5 induction. Blood 112:4991–4998

Rasmusson I (2006) Immune modulation by mesenchymal stem cells. Exp Cell Res 312: 2169–2179

Sackstein R, Merzaban JS, Cain DW et al (2008) Ex vivo glycan engineering of CD44 programs human multipotent mesenchymal stromal cell trafficking to bone. Nat Med 14:181–187

Sakaida I, Terai S, Yamamoto N et al (2004) Transplantation of bone marrow cells reduces CCl4-induced liver fibrosis in mice. Hepatology 40:1304–1311

Salem HK, Thiemermann C (2010) Mesenchymal stromal cells: current understanding and clinical status. Stem Cells 28:585–596

Sato Y, Araki H, Kato J et al (2005) Human mesenchymal stem cells xenografted directly to rat liver are differentiated into human hepatocytes without fusion. Blood 106:756–763

Segers VF, Van Riet I, Andries LJ et al (2006) Mesenchymal stem cell adhesion to cardiac microvascular endothelium: activators and mechanisms. Am J Physiol Heart Circ Physiol 290:H1370–H1377

Serralta A, Donato MT, Orbis F et al (2003) Functionality of cultured human hepatocytes from elective samples, cadaveric grafts and hepatectomies. Toxicol In Vitro 17:769–774

Serralta A, Donato MT, Martinez A et al (2005) Influence of preservation solution on the isolation and culture of human hepatocytes from liver grafts. Cell Transplant 14:837–843

Shake JG, Gruber PJ, Baumgartner WA et al (2002) Mesenchymal stem cell implantation in a swine myocardial infarct model: engraftment and functional effects. Ann Thorac Surg 73: 1919–1925

Shi M, Li J, Liao L et al (2007) Regulation of CXCR4 expression in human mesenchymal stem cells by cytokine treatment: role in homing efficiency in NOD/SCID mice. Haematologica 92:897–904

Silva GV, Litovsky S, Assad JA et al (2005) Mesenchymal stem cells differentiate into an endothelial phenotype, enhance vascular density, and improve heart function in a canine chronic ischemia model. Circulation 111:150–156

Sotiropoulou PA, Perez SA, Gritzapis AD et al (2006) Interactions between human mesenchymal stem cells and natural killer cells. Stem Cells 24:74–85

Tang J, Xie Q, Pan G et al (2006) Mesenchymal stem cells participate in angiogenesis and improve heart function in rat model of myocardial ischemia with reperfusion. Eur J Cardiothorac Surg 30:353–361

Till JE, Mcculloch EA (1961) A direct measurement of the radiation sensitivity of normal mouse bone marrow cells. Radiat Res 14:213–222

Tögel F, Weiss K, Yang Y et al (2007) Vasculotropic, paracrine actions of infused mesenchymal stem cells are important to the recovery from acute kidney injury. Am J Physiol Renal Physiol 292:F1626–F1635

Tokcaer-Keskin Z, Akar AR, Ayaloglu-Butun F et al (2009) Timing of induction of cardiomyocyte differentiation for in vitro cultured mesenchymal stem cells: a perspective for emergencies. Can J Physiol Pharmacol 87:143–150

Tomchuck SL, Zwezdaryk KJ, Coffelt SB (2008) Toll-like receptors on human mesenchymal stem cells drive their migration and immunomodulating responses. Stem Cells 26:99–107

Urbanek K, Rota M, Cascapera S et al (2005) Cardiac stem cells possess growth factor-receptor systems that after activation regenerate the infarcted myocardium, improving ventricular function and long-term survival. Circ Res 97:663–673

Waterman RS, Tomchuck SL, Henkle SL et al (2010) A new mesenchymal stem cell (MSC) paradigm: polarization into a pro-inflammatory MSC1 or an immunosuppressive MSC2 phenotype. PLoS One 5:e10088

Weissman IL (2000) Stem cells: units of development, units of regeneration, and units in evolution. Cell 100:157–168

Wollert KC, Meyer GP, Lotz J et al (2004) Intracoronary autologous bone-marrow cell transfer after myocardial infarction: the BOOST randomised controlled clinical trial. Lancet 364:141–148

Zhao DC, Lei JX, Chen R et al (2005) Bone marrow-derived mesenchymal stem cells protect against experimental liver fibrosis in rats. World J Gastroenterol 11:3431–3440

Chapter 7
Tissue Engineering Based on the Importance of Collaboration Between Clinicians and Basic Scientists Regarding Mesenchymal Stromal Cells

Aysel Yurtsever

Abstract Tissue engineering is an interdisciplinary field that applies the principles of biology and engineering to developing tissue substitutes to restore, maintain, or improve the function of diseased or damaged human tissues. Autologous mesenchymal stromal cells (MSCs) are good candidates for tissue engineering and regenerative medicine in that they can replace damaged tissues in the human body owing to their self-renewal, plasticity, engraftment, and homing capacity. MSCs can easily differentiate into adipocytes, osteoblasts, and fibroblasts using various transcription factors and hormones. For cell treatments, nanotechnological scaffolds in various structures are needed for differentiation of stem cells. Biodegradable polymeric constructs for bone tissue engineering, are three-dimensional structures that allow bone cells to attach and reproduce on them. Because of biodegradability properties, they are not permanent in the body and are degraded slowly while bone cells are reproducing. Thus, bone cells replace the scaffold in time, which means healing of the defective site.

7.1 Introduction

Tissue engineering is involved with the creation of organs and tissues under laboratory conditions with the aim of transplantation to patients. It is at the crossroads of biotechnology, nanobiotechnology, molecular biology and genetics, bioinformatics, medicine, and engineering specialties and is a rapidly developing area.

Among adult stem cells, mesenchymal stromal cells (MSCs) are those best suited to tissue engineering because of their mesodermal origin (Fig. 7.1). Human MSCs are spindle-shaped, fibroblast-like cells. Many MSCs (91%) stay in the G_0/G_1

A. Yurtsever, M.D. (✉)
Ege Üniversity Cancer Research Center (consultant), İzmir, Turkey
e-mail: aysel.yurtsever@gmail.com

K. Turksen (ed.), *Adult and Embryonic Stem Cells*, Stem Cell Biology and Regenerative Medicine, DOI 10.1007/978-1-61779-630-2_7,

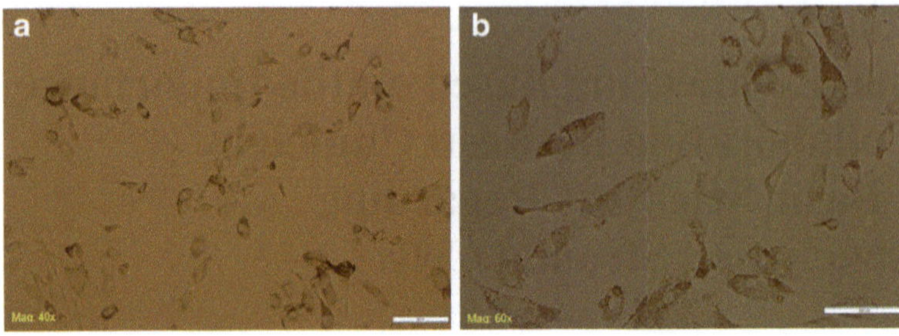

Fig. 7.1 Nonmagnetic enrichment of mesenchymal stem cells (MSCs) from umbilical cord blood (UCB) with 20 h of incubation prior to standard centrifugation with Ficoll. (**a**) ×40. (**b**) ×60

phase for most of their life. They have various specific characteristics—self renewal, plasticity, engraftment, homing—that enable them to replaced damaged tissues in the human body. MSCs can easily differentiate into adipocytes, osteoblasts, and fibroblasts using various transcription factors and hormones (Patel et al. 2008a; Carlo-Stella and Gianni 2005). Tumor tracking properties of MSC provide an attractive opportunity for targeted transgene delivery into the sites of tumor formation (Kucerova et al. 2007).

7.2 Approach and Experimental Methods

We obtained umbilical cord blood (UCB) with informed consent from voluntary donors and used a fully automated method (Sepax UCB-HES method) to isolate the mononuclear stem cells (SCs) from the UCB. This method involves a closed system that uses gradient density centrifugation as the isolation technique. Then the SCs are tested for their number and viability using a flow cytometer (ViCell Beckman Coulter Automated Cell Viability Analyzer, USA) and microscopic trypan blue staining methods. MSCs were obtained from UCB using RosetteSep Human Mesenchymal Stem Cell Enrichment Cocktail (Stemcell Technologies Vancouer Canada). RosetteSep antibody cocktail was added to each UCB sample, and the samples were incubated at room temperature for 20 min. The samples were then diluted with Hank's balanced salt solution (HBSS) and layered over an equal volume of Ficoll-Paque (1.077 g/ml). The enriched cells were recovered from the gradient interface after centrifugation at 400 g for 25 min and washed twice with HBSS after centrifugation at 550 g for 5 min.

These enriched mesenchymal precursor cells were suspended in MesenCult MSC Basal Medium (Human) (Stemcell Technologies Vancouer Canada) supplemented with MesenCult Mesenchymal Stem Cell Stimulatory Supplements (Human) and plated in 25 cm^2 culture flasks at a density of 1×10^5 cells/ml. The cultures were maintained at 37°C in 5% CO_2 in fully humidified air. On day 5, the medium was changed for the first time and then changed every 2 days thereafter (Fig. 7.2).

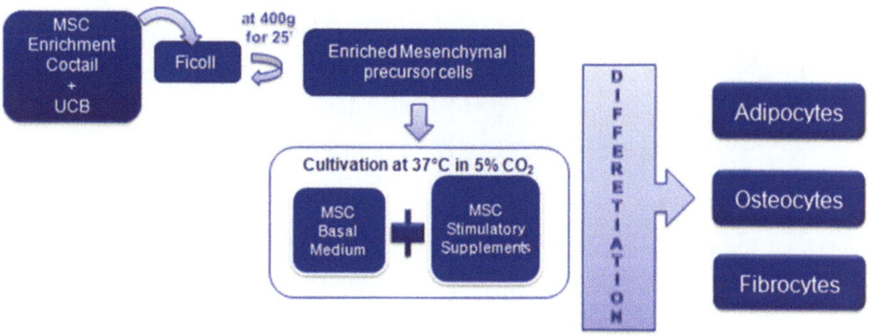

Fig. 7.2 Experimental methods

7.3 Results and Analysis

The adherent cells were observed on the fifth day of cultivation. The spindle-shaped cells appeared at the bottom of culture flasks. Through continuous changes of the medium, the suspended cells in the culture flask became fewer. When the medium had been changed twice, the suspended cells were completely removed from the medium. The adherent cells were fibroblast-like and grew as a whirlpool.

Fibroblasts were observed on the tenth day of culture (Fig. 7.3b). Adipocytes were seen after day 10, and they appeared as mature adipocytes on day 18 (Fig. 7.3c). Osteocytes were observed starting from day 15 of culture. The alkaline phosphatase levels in the culture medium, upper phase, were evaluated between days 16 and 28 (Fig. 7.4). It was observed that the osteocytes were active until day 28 (Fig. 7.3d). An Olympus IX71 camera was used for observations and recordings.

The MSC cultures can be characterized by CD90 and CD105 expression (Fig. 7.5). The total number of mononuclear cells in UCB are 13.67×10^6 cells/ml, and the viability of the cells is 98.83% (trypan blue) and 98.71% (flow cytometer). The total CD90- and CD105-expressing cell count is 0.13×10^6 cells/ml (0.13×10^3 cells/μl) (Fig. 7.5b–d).

7.3.1 *Sources of Mesenchymal Stem Cells*

Mesenchymal stem cells are found in various tissue types. The main source of MSCs is adipose tissue. The other source of MSCs are UBC, muscle tissue, bone marrow, cartilage, tendons, vascular tissue, dental pulp, and ligaments. These cells constitute, however, only a small percentage of the total number of MSCs in bone marrow. Pittenger et al. showed that only 0.01–0.001% of mononuclear cells isolated on a density gradient (Ficoll/Percoll) give rise to plastic adherent fibroblast-like colonies (Pittenger et al. 1999). In addition to bone marrow, MSCs are located

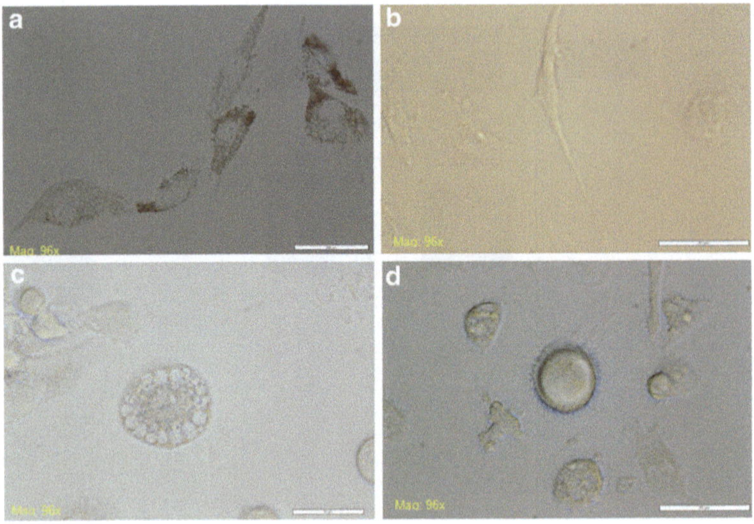

Fig. 7.3 (**a**) MSCs (×96). (**b**) Fibrocyte (×96). (**c**) Adipocyte (×96). (**d**) Osteocyte (×96)

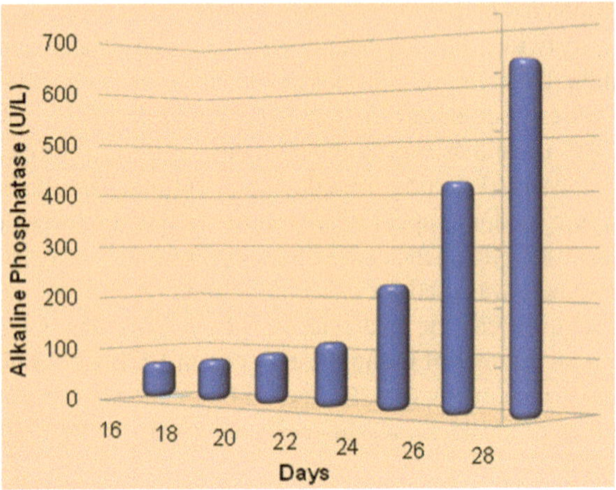

Fig. 7.4 Alkaline phosphatase levels (U/L) at 16–28 days of MSC culture at 37°C in 5% CO_2

in other tissues of the human body. There is an increasing number of reports describing their presence in adipose tissue (Gronthos et al. 2001), UCB, chorionic villi of the placenta (Igura et al. 2004), amniotic fluid (Tsai et al. 2004), peripheral blood (Zvaifler et al. 2000), and even in exfoliated deciduous teeth (Miura et al. 2003). Studies during the last decade showed that human UCB contains hematopoietic stem cells and MSCs, both of which can be used as alternative sources to bone marrow for cell transplantation and therapy. The hematopoietic stem cells in UCB have already been proven useful for treating various hematological disorders.

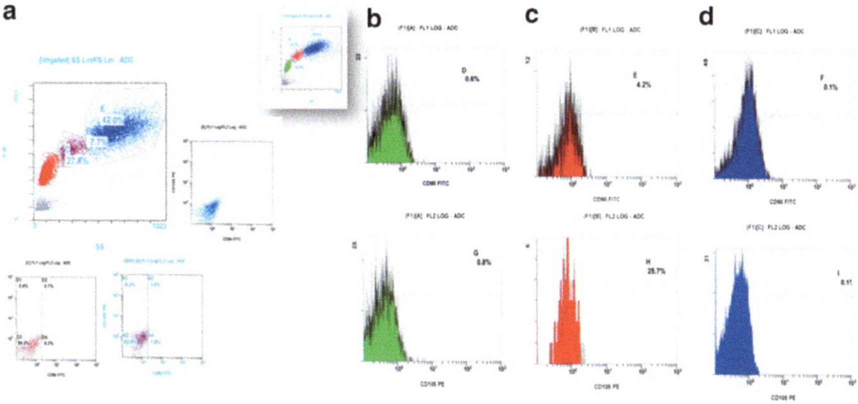

Fig. 7.5 Flow cytometric analysis of UCB for MSC markers CD90 and CD105. (**a**) Gate C (*red*) lymphocytes; gate B (*purple*) monocytes; gate E (*blue*) granulocytes. (**b**) CD90+/CD105+ lymphocytes. (**c**) CD90+/CD105+ monocytes. (**d**) CD90+/CD105+ granulocytes

7.3.2 Surface Markers on Mesenchymal Stem Cells

The MSCs constitute a heterogeneous population of cells in terms of their morphology, physiology, and expression of surface antigens. Up to now, no single specific marker has been identified. MSCs express a large number of adhesion molecules, extracellular matrix proteins, cytokines, and growth factor receptors associated with their function and cell interactions within the bone marrow stroma (Bobis et al. 2006; Devine and Hoffman 2000). The population of MSCs isolated from bone marrow express CD44, CD105 (SH2, or endoglin), CD106 (vascular cell adhesion molecule, or VCAM-1), CD166, CD29, CD73 (SH3 and SH4), CD90 (Thy-1), CD117, STRO-1, and Sca-1 (Baddoo et al. 2003; Boiret et al. 2005; Cognet and Minguell 1999; Dennis et al. 2002; Gronthos et al. 2003).

The MSCs have multipotential differentiation capacity. MSCs are homogeneous, fibroblast-like cells. Population of cells in G_0/G_1 phase are in large quantities, which means that MSCs have very high potential for differentiation. MSC subgroups vary in their RNA and DNA content, size, and granule content. At rest, MSCs are immobile. They can be isolated from many sources, including bone marrow, tendons, blood vessels, dental pulp, periodontal ligaments, adipose tissue, cartilage tissue, and cord blood. MSCs serve as "cellular vehicles" for reaching the target tissue.

The MSCs can act as local production sites for soluble biological agents, considered an important therapeutic strategy (Klingemann et al. 2008). MSCs with directed migration ability to tumor sites can be used as potential vehicles to deliver anticancer agents to malignant cells (Klingemann et al. 2008). MSCs can thus be used for delivery of therapeutic agents to appropriate targets, making them attractive cellular therapy devices.

Some MSCs have been used in preclinical models for tissue engineering of bone, cartilage, muscle, marrow stroma, tendon, fat, and other connective tissues. These tissue-engineered materials show considerable promise for use in rebuilding damaged or diseased mesenchymal tissue (Caplan 2007).

Bone regeneration is achieved by ex vivo diffusion of bone marrow-derived stem cells and attachment of these cells to hydroxyapatite/tricalcium phosphate ceramic three-dimensional (3D) scaffolds. Such scaffolds have been transplanted into segmental defects that were experimentally opened in os longum cells. An appropriate in vivo 3D structure can be obtained with this technique. Successful applications of cartilage and bone tissue engineering in rabbits, horses, and pigs led to promising results in cartilage and bone tissue repair. These applications were then entered into clinical trials with autologous 3D chondrocyte and bone grafts, respectively. In 2003, "proof of market" could be shown for two new engineering products: BioSeed-C and BioSeed-Oral Bone (Bio tissue, Freiburg, Germany). These molecules are immunosuppressive, especially for T cells.

Thus, allogeneic MSCs can be considered for therapeutic use. In this context, the secreted bioactive molecules provide a regenerative microenvironment for a variety of injured adult tissues to limit the area of damage and to mount a self-regulated regenerative response. This regenerative microenvironment is referred to as trophic activity. MSCs appear to be valuable mediators for tissue repair and regeneration. The natural titers of MSCs that are drawn to sites of tissue injury can be augmented by allogeneic MSCs delivered via the bloodstream. Indeed, human clinical trials are now under way to use allogeneic MSCs for treatment of myocardial infarcts, graft-versus-host disease, Crohn's disease, cartilage and meniscus repair, stroke, and spinal cord injury (Caplan 2007).

7.3.3 Tissue Engineering

In tissue engineering, various branches of medicine and engineering involved with growing molecules, cells, tissues, and organs under laboratory conditions work together for the support, development, or regeneration of nonfunctioning, damaged, or congenitally abnormal tissues and organs. MSCs are being utilized in bone tissue, cartilage tissue, and adipose tissue engineering.

Tissue engineering research is making rapid progress because of the collaboration of basic scientists and clinicians. Scaffolds with different structures are needed to differentiate the cells grown on a 3D tissue culture basis in a desired direction. In accordance with the interdisciplinary division of labor, the basic scientist prepares the material to be applied and the clinician carries out the application procedure.

Adipose tissue is important for breast enlargement by mammoplasty, other clinical applications, and even treatment of type II diabetes. Adipose cells extract lipids and fatty acids from the bloodstream, and they have an important role in converting glucose into energy via insulin. Polietilen tereftalat (PET), or Dacron fibrils, are used as the scaffold.

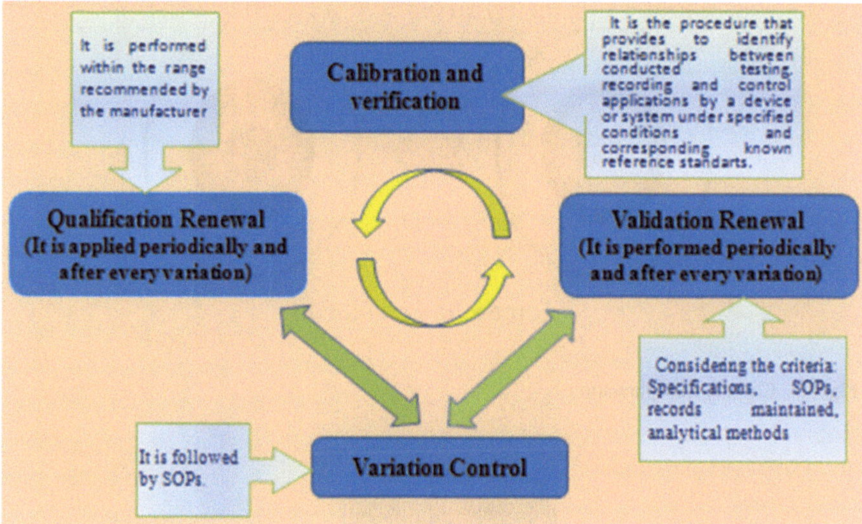

Fig. 7.6 Good manufacturing practice: maintenance, variation, renewal

Basic scientists have conducted studies on the proliferation and differentiation of cells, characterization and validation, designing biomaterials to direct the growth and differentiation of cells during functional tissue formation, bioreactors supporting cell growth, gene and protein arrays, automated quality systems, and Petri nets. These studies can be conducted only under laboratory conditions that comply with Good Manufacturing Practices (GMPs) standards (Fig. 7.6).

An example of collaboration between basic scientist and clinician is the following. The cultivation of cells needed for corneal transplantation is performed by basic scientists, but application to the patient is performed by surgeons specialized in ophthalmology. Tissue engineering of the cornea represents a paradigm shift in medical treatment to overcome the present disadvantages of corneal transplantation, primarily immune rejection and the shortage of donor corneas. Transplantation of cultivated corneal epithelial cells expanded ex vivo from corneal epithelial stem cells has been developed and has already entered the clinical realm (Fig. 7.7). However, there are still many hurdles to overcome. The author and colleagues are developing a method to transplant cultivated cell sheets that uses a temperature-responsive culture dish. Nishida reviewed the present situation regarding tissue-engineered corneal epithelium and introduced the results of this program (Nishida 2003).

In the area of cartilage and bone repair in the field of orthopedics, whereas the 3D collagen cultures and allogenic bone tissue suitable for transplantation are prepared by the basic scientist, the orthopedic surgeon carries out the surgical procedure. Basic scientists and clinicians also collaborate on the treatment of peripheral vascular diseases (e.g., diabetes mellitus, thromboangiitis obliterans). A final example lies in the area of heart disease, where heart valves are manufactured via tissue engineering and their application by the medical profession has become common practice (2008).

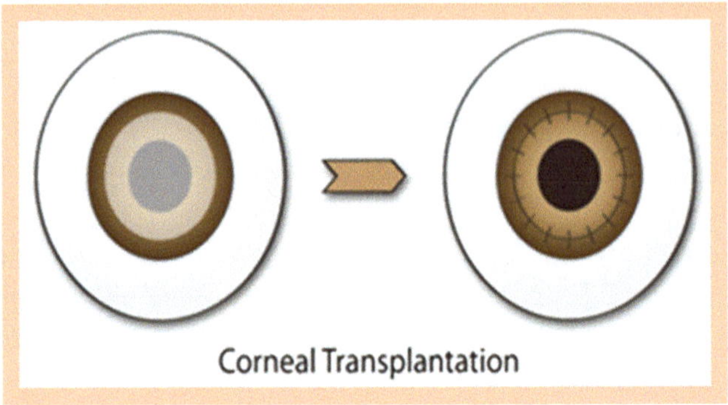

Fig. 7.7 Corneal transplantation

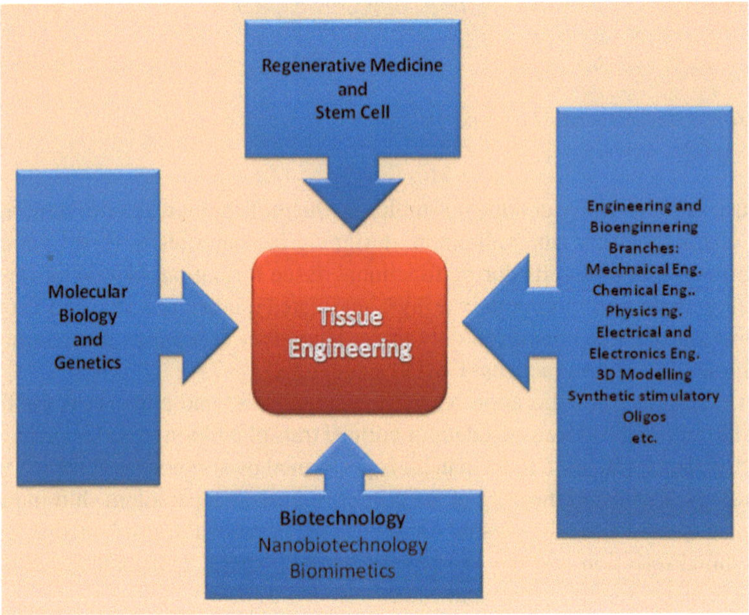

Fig. 7.8 Tissue engineering at the molecular level for mesenchymal cellular therapy

Tissue engineering will reach its highest level in centers of excellence conducting interdisciplinary studies involving genes, nanotechnology, and tissue stem cells (Fig. 7.8). Valvular heart disease is a significant cause of morbidity and mortality worldwide. Classic replacement surgery involves the implantation of mechanical valves or biological valves (xenografts or homografts). Tissue engineering of heart valves represents a new experimental concept for improving current modes of therapy in valvular heart surgery (Fig. 7.9). Various approaches have been developed that differ either in the choice of scaffolds (synthetic biodegradable polymers, decellularized xenografts or homografts) or the cell sources for producing living tissue like vascularly derived cells, bone marrow cells, progenitor cells from peripheral blood (Sacks et al. 2009).

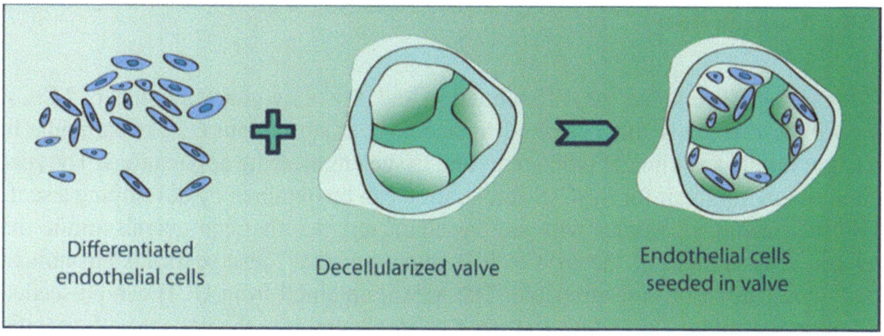

Fig. 7.9 Heart valve production via tissue engineering

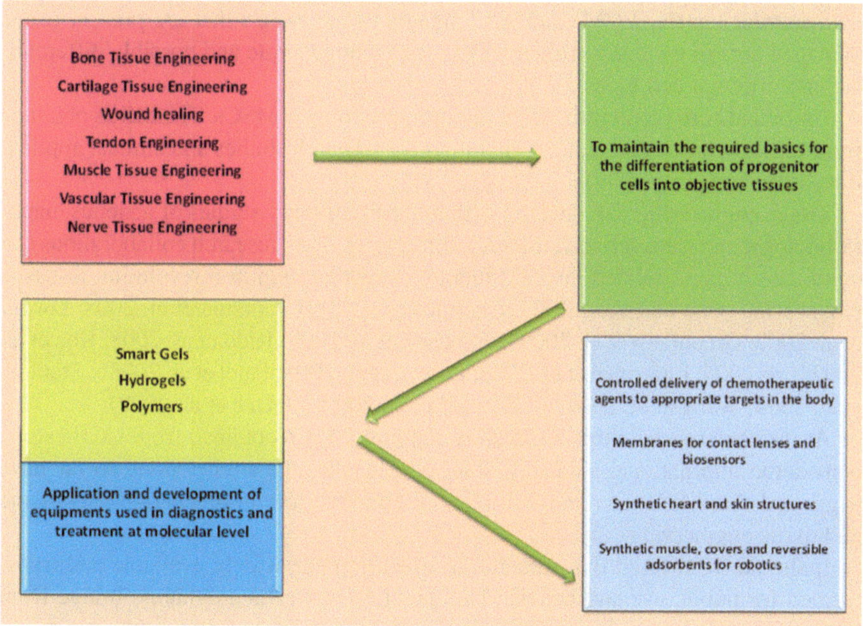

Fig. 7.10 Interdisciplinary research collaborations in tissue engineering

The use of autologous bone marrow cells in combination with synthetic biodegradable scaffolds has advantages over other tissue engineering approaches. It is safe; it leads to completely autologous prostheses; and the cells are more easily obtained. Clinical trials with stem cells for liver diseases provide renewed hope for the future. They are certainly timely in the current climate of increasing morbidity and mortality associated with transplant waiting lists around the world (Neuenschwander and Hoerstrup 2004). Interdisciplinary research collaboration in tissue engineering is creating centers of excellence that bring together unique teams of cell and molecular biologists, biomaterials scentists and engineers, and surgeons (Fig. 7.10).

7.4 Conclusion

By learning more about the differentiation and plasticity characteristics of MSCs, we will be able to use them more advantageously as autologous cells in particular in a larger number of clinical indications and tissue engineering applications. The specialized environment for MSC differentiation can be obtained by developing a scaffold that has nanoparticles with similar characteristics that can overall mimic the natural environment of the MSCs. It is important that these scaffolds are human body-compatible or biodegradable. The MSCs obtained from UCB can be seeded on these scaffolds. These constructs can be implanted into the patient with specific growth factors to induce differentiation into various cell types such as adipocytes, osteoblasts, and fibroblasts. These nanotechnological scaffolds can also be designed by tissue engineering methods in such a manner that they can replace the natural extracellular matrix (ECM) and send signals to the MSCs that are present in the damaged area in the body, triggering the cells to regenerate and possibly differentiate into different cell types.

Umbilical cord blood is the most important source for MSCs. The MSCs obtained from UCB do not lead to ethical dilemmas for research studies and clinical applications (Cao and Feng 2009).

Tissue engineering will develop with multidisciplinary studies of genetics, nanotechnology, and stem cell technologies. Interdisciplinary research collaborations can culminate in personalized targeted therapy—another step in a revolution in tissue engineering (Cao and Feng 2009; Choppes et al. 2009; Corsten et al. 2008; Dwyer et al. 2007; Greenberg et al. 2009; Jorgensen et al. 2009; Kidd et al. 2008; Koç et al. 2002; Lee 2005; Lazennec et al. 2008; Ozawa et al. 2008; Patel et al. 2008b; Studeny et al. 2004; Wang et al. 2004; Woodward et al. 2009; Zielske et al. 2009).

At the centers where this research is ongoing, MSCs obtained from UCB can be considered international stem cell sources. Many of the studies done so far have suggested that MSCs can and will play an effective role in regenerative medicine and tissue engineering.

It should be ensured that research results are made widely available and transformed for public use and benefit. This practice serves as a dynamic bridge from laboratory to industry to make certain that promising new technologies are translated into products and services that benefit society and the world.

Human adipose tissue-derived mesenchymal stem cells (ATMSC) are a promising source of autologous stem cells to be used in personalized cell-based therapy. Tumor tracking properties of MSCs provide an attractive opportunity for targeted transgene delivery to sites of tumor formation.

References

Baddoo M, Hill K, Wilkinson R, Gaupp D, Hughes C, Kopen GC, Phinney DG (2003) Characterization of mesenchymal stem cells isolated from murine bone marrow by negative selection. J Cell Biochem 89:1235–1249

Bobis S, Jarocha D, Majka M (2006) Mesenchymal stem cells: characteristics and clinical applications. Folia Histochem Cytobiol 44(4):215–230

Boiret N, Rapatel C, Veyrat-Masson R, Guillouard L, Guérin J-J, Pigeon P, Descamps S, Boisgard S, Berger MG (2005) Characterization of nonexpanded mesenchymal progenitor cells from normal adult human bone marrow. Exp Hematol 33:219–225

Cao F-J, Feng S-Q (2009) Human umbilical cord mesenchymal stem cells and the treatment of spinal cord injury. Chin Med J (Engl) 122(2):225–231

Caplan AI (2007) Adult mesenchymal stem cells for tissue engineering versus regenerative medicine. J Cell Physiol 213(2):341–347

Carlo-Stella C, Gianni MA (2005) Biology and clinical applications of marrow mesenchymal stem cells. Pathol Biol 53:162–164

Choppes RP et al (2009) Stem cell therapy to reduce radiation-induced normal tissue damage. Semin Radiat Oncol 19:112–121

Cognet PA, Minguell JJ (1999) Phenotypical and functional properties of human bone marrow mesenchymal progenitor cells. J Cell Physiol 181:67–73

Corsten MF et al (2008) Therapeutic stem-cells for cancer treatment: hopes and hurdles in tactical warfare. Lancet Oncol 9:376–384

Dennis JE, Carbillet JP, Caplan AI, Charbord P (2002) The STRO-1+ marrow cell population is multipotential. Cells Tissues Organs 170:73–82

Devine SM, Hoffman R (2000) Role of mesenchymal stemcell in hematopoietic stem cell transplantation. Curr Opin Hematol 7:358–363

Dwyer RM et al (2007) Monocyte chemotactic proteion-1 secreted by primary breast tumors stimulates migration of mesenchymal stem cells. Clin Cancer Res 13(17):5020–5027

Greenberg JS et al (2009) Bone marrow-derived stem cells and radiation response. Semin Radiat Oncol 19:133–139

Gronthos S, Franklin DM, Leddy HA, Robey PG, Storms RW, Gimble JM (2001) Surface protein characterization of human adipose tissue-derived stromal cells. J Cell Physiol 189:54–63

Gronthos S, Zannettino AC, Hay DJ, Shi S, Graves SE, Kortesidis A, Simmonos PJ (2003) Molecular and cellular characterization of highly purified stromal stems derived from human bone marrow. J Cell Sci 116:1827–1835

Igura K, Zhang X, Takahashi K, Mitsuru A, Yamaguchi S, Takashi TA (2004) Isolation and characterization of mesenchymal progenitor cells from chorionic villi of human placenta. Cytotherapy 6:543–553

Jorgensen C et al (2009) Link between cancer stem cells and adult mesenchymal stromal cells: implications for cancer therapy. Regen Med 4(2):149–152

Kidd S et al (2008) The auspicious role of mesenchymal stromal cells in cancer: be it friend or foe. Cytotherapy 10(7):657–667

Klingemann H et al (2008) Mesenchymal stem cells: sources and clinical applications. Transfus Med Hemother 35:272–277

Koç ON et al (2002) Allogeneic mesenchymal stem cell infusion for treatment of metachromatic leukodystrophy (MLD) and Hurler syndrome (MPS-IH). Bone Marrow Transplant 30:215–222

Kucerova L, Altanerova V, Matuskova M, Tyciakova S, Altaner C (2007) Adipose tissue-derived human mesenchymal stem cells mediated prodrug cancer gene therapy. Cancer Res 67(13): 6304–6313

Lazennec G et al (2008) Concise review: adult multipotent stromal cells and cancer: risk or benefit? Stem Cells 26(6):1387–1394

Lee K-D (2005) Mesenchymal stem cells. Chang Gung Med J 31(3):228–235

Miura M, Gronthos S, Zhao M, Lu B, Fisher LW, Robey PG, Shi S (2003) SHED: stem cells from human exfoliated deciduous teeth. Proc Natl Acad Sci USA 100:5807–5812

Neuenschwander S, Hoerstrup SP (2004) Organogenesis and tissue engineering in transplantation medicine. Transpl Immunol 12(3–4):359–365

Nishida K (2003) Tissue engineering of the cornea. Cornea 22:S28–S34

Ozawa K et al (2008) Cell and gene therapy using mesenchymal stem cells (MSCs). J Autoimmun 30:121–127

Patel SA, Sherman L, Munoz J, Rameshwar P (2008a) Immunological properties of mesenchymal stem cells and clinical implications. Arch Immunol Ther Exp 56:1–8

Patel SA et al (2008b) Immunological properties of mesenchymal stem cells and clinical implications. Arch Immunol Ther Exp (Warsz) 56:1–8

Pittenger MF, Mackay AM, Beck SC, Jaiswal RK, Douglas R, Mosca JD, Moorman MA, Simoneti DW, Craig S, Marshak DR (1999) Multilineage potential of adult human mesenchymal stem cells. Science 284:143–147

Studeny M et al (2004) Mesenchymal stem cells: potential precursors for tumor stroma and targeted-delivery vehicles for anticancer agents. J Natl Cancer Inst 96(21):1593–1603

Sacks MS, Schoen FJ, Mayer JE (2009) Bioengineering challenges for heart valve tissue engineering. Ann Rev Biomed Eng 11:289–313

Tsai MS, Lee JL, Chang YJ, Hwang SM (2004) Isolation of human multipotent mesenchymal stem cells from second trimester amniotic fluid using a novel two-stage culture protocol. Hum Reprod 19:1450–1456

Wang HS et al (2004) Mesenchymal stem cells in the Wharton's jelly of the human umbilical cord blood. Stem Cells 22:1330–1337

Woodward WA et al (2009) Radiosensitivity of cancer-initiated cells and normal stem cells (or what the Heisenberg uncertainly principle has to do with biology). Semin Radiat Oncol 19:87–95

Zielske SP et al (2009) Radiation increases invasion of gene-modified mesenchymal stem cells into tumors. Int J Radiat Oncol Biol Phys 75(3):843–853

Zvaifler NJ, Marinova-Mutafchieva L, Adams G, Edwards CJ, Moss J, Burger JA, Maini RN (2000) Mesenchymal precursor, cells in the blood of normal individuals. Arthritis Res 2:477–488

Chapter 8
Synchroton Radiation and Nanotechnology for Stem Cell Researchers

F. Fiori, A. Giuliani, A. Manescu, C. Renghini, and F. Rustichelli

Abstract Stem cell-based tissue engineering therapies involve the administration of ex vivo manipulated stem cell populations for the purpose of repairing and regenerating damaged or diseased tissue. Currently available methods for monitoring transplanted cells are limited. Monitoring stem cell therapy outcomes requires the development of nondestructive strategies capable to identify the location, magnitude, and duration of cellular survival and fate. The recent development of imaging techniques offers great potential for addressing these critical issues by noninvasively tracking the fate of the transplanted cells. We offer a focused presentation of some examples of the use of imaging techniques connected to the nanotechnological world in research areas related to stem cells. In particular, investigations concerning human stem cell treatment of Duchenne muscular dystrophy in animal models, bioscaffolds for cell proliferation to form muscular fibers, and bone tissue engineering are discussed.

8.1 Introduction

Tissue engineering and regenerative medicine are an emerging research area that promises new therapeutic techniques for the repair and replacement of tissues and organs that have lost functions due to aging, disease, damage, or congenital defects (Langer and Vacanti 1999; Atala 2005; Jones and Hench 2003). Clinical applications have already begun to repair a wide variety of tissues, such as blood, skin, cornea, cartilage, and bone.

F. Fiori (✉) • A. Giuliani • A. Manescu • C. Renghini • F. Rustichelli
Dipartimento Di.S.C.O. - Sezione Biochimica Biologia e Fisica, Università Politecnica delle Marche, Via Brecce Bianche, 60131 Ancona, Italy
e-mail: f.fiori@univpm.it

K. Turksen (ed.), *Adult and Embryonic Stem Cells*, Stem Cell Biology and Regenerative Medicine, DOI 10.1007/978-1-61779-630-2_8,

Imaging techniques are playing an increasingly important role in the rigorous characterization of biomaterial properties and function. Sophisticated two-dimensional (2D) imaging technologies have been developed to complement histological evaluation and probe complex biological events occurring at the interface between tissues and biomaterials (Boskey and Pleshko Camacho 2007; Campbell and Kim 2007; Huebsch and Mooney 2007). However, there is a clear need for high-resolution three-dimensional (3D) imaging technologies that reveal the spatial distribution of tissues forming within porous biomaterials in vitro and in vivo.

For regeneration of vascularized tissues such as bone or muscle, the ability to quantify 3D vascular ingrowth would be tremendously valuable, particularly for studies exploring the potential to enhance regeneration via therapeutic angiogenesis strategies (Silva and Mooney 2007). The imaging modality that has been applied most extensively for this purpose, particularly for bone tissue engineering studies (Mastrogiacomo et al. 2004; Komlev et al. 2006; Papadimitropoulos et al. 2007; Eniwumide et al. 2007), is high-resolution X-ray computed tomography (CT). CT provides rapid reconstruction of 3D images and quantitative volumetric analysis of X-ray attenuating materials or tissues. In the perspective of clinical translation of stem cell research, it would be advantageous to develop new techniques to detect donor cells after transplantation to track their fate and thus better understand their role in the regeneration of damaged and diseased tissues.

Several groups have reported successful labeling of mesenchymal pig (Hill et al. 2003) and mouse (Hoehn et al. 2002) embryonic stem cells with nanoparticles of iron oxide (SPIO). These particles are used as contrast agents for magnetic resonance imaging (MRI) (Arbab et al. 2003; Frank et al. 2003; Wang et al. 2001). It appears that cells that are able to incorporate SPIO intracellularly are readily detectable with MRI, allowing in vivo tracking of such "tagged" cells (Bulte et al. 2002). MRI provides a noninvasive and repeated 3D visualization of transplanted "tagged" stem cells in organs, making it particularly attractive for imaging studies (Nuzzo et al. 2002). The aim of this chapter, therefore, is to present some of the recent progress obtained using innovative, noninvasive imaging techniques and nanodiffraction involving nanotechnologies in research areas related to stem cells. In particular, we provide some examples of studies concerning human stem cell treatment of Duchenne muscular dystrophy in animal models, bioscaffolds for cell proliferation striving to form muscular fibers, and bone tissue engineering.

8.2 Synchrotron Radiation X-ray Computed Microtomography

X-ray computed microtomography (micro-CT) is similar to conventional CT systems usually employed in medical diagnostics. The main difference is that with micro-CT a spatial resolution of the order of a few hundred nanometers can be achieved (in contrast to about 0.5 mm for standard CT). Of course, such high spatial resolution can be obtained only for samples extremely reduced in size (a few cubic

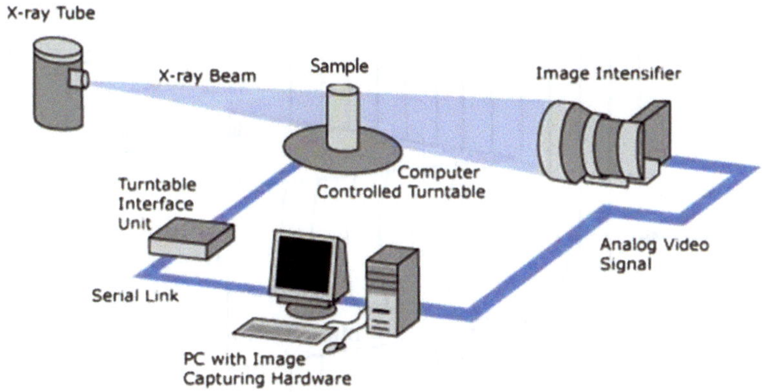

Fig. 8.1 X-ray computed microtomography setup

millimeters). In particular, with synchrotron radiation—available at some European large-scale facilities such as ESRF in Grenoble, PSI/SLS in Zurich, BESSY in Berlin, HASYLAB in Hamburg, and ELETTRA in Trieste—it is possible to couple high spatial resolution to a high signal-to-noise ratio (Nuzzo et al. 2002; Salomé et al. 1999). Furthermore, with respect to conventional laboratory sources, the advantages of X-rays produced at synchrotron radiation sources also include a very high photon flux and a tunable-energy monochromatic beam with high coherency in parallel beam geometry.

In a typical micro-CT set-up (Fig. 8.1), the sample is fixed on a turntable and is rotated through 180° or 360° with an angular step of the order of some tenths of a degree. The detecting device is usually comprised of a scintillator followed by a light intensifier and a CCD camera (e.g., 2,048 × 2,048 pixels). The pixel size in the image, and thus the resolution, depends on the field of view (FOV) to be explored: the wider the FOV, the worse the resolution.

Hundreds of radiographs (projections) are obtained at different rotation angles of the sample (Fig. 8.1). The information stored in each of them is a 2D projection of the X-ray absorption coefficient map, related to the various materials encountered by the beam along its path in the plane perpendicular to the beam direction. By means of suitable algorithms (filtered back-projection) based on the radon transform theory (Kak and Slaney 1988), the 2D projections are used to reconstruct the 3D volume morphology of the sample.

In the pure absorption experimental setup, the sample is as close as possible to the detector, as in this way image blurring is minimized. Anyway, by choosing a suitable sample-to-detector distance, the contrast is no longer given by pure absorption only but also by the phase differences among the scattered X-ray waves. In particular, this phase-contrast effect puts into evidence the interface and edges between two materials (Fig. 8.2), and it is particularly useful when media with similar absorption coefficients should be discriminated.

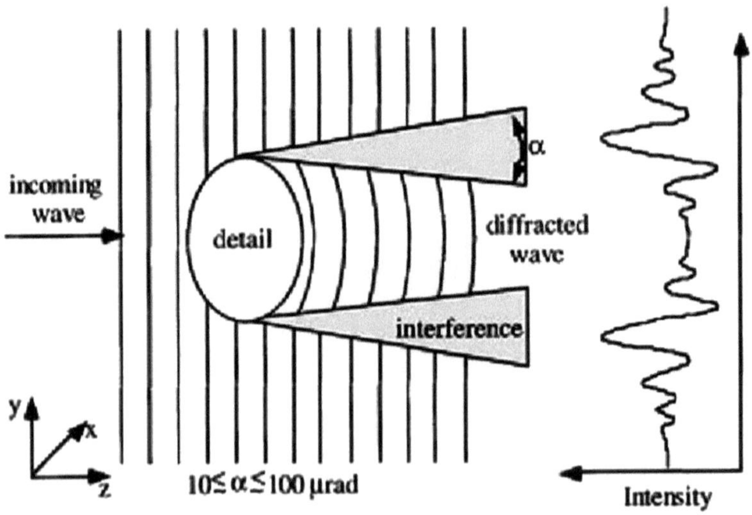

Fig. 8.2 Phase-contrast principle

8.3 Applications

8.3.1 Study of Stem Cell Fate in Dystrophic Muscular Tissue

This work (Torrente et al. 2006) was carried out in the framework of a more general research program aiming to use human stem cells to repair muscle damage in Duchenne muscular dystrophy. In previous studies (Torrente et al. 2004; Gavina et al. 2006) it was shown that, after intraarterial delivery to murine dystrophic muscle human blood-derived CD133+ cells localize under the basal lamina and expressed the satellite cells markers M-cadherin and Myf5, differentiating into human muscle fibers and causing significant amelioration of skeletal muscle structure. Elucidation of the mechanisms involved in muscle homing of stem cells can aid in improving potential therapy for muscular dystrophy based on the systemic delivery of such stem cells. Iron oxide nanoparticle (Endorem; Guerbet, Sulzback, Germany) labeling is a promising approach to visualize stem cells in vivo and thus can help us understand the basic processes involved in stem cell homing and migration (Gupta and Gupta 2005; Reimer and Weissleder 1996).

Human blood-derived CD133+ cells were isolated from mononucleated cells collected by centrifugation (Ficoll-Hypaque; Pharmacia Biotech, Uppsala, Sweden) of several buffy coats, diluted 1:2 in RPMI 1640 medium (GIBCO, Invitrogen Life Technologies, Grand Island, NY, USA), incubated with CD133-phycoerythrin (CD133PE; Miltenyi Biotech, Bergisch-Gladbach, Germany), and sorted to obtain purified CD133+ cells. Stem cells were labeled with Fe_3O_4 nanoparticles (Endorem)

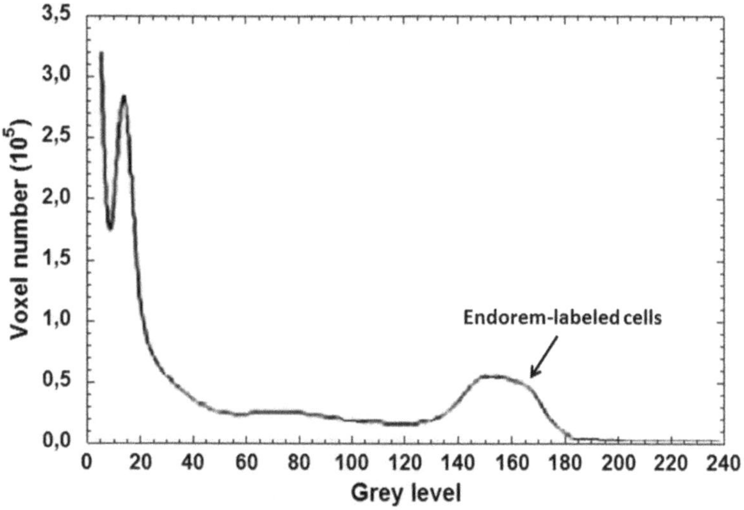

Fig. 8.3 *Grey level* histogram showing the peaks corresponding to different materials in the biopsies. The peak corresponding to the Endorem-labeled cells is evidenced

(250 mg/ml). Endorem has been approved for human use and is commercially available in the form of an aqueous colloid. It is a magnetic contrast agent based on dextran-coated iron oxide nanoparticles, with an average size of 150 nm. Labeling was performed in RPMI 1640 medium enriched with epidermal growth factor (EGF) (20 ng/ml) and basic fibroblastic growth factor (bFGF) (10 ng/ml) for 24 h. The mean iron concentration in a 2 ml sample containing one million cells was 88.5 mg/ml, corresponding to an average iron content of 177 pg/cell. The labeled CD133+ cells were injected into the femoral artery of scid/mdx mice, a dystrophic animal model that allows transplantation of human cells. Different stem cell numbers (5×10^4, 1×10^5, 5×10^5) were considered at different times (0, 2, 12 and 24 h) after injection.

Ex vivo measurements were carried out at the BM05 beamline of the European Synchrotron Radiation Facility (ESRF) in Grenoble, France. A sample-to-detector distance of 15 mm was used, and a few preliminary measurements were performed, varying the X-ray energy values between 18 and 27 keV to obtain optimal conditions for the X-ray absorption contrast among the different phases contained in the samples under investigation. A total of 1,000 projections were obtained from each sample over 180°, with an exposure time of 1 s per projection. A Gadox scintillator associated with a FReLoN 2,048×2,048 pixel CCD camera was used as a detector, with the pixel size set to 1.65 mm, giving an FOV of about 3 mm. Tibialis anterior biopsies ($2\times2\times2$ mm^3) were isolated from injected legs and analyzed for different numbers of initially injected cells (5×10^5, 1×10^5, 5×10^4) and different times after the injection (2, 12, and 24 h).

The different absorption coefficients of the materials in the samples give rise to different peaks in the gray level scale (Fig. 8.3). In particular, the absorption

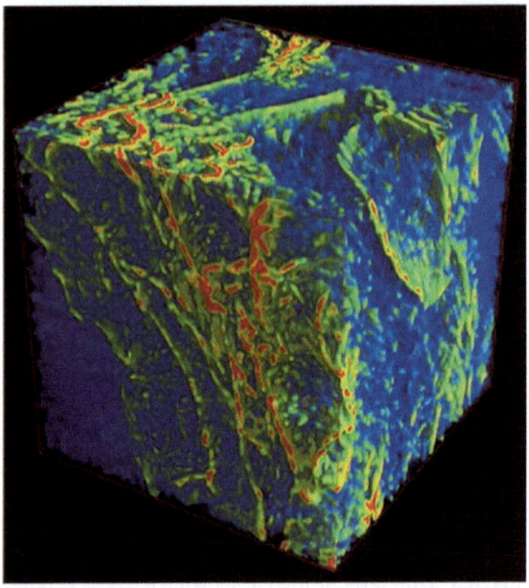

Fig. 8.4 3D display showing the distribution of labeled stem cells (5×10^5 injected cells, 24 h after injection; *red*: labeled cells, *green*: vessels, *blue*: muscular tissue)

coefficient of the Endorem-labeled stem cells is higher than the one of other tissues, and in the reconstructed 3D volumes the labeled cells are visualized as red spots (Fig. 8.4). Furthermore, it is possible to use the 3D image processing to "cancel" a phase to allow more accurate observation of the spatial distribution of each phase (Fig. 8.5). The signal of labeled cells was clear at all concentrations higher than 5×10^4 cells. No difference in the location of stem cells was observed at different times after injection, and stem cells appeared to be distributed along the vessels. The volume fraction of migrated labeled stem cells was calculated by counting their corresponding pixels using an algorithm that automatically separates them from other tissues (Fig. 8.6).

Therefore, ex vivo experiments showed the feasibility of the technique for visualizing Endorem-labeled stem cells for different numbers of injected cells and at different times after injection as well as its capability to determine the cell distribution in the tissue. It was also possible to extract quantitative parameters such as the volume fraction of migrated cells.

8.3.2 Phase-Contrast micro-CT for Analysis of Extracellular Matrix Fibers Organization in Bioscaffolds

Spatiotemporal organized patterns of cell surface-associated and extracellular matrix (ECM)-embedded molecules play important roles in the development and functioning of tissues. ECM proteins interact with the surface of the bioscaffold and influence the material-driven control of cell differentiation. In fact, cells continually

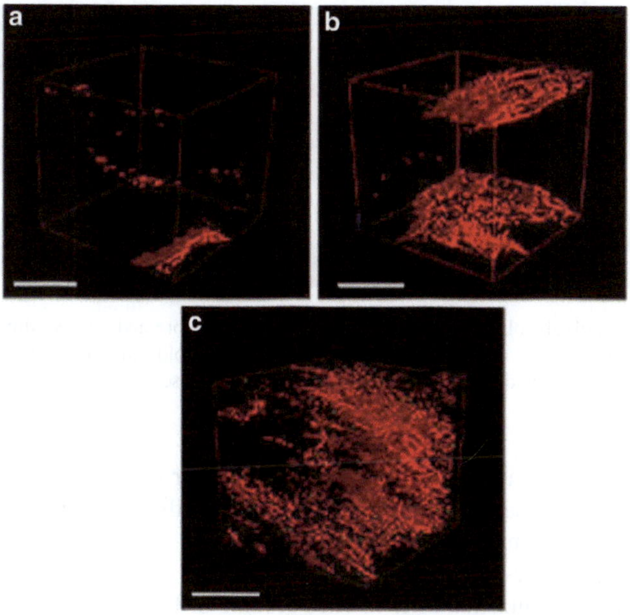

Fig. 8.5 3D distribution of labeled stem cells (in *red*) within the muscle biopsies, 12 h after injection; (**a**) 5×10^4, (**b**) 1×10^5, (**c**) 5×10^5 injected cells (the markers correspond to 700 mm)

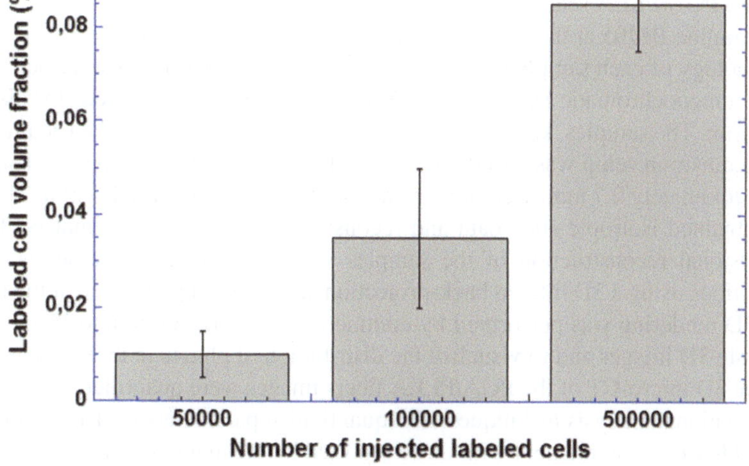

Fig. 8.6 Volume fraction of labeled cells in the muscle biopsies, 12 h after injection

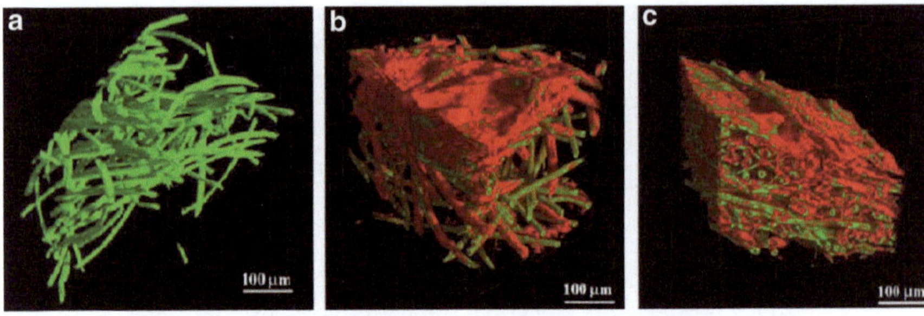

Fig. 8.7 Fiber polyglycolic acid–polylactic acid scaffolds before and after seeding with mesenchymal stem cells (MSCs). 3D display of subvolumes of scaffolds cultured without cells (**a**) and with human MSCs (**b**) and mouse MSCs (**c**) for 15 days: giber scaffold (*green*), thin extracellular matrix (*red*)

secrete complex mixtures of ECM proteins and other regulators of cell behavior, which may affect what happens when exogenous cells or material are implanted (Yamada and Cukierman 2007).

Using X-ray phase-contrast micro-CT, Albertini et al. (2009) visualized the 3D image of ECM organization after in vitro seeding of bone marrow-derived human and murine mesenchymal stem cells (hMSCs and mMSCs, respectively) induced to myogenic differentiation, labeled with iron oxide nanoparticles, and seeded onto polyglycolic acid–polylactic acid scaffolds.

Briefly, hMSCs were obtained from iliac crest marrow aspirates of healthy donors (ages 31–42 years) after obtaining informed consent. The mMSCs were obtained from C57Bl=6 mice as previously described (Augello et al. 2005). The scaffold consisted of nonwoven fibers of PGA and PLLA each present at a percentage of 50% (BIOFELT; Concordia Fibers, Coventry, RI, USA). After incubation and treatment with 5-Aza, 50×10^4 Endorem-labeled MSCs were seeded using gravity onto PGA/PLLA scaffolds.

Beamline BM05 at the ESRF was used to image and quantify the 3D structural morphology of each sample noninvasively. Micro-CT experiments were performed using a monochromatic beam of 20 keV energy and a sample-to-detector distance of 20 mm. The samples, kept in 70% ethanol, were air-dried before data acquisition. The acquisition setup was based on 3D parallel tomography with the pixel size set at approximately 0.7 mm, yielding a visual field of approximately 1.5 mm. The system obtained isotropic slice data and reconstructed them into 2D images. Three-dimensional reconstruction of the samples was obtained from the series of 2D projections using a 3D filtered back-projection algorithm implemented at the ESRF. The 3D rendering was performed by commercial software VGStudio MAX 1.1 to generate 3D images and to visualize the distribution of phases in three dimensions.

The 3D micro-CT of the PGA/PLLA fibers images were quantified using spatial computational analysis techniques. The quantitative parameters measured for these scaffolds were in agreement with those reported by the manufacturer.

The 3D micro-CT analysis was easily able to distinguish empty PGA/PLLA structures from cell-loaded PGA/PLLA scaffolds (Fig. 8.7). Cell–scaffold interactions produced modification of the PGA/PLLA structure, producing images in

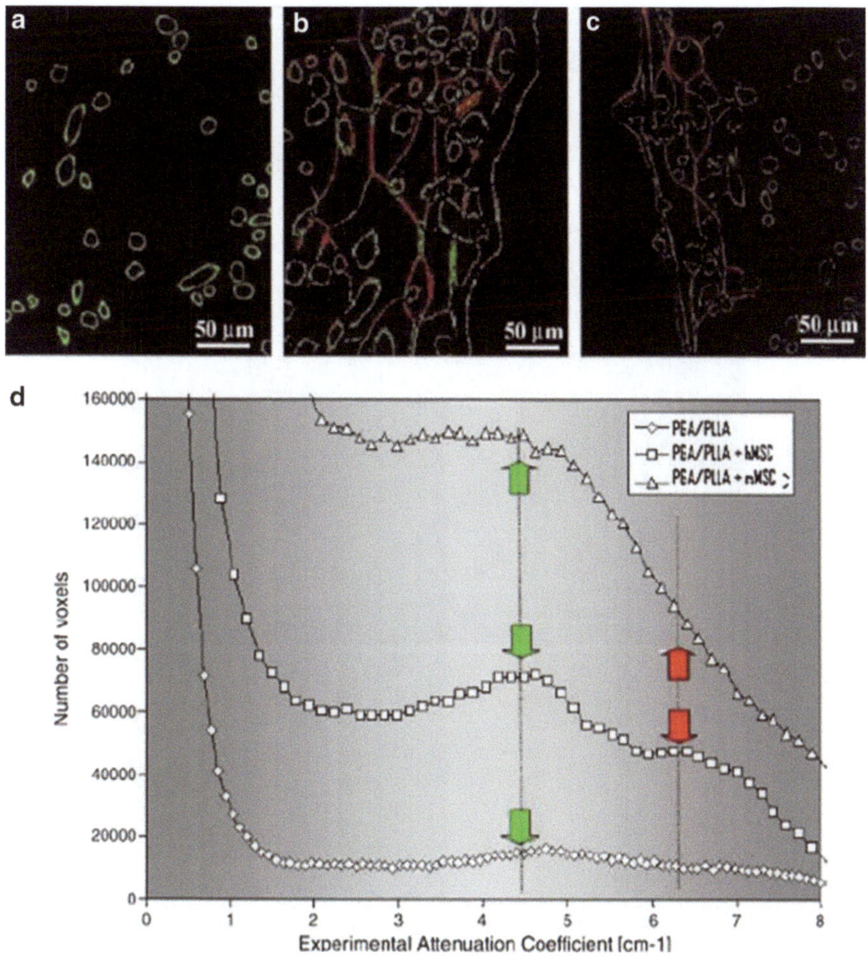

Fig. 8.8 Evidences of new extracellular matrix formation. 2D (*frontal section*) display of subvolumes of scaffolds cultured without cells (**a**), with human mesenchymal stem cells (hMSCs) (**b**) and murine mesenchymal stem cells (mMSCs) (**c**) for 15 days: fiber scaffold (*green*), thin layered matrix (*red*). Experimental attenuation coefficient histograms of scaffolds (**d**) without cells (◊) and with hMSCs (□) and mMSCs (Δ). The *green arrows* indicate the polyglycolic acid–polylactic acid fibers; the *red arrows* indicate the newly formed fibrillar matrix

which two phases with different attenuation coefficients were evident. The different phases were colored using 3D display software to make them more easily recognizable. Unmodified PGA/PLLA fibers were shown in green, and thin layers of fibrillar matrix produced by cells grown on the bioscaffold were depicted in red. The layers were partially connected to fibers and covered some regions in the network of the fiber itself (Fig. 8.7b, c). The measured average thicknesses of the layers were 5.4 ± 1.3 µm and 2.7 ± 0.7 µm for the scaffolds seeded with mMSCs and hMSCs, both labeled with iron nanoparticles, respectively (Fig. 8.7b, c). Frontal sections of scaffolds reinforced the information given by the 3D imaging (Fig. 8.8a–c).

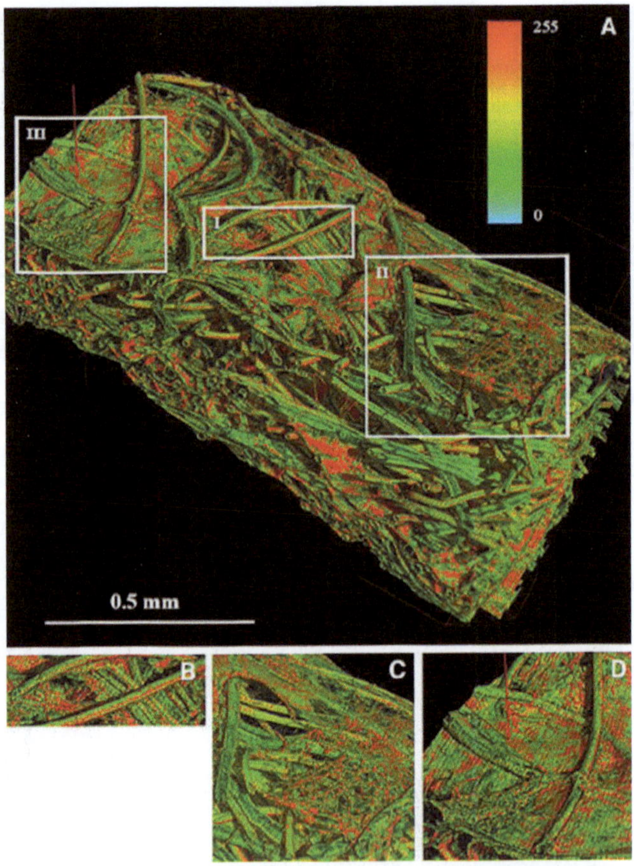

Fig. 8.9 Extracellular matrix detection using X-ray micro computed tomography. 3D display of the scaffold fibers and of the deposited matrix (**a**). The "stages" were zoomed in. The matrix fibrils were initially deposited on the scaffold fibers (Stage I) (**b**). At Stage II, they appear to be organized in chains at different sites (**c**). At Stage III, chains appear to be organized as networks at different sites, indicating that the aggregation process contributes to developing matrix layers (**d**)

It is easily recognizable that although in the scaffold cultured without cells there was no evidence of the red-layered phase (Fig. 8.8a) this phase was visible in both of the scaffolds cultured with hMSCs (Fig. 8.8b) and mMSCs (Fig. 8.8c). To make such evidence clearer, the profile of the "number of voxels versus the experimental attenuation coefficient" was drawn (Fig. 8.8d).

Elaboration of the 3D image (Fig. 8.9) revealed the presence of thin layers of fibrillar matrix, possibly corresponding to specific spatiotemporal patterns of fibril organization. Initially, a few filaments were deposited onto the scaffold fibers, creating a sparse structure (stage I); they appeared to be organized in chains (stage II). Finally, layers of filament chains appeared to be organized in the form of a thicker structure (stage III). The fiber organization and its connection with the scaffold were similar in samples loaded with hMSCs and mMSCs.

In conclusion, despite the fact that micro-CT alone is not recommended for analysis of soft tissues present in the samples because of such low X-ray absorption (Potter et al. 2006) X-ray micro-CT enabled the authors to detect with high spatial resolution the 3D structural organization of ECM in the bioscaffold and how the presence of cells modified the construct arrangement. The use of synchrotron X-rays has in fact several advantages over laboratory or industrial X-ray sources, including (1) a high photon flux, which permits measurements at high spatial resolution; (2) a tunable X-ray source, allowing measurements at different energies; (3) the use of monochromatic X-ray radiation, which eliminates beam hardening effects; and (4) parallel beam acquisition, which allows the use of exact tomographic reconstruction algorithms. Furthermore, the authors performed the experiment with a semiphase contrast setup that enhances the possibility of visualizing the interfaces between different phases, highlighting the organization of the ECM.

The images extrapolated from the synchrotron analysis indicated that ECM fibers aggregated according to a spatiotemporal pattern. Isolated collagen-like fibers that in other portions of the scaffolds appeared aggregated in progressively thicker layers (indicating continuous secretion of matrix by the seeded cells) covered a few scaffold areas. Species-specific differences between the matrix produced by human and murine cells were observed. In fact, mMSCs secreted a larger amount of ECM proteins than hMSCs, confirming the species-specific behavior of MSCs (Augello et al. 2005; Kuznetsov et al. 2001). In this context, noninvasive and quantitative X-ray micro-CT can be considered a potentially important tool for challenging new applications in tissue engineering research. Current microscopy techniques are limited to 2D local information or otherwise require laborious 3D reconstruction of serial sections. Here, the authors determined the feasibility of using synchrotron analysis to depict the fine spatiotemporal organization of the net of matrix fibers layered by MSCs in contact with PGA/PLLA bioscaffolds.

8.3.3 Scaffolds for Bone Tissue Engineering

One of the most critical issues in tissue engineering is the fabrication of scaffolds with tailored physical, mechanical and biological properties that act as substrates for cellular in growth and proliferation, and support new tissue formation (Causa et al. 2007). Scaffolds able to mimic the architecture and biological functions of ECM are very promising substitutes since they might provide mechanical support, carry inductive molecules or cells, and supply signals to control structure and function of newly formed tissue. In recent years, biomaterials design has evolved from the classical, first-generation, biofunctional materials that seek to incorporate instructive signals into scaffolds to modulate cellular functions such as proliferation, differentiation, and morphogenesis. Adult stem cells are defined by two major functions: multilineage differentiation and self-renewal. These functions are evident in the key role that the stem cells play in development and regeneration of specific tissues, in fact stem cells take part increasingly in tissue engineering.

Progress in the understanding of the molecular mechanisms of self-renewal and of directed differentiation of stem cells growing on bio-mimetic materials will lead to the possibility of cell-based therapies, and the possible use of stem cells in tissue engineering (Liao et al. 2008). Several works (Gerecht-Nir et al. 2004; Levenberg et al. 2003) have described culturing stem cells within 3D scaffolds, with the general aim of inducing stem cell proliferation and differentiation. Stem cells have been recognized as a promising alternative to somatic cells for cell therapy owing to their potential to renew themselves through cell division and to differentiate into a wide range of specialized cell types. In recent years, stem cells have shown significant promise for their potential to provide a source of undifferentiated progenitor cells for therapeutic applications in tissue or organ repair. Significant questions still remain, however, as to the genetic and epigenetic signals that regulate the fate of stem cells. It is now well accepted that the micro-environment of the stem cell can have a significant influence on its differentiation and phenotypic expression. Stem cells have great potential as cell sources for regenerative medicine due to both their self-renewal and multi-lineage differentiation capacity. One challenge is to develop reproducible methods to control stem cell growth and differentiation. Stem cells can be encouraged to differentiate to the required phenotype by manipulating the culture conditions under which they are maintained. In this way, it is possible to control or restrict the available differentiation pathways and to generate selectively cultures enriched with a particular phenotype. Such manipulations include stimulation of cells with particular cytokines, growth factors, amino acids, other proteins and active ions and co-culture with a relevant cell/tissue type.

Different biomaterials have been proposed as scaffolds for the delivery of cells and/or biological molecules to repair or regenerate damaged or diseased bone tissues. Imaging techniques are serving an increasingly important role in the rigorous characterization of biomaterial properties and function. Chemical composition, density, pore shape, pore size, and pore interconnection are elements that have to be considered to improve the efficiency of the biomaterials.

In the study reported in (Renghini et al. 2009) an accurate analysis of the structure was performed in order to confirm and extend the promising results from previous works (Vitale-Brovarone et al. 2005, 2008), concerning the use of CEL2 glass–ceramic as effective biomaterial for scaffolding. 3-D highly porous scaffolds, with a trabecular texture similar to cancellous bone, were fabricated via a sponge-replication method. The prepared scaffolds were soaked in a simulated body fluid (SBF) and in a buffer solution (Tris–HCl) for different time intervals and then investigated by means of m-CT. In particular, micro-CT analysis was used to study the new phase 3-D distribution in the bulk material and its evolution as a function of the soaking time in SBF and Tris–HCl medium. It was observed that a hydroxyapatite (HA) layer grew onto the samples soaked in SBF (Fig. 8.10), showing the high bioactivity and biocompatibility of the glass–ceramic scaffolds. Moreover, the decrease in the mean thickness of the walls with immersion time in Tris showed the bioresorbability of the scaffolds. Finally, the porosity determined by micro-CT demonstrated that all the scaffolds, before and after the soaking in SBF and Tris, exhibited a high percentage of porosity (50–60 vol.%). In particular, microstructural observations (Fig. 8.11) revealed that the samples were characterized by a

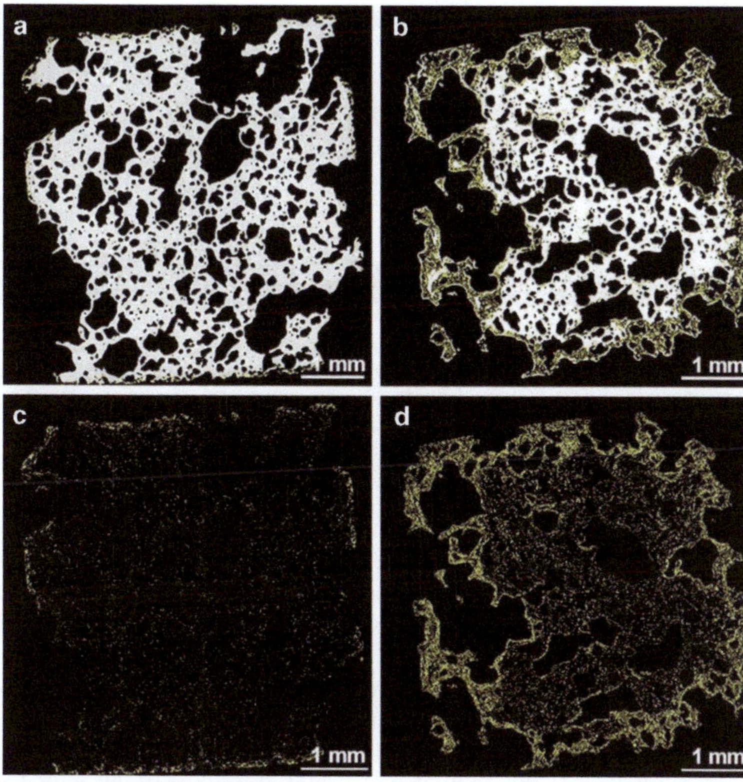

Fig. 8.10 Example of a central virtual slice obtained by micro-CT with the material scaffolds (*white*) and the new phase (*yellow*): slice of sample after treatment in SBF for (**a**) 1 week and (**b**) 4 weeks; image of the same slice after "cancelling" the scaffold material: slice of sample after treatment in SBF for (**c**) 1 week and (**d**) for 4 weeks

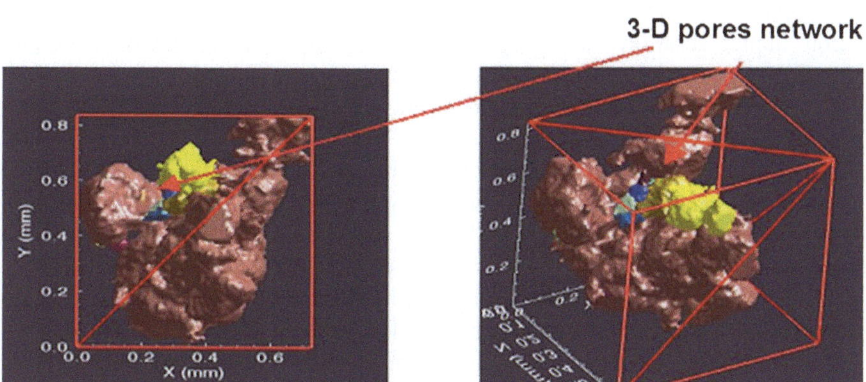

Fig. 8.11 The 3-D pore network showing the bimodal porous structure of the tissue-engineering scaffold as obtained by micro-CT

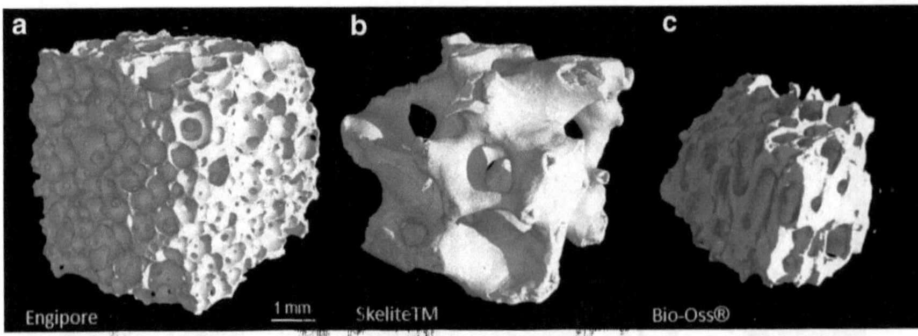

Fig. 8.12 3D display of different scaffolds before implantation: (**a**) Engipore (hydroxyapatite) produced by FinCeramica, Faenza, Italy; (**b**) Skelite™ (silicon-stabilized tricalcium phosphate) produced by Millenium Biologics Kingston, Ontario, Canada; (**c**) Bio-Oss® (natural bone mineral) produced by Geistlich Pharma AG, Wolhusen, Switzerland

bimodal porous structure comprising macropores, necessary for the growth of new bone and the vascularization of the implant, and micropores, important for cells adhesion and proliferation. Therefore, the micro-CT analysis shows that the proposed CEL2 scaffolds are very interesting candidates for bone tissue engineering applications.

Recently, different ceramic scaffolds with high porosity were characterized by Komlev et al. (2010) and 3D bone growth into tissue engineering constructs was evaluated *in vivo* at different implantation times by using micro-CT associated with synchrotron radiation. In this study three types of ceramic scaffolds with different composition and structure [namely synthetic 100% hydroxyapatite (HA; Engipore), synthetic calcium phosphate multiphase biomaterial containing 67% silicon stabilized tricalcium phosphate (Si-TCP; Skelite™) and natural bone mineral derived scaffolds (Bio-oss®)] were seeded with mesenchymal stem cells (MSC) and ectopically implanted for 8 and 16 weeks in immunodeficient mice. X-ray synchrotron radiation microtomography was used to derive 3D structural information on the same scaffolds both before and after implantation. The images of the three scaffolds before implantation revealed an appreciable difference among their morphologies (Fig. 8.12). In particular the Bio-Oss®, which was investigated by us for the first time, contained elongated ellipsoidal pores, whereas the HA scaffold contained roughly spherical pores.

The histograms of the distribution of the thickness of the scaffold wall of Fig. 8.13 (A) and 4(B), confirmed the results previously obtained in (Papadimitropoulos et al. 2007), namely biodegradation for the Si-TCP scaffold and lack of it for the HA scaffold. The newly investigated Bio- Oss® showed a very little decrease of the scaffold wall thickness; the decrease was at the limit of detectability, and needs to be confirmed by additional experiments. Figure 8.14 (Panels A1-C1), obtained by an innovative imaging procedure, gives an instantaneous pictorial view of the variation in scaffold wall thickness and confirm in a rather impressive way the uniqueness of the biodegradation process in Skelite™.

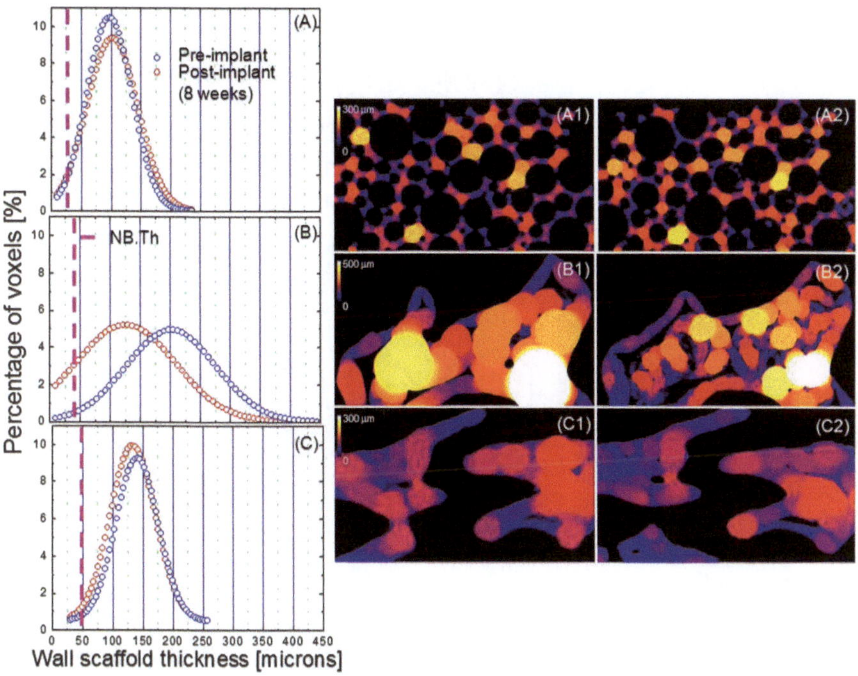

Fig. 8.13 Histograms of the distribution of wall thickness before and after scaffold implantation: (**a**) Engipore (hydroxyapatite); (**b**) Skelite™ (silicon-stabilized tricalcium phosphate); (**c**) Bio-Oss® (natural bone mineral). NbTh = New Bone Thickness. (**A1-C2**) Examples of central slices through the samples within the 3D local wall thickness map before (**A1-C1**) and after implantation (**A2-C2**). The thickness in each point is coded according to the color map included in panels **A1**, **B1**, **C1**

Based on these findings, the scaffold degradation in the tissue engineered implanted constructs was investigated after 16 weeks implantation only for the SkeliteTM. 3D displays of registered images of pre- and post implantation SkeliteTM samples implanted for 8 (a) and 16 (b) weeks are presented in Fig. 8.15 (panels A-B), respectively. As in the registered images of Fig. 8.14 (panel B2) blue and yellow correspond to totally or partially resorbed scaffold. The volume percentage distribution of the different phases is presented in panels A1 and B1. An increase in the percentage of the resorbed scaffold was observed with the increased implantation time. The analysis proposed in this work is a major improvement as compared to the imaging procedure adopted in previous works (Papadimitropoulos et al. 2007), where only a comparison between different subvolumes of the implants before and after implantation was made.

Finally, a high content of innovation is associated to the detailed kinetics studies on the Skelite™ scaffolds implanted for different times, not only due to the large number of the implantation times investigated, but also to the recording in the X-ray absorption histograms of separate peaks associated to HA and TCP in the same scaffold (Fig. 8.16).

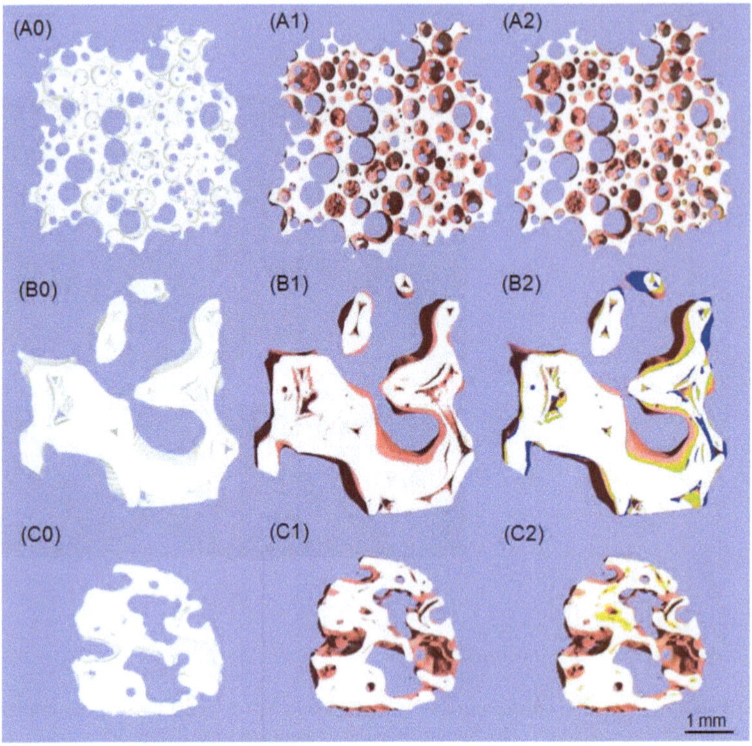

Fig. 8.14 3D display of subvolumes of scaffolds before and after implantation: (**a**) Engipore; (**b**) Skelite™; (**c**) Bio-Oss®; (**A1-C1**) Subvolumes of implanted samples before (**A0-C0**) and after 8 weeks (**A1-C1**) implantation. The images show the new bone (*pink*) on the surface of the scaffolds (*white*). In the panels **A2-C2** are presented the images obtained by combining (registering) the data of panels **A0-C0** with those of panels **A1-C1**. Blue volumes indicate portions of scaffolds present in panels **A0-C0** (pre-implant) and absent in panels **A1-C1** (after implantation) and correspond to completely resorbed scaffold. Yellow volumes indicate virgin scaffold volume in which after implantation a reduction of the sample density is observed

It is therefore possible to observe that the progressive biodegradation of Skelite™ scaffold is eventually due to the TCP component. It should be noted that when we investigated by microdiffraction studies the interfaces between the newly formed bone and the Skelite™ scaffold, the local structural study at the interface indicated that scaffold biodegradation was mainly due to TCP depletion (Papadimitropoulos et al. 2007). Moreover, saturation in the TCP resorption occurred at an implantation time of about 10 weeks, whereas saturation in the tissue engineered bone occurred at an implantation time of about 22 weeks. This could indicate that the bone growth did not occur only in the scaffold volume that was resorbed, but also in the inward direction with respect to the pore surface. This finding is in agreement with previous results (Papadimitropoulos et al. 2007; Mastrogiacomo et al. 2007). From these examples it appears that non-destructive

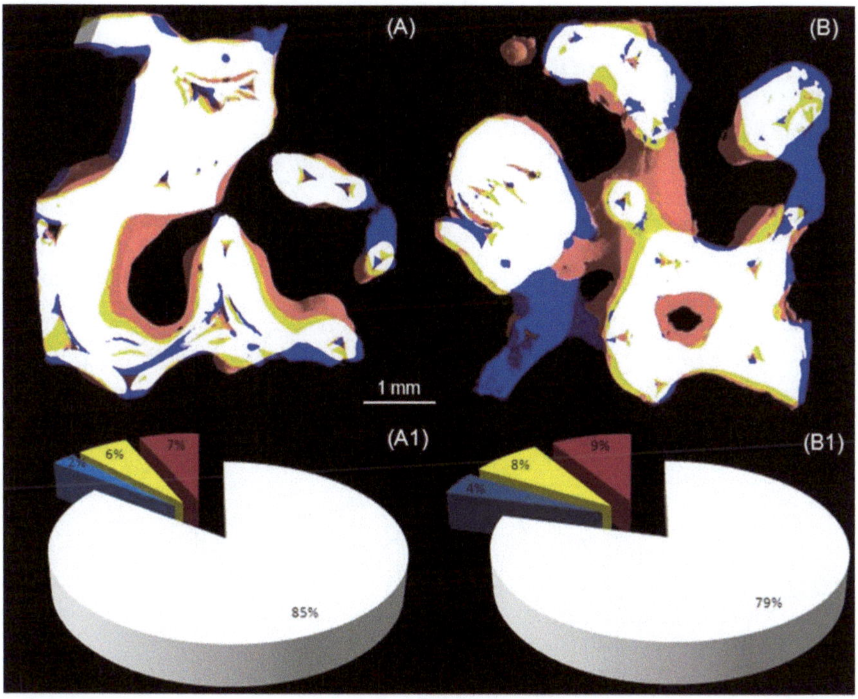

Fig. 8.15 Display based on a combination of the 3D structure of pre- and post implanted Skelite™ samples for 8 (**a**) and 16 (**b**) weeks, respectively (*white*—scaffold; *pink*—new bone; *blue*—total resorption; *yellow*—partial resorption (see caption of Fig. 8.3)). (**A1-B1**) volume percentage distribution of the different phases

3D imaging techniques, such as micro-CT can increasingly provide also a powerful set of quantitative data to aid in the development and evaluation of porous biomaterials and of engineered tissues and organs.

In a recent study (Komlev et al. 2009) it was proposed to use X-ray synchrotron radiation pseudo-holotomography to visualize, at three-dimensional (3D) level, microvascular networks for the first time with no need for contrast agents, and to extract quantitative structural data in a Bioceramic/MSC composite implanted for 24 weeks in a mouse.

The pseudo-holotomography technique is a new imaging method based on classical micro-CT and recently developed technique holotomography. In classical tomography the detector, set directly behind the sample, measures the attenuation, which allows calculation of the integral of attenuation coefficient along the transmitted path. By repeating this measurement for a large number of angular positions of the sample and by using a tomographic reconstruction algorithm, it is possible to reconstruct the attenuation map. On the other hand, holotomography allows to reconstruct the phase map by knowledge of the phase distribution for each angular setting of the sample. In fact, with a coherent X-ray beam, phase contrast may be

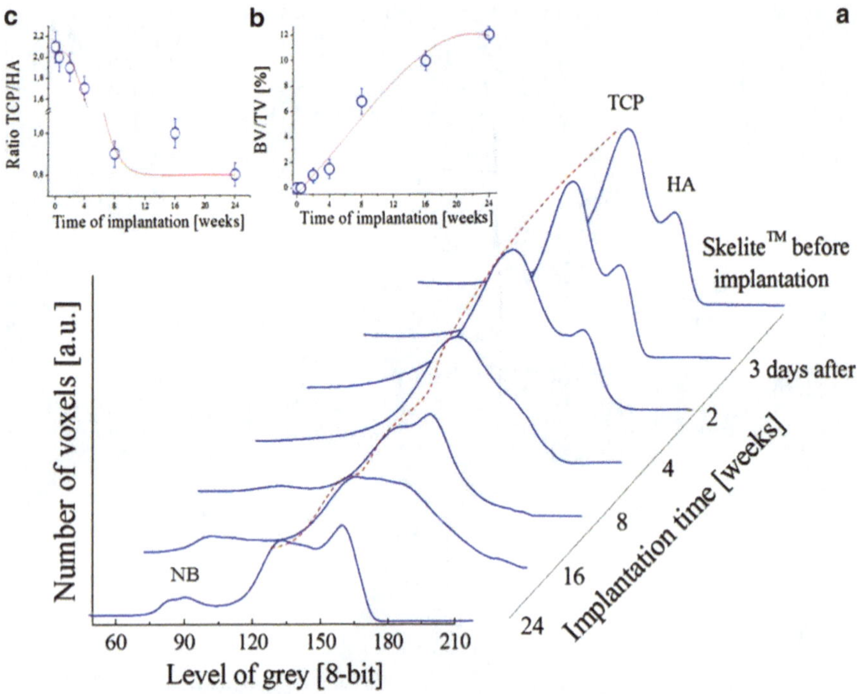

Fig. 8.16 Volume histograms of Skelite™ scaffolds implanted from 3 days to 24 weeks (NB new bone, TCP tricalcium phosphate, HA hydroxyapatite). (**a**) New bone deposition kinetics; (**b**) Percentage of bone volume/total volume; (**c**) TCP/HA mean ratio as a function of the implantation time

simply obtained by free space propagation (i.e., by positioning the detector at some distance from the sample), while a 2-D projection of the phase map can be obtained from three or four series of images, each series being recorded at different distance from the object at each of the different angles of rotation. Then, the 3-D phase map is reconstructed with the same algorithm as in classical tomography. Anyway, through the weighted superposition of both attenuation and phase maps (pseudo-holotomography), it is possible to generate better images.

Figure 8.17a and b is a 3D reconstruction of a bone tissue- engineered construct 24 weeks after the implantation. Three phases are clearly distinguishable: the scaffold (white), the engineered bone (light brown), and the vessel networks within the pores (green). In Fig. 8.8b, the engineered bone was removed by digital processing to obtain a clearer evidence of vessel network structure. Three-dimensional representation of the structure within one single pore is illustrated in Fig. 8.17c, d. Vessels are easy to see also in 2D microCT images (Fig. 8.18c). Figure 8.18d shows a histogram of the 3D vessel diameter distribution measured within the full volume of the sample implanted for 24 weeks. The mean vessel diameter measured from pseudo-holotomography data was 49 ± 25 mm. This value was comparable to the 47 ± 18 mm measured in control histology sections.

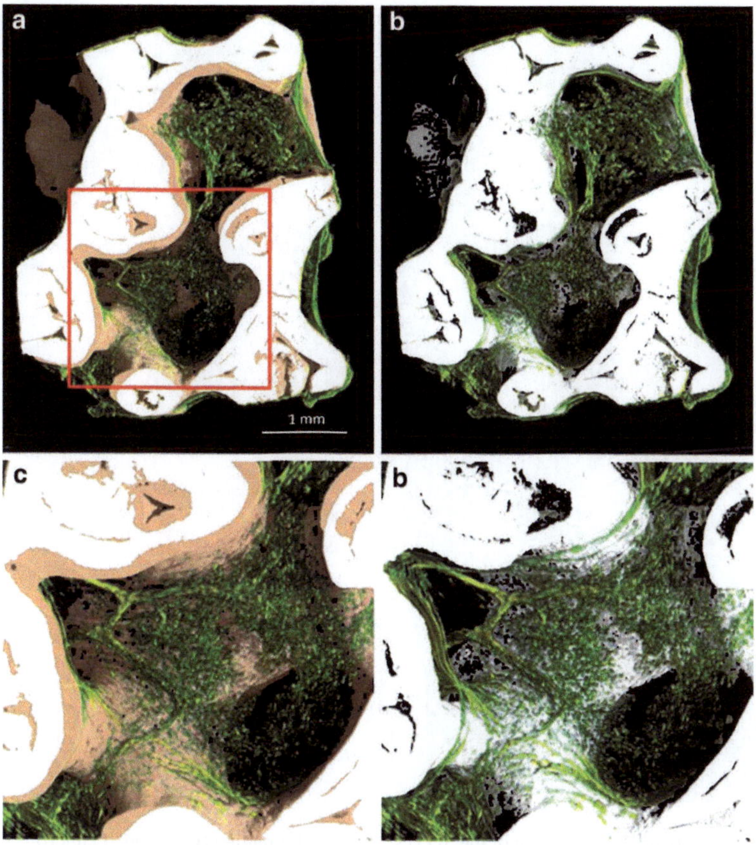

Fig. 8.17 3D Pseudo-holotomographic images of the tissue-engineered construct after 24 weeks of implantation in an immunocompromised mouse. The images show the vessel in-growth inside the scaffold: vessel growth occurred both in the presence (**a**, *green*) and in the absence of newly formed bone (**b**, *brown=pink*). (**b–d**) Details of 3D spatial distribution of the phases into scaffolds within one single pore

Pseudo-holotomography was applied to the study of the microvasculature in bone tissue-engineered constructs implanted in a small animal model, but the potential methodological development due to this study is likely to be of a much more general interest, as compared to the chosen stem cell therapeutic approach (bone tissue engineering). In particular, the progress associated to the present study could be extrapolated to different biomedical research areas where angiogenesis and microvasculogenesis play an important role, as for the development of tissues such as bone, in regenerative medicine, or in pathologies characterized by inflammation and tissue damage such as diabetes, osteoarthritis, and muscular dystrophy. Of great interest could also be the application of the pseudo-holotomography to investigations of therapeutical roadmaps for tumor treatment involving the suppression of vascularization.

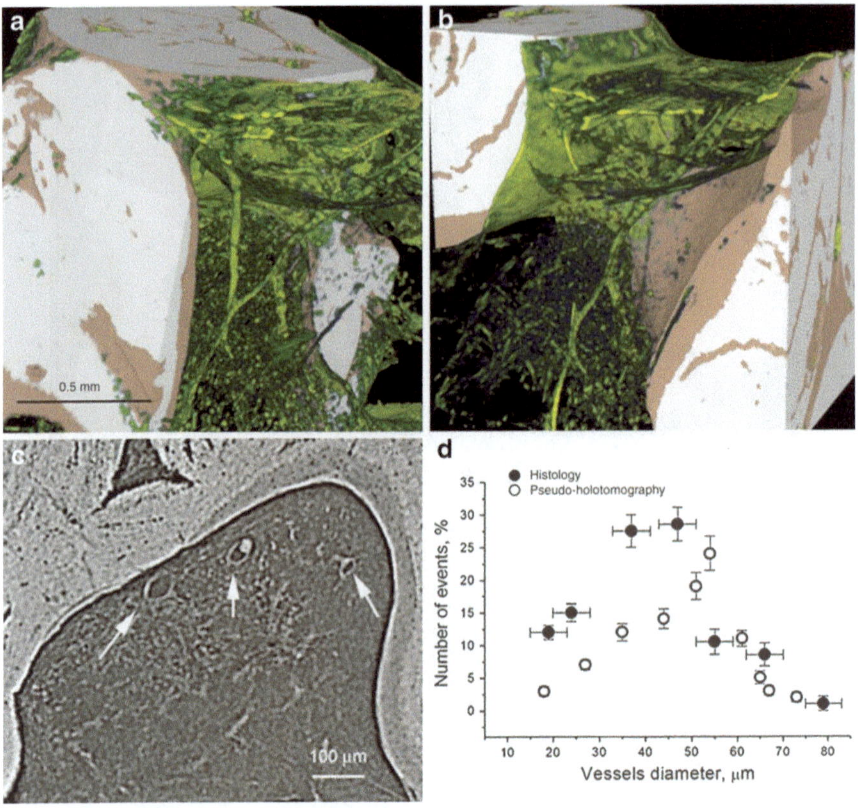

Fig. 8.18 Statistical information of the 3D vessel network imaging (**a**, **b**) and 2D micro-CT image (**c**). (**d**) Histogram of the vessel diameter distribution measured for pseudo-holotomography data (*open circles*) and histology

8.4 Conclusion

Stem cell based tissue engineering therapies involve the administration of ex vivo manipulated stem cell populations with the purpose of repairing and regenerating damaged or diseased tissue.

Non-destructive 3D imaging techniques such as micro-CT are increasingly providing a powerful set of quantitative tools to aid in the development and evaluation of porous biomaterials and new approaches to engineering tissues and organs. A key advantage of micro-CT imaging is that this method, as well as MRI, may be applicable to monitoring the stem cell homing, after cell labeling with iron oxide nanoparticles. When working on biopsies of small sizes (few millimeters) or small animals, micro-CT has an appreciably higher spatial resolution as compared to magnetic resonance imaging, which on the other hand has the advantage to be applicable to human body.

Micro-CT has been also shown to be feasible for 3D studies of bioscaffolds for tissue engineering, also allowing the 3D visualization, as well as quantitative evaluations, of features very difficult to be detected by other imaging techniques, such as the vascularization network in engineered bone tissue.

References

Albertini G, Giuliani A, Komlev V, Moroncini F, Pugnaloni A, Pennesi G, Belicchi M, Rubini C, Rustichelli F, Tasso R, Torrente Y (2009) Organization of extracellular matrix fibers within polyglycolic acid-polylactic acid scaffolds analyzed using X-ray synchrotron-radiation phase-contrast micro computed tomography. Tissue Eng Part C Methods 15(3):403–411

Arbab AS, Bashaw LA, Miller BR, Jordan EK, Bulte JW, Frank JA (2003) Intracytoplasmic tagging of cells with ferumoxides and transfection agent for cellular magnetic resonance imaging after cell transplantation: methods and techniques. Transplantation 76(7):1123–1130

Atala A (2005) Tissue engineering, stem cells and cloning: current concepts and changing trends. Expert Opin Biol Ther 5:879–892

Augello A, Tasso R, Negrini SM, Amateis A, Indiveri F, Cancedda R, Pennesi G (2005) Bone marrow mesenchymal progenitor cells inhibit lymphocyte proliferation by activation of the programmed death 1 pathway. Eur J Immunol 35(5):1482–1490

Boskey A, Pleshko Camacho N (2007) Infrared imaging of native and tissue-engineered bone and cartilage. Biomaterials 28(15):2465–2478

Bulte JW, Duncan ID, Frank JA (2002) In vivo magnetic resonance tracking of magnetically labeled cells after transplantation. J Cereb Blood Flow Metab 22(8):899–907

Campbell CT, Kim G (2007) SPR microscopy and its applications to high-throughput analyses of biomolecular binding events and their kinetics. Biomaterials 28:2380–2392

Causa F, Netti PA, Ambrosio L (2007) A multi-functional scaffold for tissue regeneration: the need to engineer a tissue analogue. Biomaterials 28(34):5093–5099

Eniwumide JO, Yuan H, Cartmell SH, Meijer CJ, de Bruijn JD (2007) Ectopic bone formation in bone marrow stem cell seeded calcium phosphate scaffolds as compared to autograft and (cell seeded) allograft. Eur Cell Mater 14:30–38

Frank JA, Miller BR, Arbab AS, Zywicke HA, Jordan EK, Lewis BK, Bryant LH Jr, Bulte JW (2003) Clinically applicable labeling of mammalian and stem cells by combining superparamagnetic iron oxides and transfection agents. Radiology 228:480

Gavina M et al (2006) VCAM-1 expression on dystrophic muscle vessels has a critical role in the recruitment of human blood-derived CD133+ stem cells after intra-arterial transplantation. Blood 108(8):2857–2866

Gerecht-Nir S, Cohen S, Itskovitz-Eldor J (2004) Bioreactor cultivation enhances the efficiency of human embryoid body (hEB) formation and differentiation. Biotechnol Bioeng 86(5):493–502

Gupta AK, Gupta M (2005) Synthesis and surface engineering of iron oxide nanoparticles for biomedical applications. Biomaterials 26:3995–4021

Hill JM, Dick AJ, Raman VK, Thompson RB, Yu ZX, Hinds KA et al (2003) Serial cardiac magnetic resonance imaging of injected mesenchymal stem cells. Circulation 108:1009–1014

Hoehn M et al (2002) Monitoring of implanted stem cell migration in vivo: A highly resolved in vivo magnetic resonance imaging investigation of experimental stroke in rat. Proc Natl Acad Sci 99:16267–16272

Huebsch ND, Mooney DJ (2007) Fluorescent resonance energy transfer: A tool for probing molecular cell-biomaterial interactions in three dimensions. Biomaterials 28:2424–2437

Jones JR, Hench LL (2003) Regeneration of trabecular bone using porous ceramics. Curr Opin Solid State Mater Sci 7:301–307

Kak AC, Slaney M (1988) Principles of computerized tomographic imaging. IEEE Press, New York

Komlev VS, Peyrin F, Mastrogiacomo M, Cedola A, Papadimitropoulos A, Rustichelli F, Cancedda R (2006) Kinetics of in vivo bone deposition by bone marrow stromal cells into porous calcium phosphate scaffolds: an X-ray computed microtomography study. Tissue Eng 12:3449–3458

Komlev VS, Mastrogiacomo M, Peyrin F, Cancedda R, Rustichelli F (2009) X-ray synchrotron radiation pseudo-holotomography as a new imaging technique to investigate angio- and micro-vasculogenesis with no usage of contrast agents. Tissue Eng 15:425–430

Komlev V, Mastrogiacomo M, Pereira RC, Peyrin F, Rustichelli F, Cancedda R (2010) Biodegradation of porous calcium phosphate scaffolds in an ectopic bone formation model studied by X-ray computed microtomography. Eur Cell Mater 19:136–146

Kuznetsov SA, Mankani MH, Gronthos S, Satomura K, Bianco P, Robey PG (2001) Circulating skeletal stem cells. J Cell Biol 153:1133

Langer RS, Vacanti JP (1999) Tissue engineering: The challenges ahead. Sci Am 280:86–89

Levenberg S, Huang NF, Lavik E, Rogers AB, Itskovitz-Eldor J, Langer R (2003) Differentiation of human embryonic stem cells on three-dimensional polymer scaffolds. Proc Natl Acad Sci USA 100:12741–12746

Liao J, Cui C, Chen S, Ren J, Chen J, Gao Y, Li H, Jia N, Cheng L, Xiao H, Xiao L (2009) Generation of induced pluripotent stem cell lines from adult rat cells. Cell Stem Cell 4(1):11–15

Mastrogiacomo M, Komlev VS, Hausard M, Peyrin F, Turquier F, Casari S et al (2004) Synchrotron radiation microtomography of bone engineered from bone marrow stromal cells. Tissue Eng 10:1767–1774

Mastrogiacomo M, Papadimitropoulos A, Cedola A, Peyrin F, Giannoni P, Pearce SG, Alini M, Giannini C, Guagliardi A, Cancedda R (2007) Engineering of bone using bone marrow stromal cells and a silicon-stabilized tricalcium phosphate bioceramic. Evidence for a coupling between bone formation and scaffold resorption. Biomaterials 28:1376–1384

Papadimitropoulos A, Mastrogiacomo M, Peyrin F, Molinari E, Komlev VS, Rustichelli F, Cancedda R (2007) Kinetics of in vivo bone deposition by bone marrow stromal cells within a resorbable porous calcium phosphate scaffold: an x-ray computed microtomography study. Biotechnol Bioeng 98:271–281

Potter K, Sweet DE, Anderson P, Davis GR, Isogai N, Asamura S, Kusuhara H, Landis WJ (2006) Non-destructive studies of tissue-engineered phalanges by magnetic resonance microscopy and X-ray microtomography. Bone 38:350–358

Reimer P, Weissleder R (1996) Development and experimental use of receptor- specific MR contrast media. Radiology 36:153–163

Renghini C, Komlev V, Fiori F, Vernè E, Baino F, Vitale-Brovarone C (2009) Micro-CT studies on 3-D bioactive glass-ceramic scaffolds for bone regeneration. Acta Biomater 5:1328–1337

Salomé M, Peyrin F, Cloetens P, Odet C, Laval-Jeantet AM, Baruchel J, Spanne P (1999) A synchrotron radiation microtomography system for the analysis of trabecular bone samples. Med Phys 26:2194–2204

Silva EA, Mooney DJ (2007) Spatiotemporal control of vascular endothelial growth factor delivery from injectable hydrogels enhances angiogenesis. J Thromb Haemost 5:590–598

Torrente Y et al (2004) Human circulating AC133(+) stem cells restore dystrophin expression and ameliorate function in dystrophic skeletal muscle. J Clin Invest 114:182–195

Torrente Y, Gavina M, Belicchi M, Fiori F, Komlev V, Bresolin N, Rustichelli F (2006) High-resolution X-ray microtomography for three-dimensional visualization of human stem cell muscle homing. FEBS Lett 580:5759–5764

Wang YX, Hussain SM, Krestin GP (2001) Superparamagnetic iron oxide contrast agents: physicochemical characteristics and applications in MR imaging. Eur Radiol 11:2319

Yamada KM, Cukierman E (2007) Modeling tissue morphogenesis and cancer in 3D. Cell 130:601–610

Chapter 9
Controversies in Corneal Epithelial Stem Cell Biology

Haifa Ali, Charles Osei-Bempong, Ani Ray-Chaudhuri, Bakiah Shaharuddin, Arianna Bianchi, Mohit Parekh, and Sajjad Ahmad

Abstract The loss or dysfunction of corneal epithelial stem cells, or limbal stem cells as they are more commonly known, results in the painful and blinding disease of limbal stem cell deficiency. In 1997, it was proposed that limbal stem cell deficiency could be treated by transplanting cultured limbal stem cells containing human limbal epithelium. The area of limbal stem cell biology therefore now encompasses not only the basic science of stem cell biology but also the area of translational research and cell therapeutics. Ranging from the laboratory to the clinic, there are still many controversies in limbal stem cell biology. In this chapter we describe and outline some of the questions that remain to be answered.

H. Ali • C. Osei-Bempong • A. Ray-Chaudhuri • A. Bianchi • M. Parekh
Institute of Human Genetics, Newcastle University, Newcastle upon Tyne, UK

North East England Stem Cell Institute, Newcastle University, Newcastle upon Tyne, UK

B. Shaharuddin
Advanced Medical and Dental Institute, Universiti Sains Malaysia, Penang, Malaysia

S. Ahmad (✉)
Institute of Human Genetics, Newcastle University, Newcastle upon Tyne, UK

North East England Stem Cell Institute, Newcastle University, Newcastle upon Tyne, UK

Royal Victoria Infirmary, Newcastle upon Tyne, UK

Department of Ophthalmology, Newcastle University, Newcastle upon Tyne, UK

Institute of Human Genetics, Newcastle University, International Centre for Life, Central Parkway, Newcastle upon Tyne, NE1 3BZ UK
e-mail: s2ahm@doctors.org.uk

K. Turksen (ed.), *Adult and Embryonic Stem Cells*, Stem Cell Biology and Regenerative Medicine, DOI 10.1007/978-1-61779-630-2_9,

9.1 Cornea, Limbus, and Limbal Stem Cells

The cornea forms the front of the eye (Fig. 9.1). It transmits and focuses light to the retina at the back of the eye, enabling visual perception. Clarity is therefore an essential property of the cornea. The cornea is a three-layered structure composed of a stratified epithelium on the outer surface, a stroma that makes up approximately 90% of the corneal thickness in the middle, and a single-layered endothelium on the inner surface. The cornea is an avascular structure nourished by the aqueous humor on the inside of the eye and the tear film on the outside. The corneal epithelium is devoid of its own stem cells but is renewed by stem cells located peripheral to the cornea in a region known as the limbus (Fig. 9.1) (Ahmad et al. 2006, 2010a). The corneal epithelial stem cells, or limbal stem cells as they are more commonly known, maintain the corneal epithelium in health and renew it after injury.

9.2 Corneal Epithelial Functions of the Limbus

The limbus has two important functions with regard to the corneal epithelium (Ahmad et al. 2006, 2010a). The first function of the limbus is to harbor stem cells for the corneal epithelium. The sclera forms the white covering of the eye, and it surrounds the cornea. The sclera is covered by the conjunctiva (Fig. 9.1). The conjunctiva is composed of a stratified epithelium and an underlying vascular stroma. It is continuous with the limbal epithelium, which in turn is continuous with the corneal epithelium. If the limbal epithelium is damaged, and hence the limbal stem cells as well, neither the limbal nor the corneal epithelia can be renewed (Puangsricharern and Tseng 1995). The surrounding conjunctival epithelium and its underlying blood vessels then begin to encroach on the corneal surface to fill the space left by the corneal epithelium. Conjunctival epithelium is phenotypically and physiologically different from the corneal epithelium, and the conjunctiva is a vascularized structure. Encroachment of both of these structures would therefore obviously result in loss of corneal clarity and function. The second function of the limbus

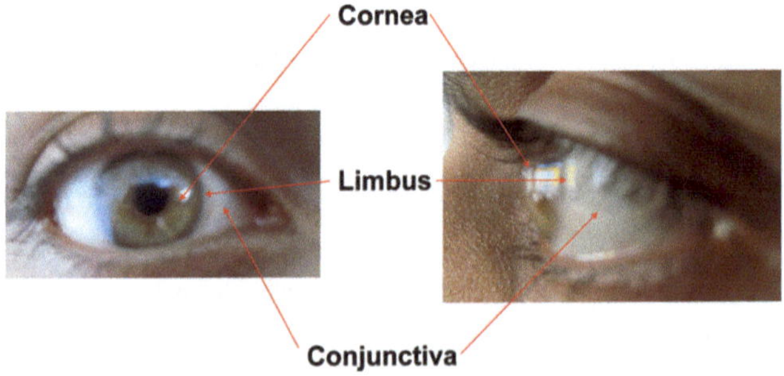

Fig. 9.1 Front and side view of the human eye showing the cornea, limbus, and conjunctiva

is therefore to act as a barrier preventing the conjunctival epithelium and its blood vessels from encroaching on to the corneal surface.

9.3 Limbal Stem Cell Deficiency

Limbal stem cell deficiency is a disease that results from the loss or dysfunction of limbal stem cells. The two functions of the limbus, described above, then fail. The corneal epithelium cannot be maintained because of the failure of limbal stem cell proliferation, and the corneal surface becomes covered by the conjunctival epithelium and its blood vessels due to loss of the barrier function of the limbus.

9.3.1 Effects of Limbal Stem Cell Deficiency

The two main effects of limbal stem cell deficiency are pain and blindness (Ahmad et al. 2006, 2010b). Recurrent corneal epithelial defects result from the inability to maintain the corneal epithelium. The corneal surface is highly innervated. A rough, irregular corneal surface therefore becomes increasingly painful especially during blinking. In addition, limbal stem cell deficiency is often accompanied by varying degrees of inflammation on the surface of the eye and even within the front of the eye, thereby exacerbating the pain. Visual impairment and even total blindness can result from the conjunctivalization in limbal stem cell deficiency and loss of corneal clarity. Glare and photophobia also result from the irregularity of the corneal surface.

9.3.2 Causes of Limbal Stem Cell Deficiency

There are many known causes of limbal stem cell deficiency (Puangsricharern and Tseng 1995), and they can be separated into primary (hereditary) and secondary (acquired) causes. Rarely, no cause can be determined, which is known as idiopathic limbal stem cell deficiency (Espana et al. 2002). Hereditary causes include ectodermal dysplasia and aniridia. With aniridia, due to a mutation in *PAX6*, it is thought that the niche for limbal stem cells is affected, which results in limbal stem cell dysfunction and dysregulation (Ramaesh et al. 2003, 2005). Acquired causes include chemical and thermal burns to the surface of the eye that result in direct damage and loss to the limbal stem cells and probably also alteration of their niche (Kolli et al. 2010). Contact lens wear can also result in limbal stem cell deficiency, either as a result of chemical damage from the solutions used, mechanical damage to the corneal and limbal surfaces, or a combination of these two processes. Inflammatory causes include Stevens-Johnson syndrome and ocular cicatricial pemphigoid and the limbal stem cell deficiency. Results from these disorders are often difficult to manage owing to prolonged and ongoing inflammation. Iatrogenic causes include the use of chemotherapeutic agents such as mitomycin C on the surface of the eye, extensive limbal surgery or cryotherapy, and irradiation.

9.3.3 Types of Limbal Stem Cell Deficiency

To decide on the management strategy for limbal stem cell deficiency, it is essential to define the type of limbal stem cell deficiency (Ahmad et al. 2006). The three main ways to define limbal stem cell deficiency are the cause, whether one or both eyes are affected, and the extent of limbal stem cell deficiency. The cause of limbal stem cell deficiency is important in determining the treatment path. Burns and iatrogenic causes are more likely to be asymmetrical and affect one eye more than the other, whereas hereditary, contact lens-related, and inflammatory causes are more likely to affect both eyes. Another way to define limbal stem cell deficiency is to define the symmetry of the disease. If one eye if affected, unilateral limbal stem cell deficiency results; and if both eyes are affected, it is termed bilateral limbal stem cell deficiency. Whether the patient has unilateral or bilateral limbal stem cell deficiency is important when determining the management strategies because with unilateral limbal stem cell deficiency there is a potential source of healthy limbal tissue on the other eye. The final way to define limbal stem cell deficiency is to determine the extent of the deficiency. If the whole limbal and corneal surface is affected, it is termed total or diffuse limbal stem cell deficiency; whereas if part of the limbus and cornea is affected, it is termed partial or focal limbal stem cell deficiency. Thus, when choosing the management option for limbal stem cell deficiency, it is important to know the cause of the disease, whether it is unilateral or bilateral, and whether it is total or partial.

9.3.4 Management Options for Limbal Stem Cell Deficiency

Management of limbal stem cell deficiency includes medical and surgical options (Fig. 9.2). Medical options include the use of lubricant eye drops to relieve discomfort, steroid eye drops to reduce inflammation, and autologous serum drops to promote epithelial healing (Geerling et al. 2004). It is important to avoid the use of preservative-containing drops if possible as they can cause epithelial toxicity or impair epithelial healing. Bandage contact lenses can also be used to reduce discomfort. Systemic steroids and immune suppression may also be used in cases of inflammatory limbal stem cell deficiency to reduce the inflammation.

Surgical options are used to provide a longer-term cure of the stem cell deficiency rather than temporary alleviation of symptoms. Prior to treating the limbal stem cell deficiency, it is vital that any problems with eyelid closure or eyelid position are addressed (DeSousa et al. 2009). Such eyelid abnormalities can be deleterious to

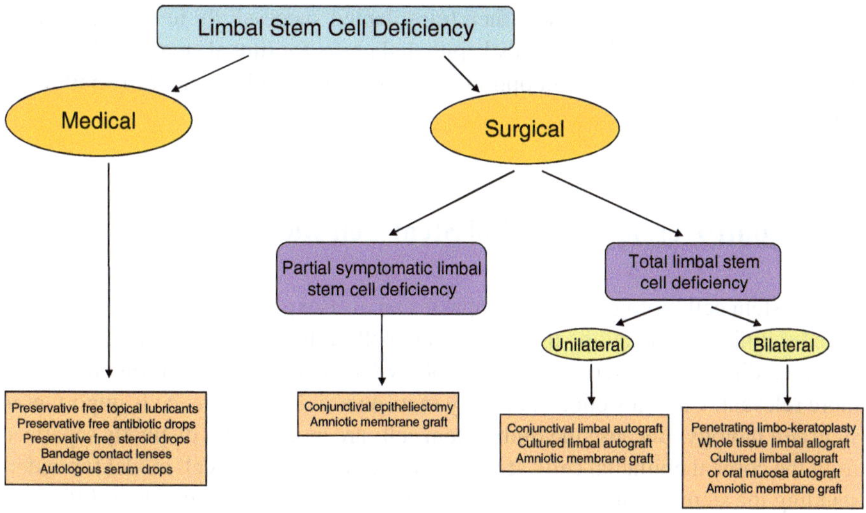

Fig. 9.2 Flow diagram shows the medical and surgical management options for limbal stem cell deficiency

any definitive surgical treatment for the stem cell deficiency. Once the eyelids are addressed, the limbal stem cell deficiency can be confronted directly. Corneal transplantation is not a viable treatment option for limbal stem cell deficiency in the first instance. Conventional corneal transplantation involves transplanting the central avascular full-thickness donor cornea. This includes the corneal epithelium, the stroma, and the endothelium. This option is beneficial for stromal and endothelial disease. For epithelial diseases, such as limbal stem cell deficiency, it is not a viable option. Once the terminal corneal epithelial cells are shed from the surface of the transplanted cornea, they need to be replaced by healthy limbal stem cells. With limbal stem cell deficiency this is not possible, and the normal corneal epithelium on the corneal transplant is not renewed, causing the transplant to fail. Corneal transplantation is, however, a viable option in limbal stem cell deficiency to treat any stromal or endothelial damage present only as a second procedure once the limbal stem cell deficiency has been rectified (Kolli et al. 2010).

Conventional surgical management of limbal stem cell deficiency involves transplantation of limbal tissue grafts (Kolli et al. 2010; Kenyon and Rapoza 1995; Kenyon and Tseng 1989). They can be autologous if the patient's other eye is healthy or allogeneic from living or cadaveric donors. In the case of allogeneic grafts, immune suppression is necessary. In all cases of limbal tissue grafts, significant amounts of limbal tissue are required to reverse the limbal stem cell deficiency. If limbal tissue is removed from a living eye (whether from the patient's healthy eye or from a donor), there is a significant risk of limbal stem cell deficiency in the donor eye because of the large amount of tissue required. For this reason, it was proposed in 1997 that a small piece of limbal stem cells containing limbal tissue could be taken and the epithelial cells then expanded by culture (Pellegrini et al. 1997).

The culture-expanded tissue was then transplanted successfully to treat the limbal stem cell deficiency. Cultured limbal epithelial grafts for limbal stem cell deficiency are contemporary management options for limbal stem cell deficiency (Shortt et al. 2007a).

9.4 Controversies in Limbal Stem Cell Biology

Limbal stem cell biology now encompasses an ever-expanding field from basic science to clinical applications for limbal stem cell biology. The clinical relevance of understanding limbal stem cell basic biology has made research in this field even more relevant. Limbal stem cell biology in its own right as a science is important in stem cell biology. Because the limbal stem cells are more or less distinctly separated from the corneal epithelial cells, it provides an important stem cell model for understanding the stem cell niche differences between the stem cells and their differentiated progeny. In addition, the translational aspects of stem cell biology in the form of cultured limbal stem cells is providing insight into how stem cell technologies and bioprocessing practices can be employed to treat diseases. Finally, the clinical application of limbal stem cell biology in the form of limbal tissue transplants or cultured limbal epithelial transplantation is providing insight into the direct application of stem cell biology to treat disease. However, in all aspects of this varied field, many answers remain unresolved. The remainder of this chapter discuss some of the controversies.

9.4.1 How Can We Identify Limbal Stem Cells?

Hematopoietic stem cells can be identified on the basis of a profile that includes factors that positively enrich for them and those that negatively enrich for them (Armstrong et al. 2004). This results in a more purified population of hematopoietic stem cells being enriched from a cell soup of differentiated hematopoietic cells and other supporting or surrounding cells in the tissue. A similar process cannot be used to enrich limbal stem cells from their more differentiated progeny, the transient amplifying cells of the corneal epithelium, and the differentiated corneal epithelial cells. An enriched population of limbal stem cells therefore cannot be obtained, and we are unable to carry out research on the limbal stem cells themselves more directly. There are, however, some markers for limbal stem cells, both negative and positive, that we do know about.

The well established and recognized negative markers for limbal stem cells are the intermediate filament proteins cytokeratins 3 and 12 (Kurpakus et al. 1990) and the gap junction protein connexion 43 (Grueterich et al. 2002; Wolosin et al. 2002). Cytokeratins 3 and 12 form a dimer that is characteristic for differentiated corneal epithelial cells; they are not found in limbal stem cells. Connexin 43 is a gap junction protein that enables cell-to-cell interaction between corneal epithelial cells. It is not found in limbal stem cells. The well recognized positive markers for limbal stem cells are the delta-N-p63 alpha isoform of the transcription factor p63 (Pellegrini et al. 2001) and the cell membrane transporter protein ATP binding cassette subfamily G member 2, ABCG2 (de Paiva et al. 2005; Budak et al. 2005). Both are well-recognized positive markers. Controversy has surrounded delta-N-p63 alpha in that many thought it could also be found in differentiated corneal epithelial cells (Di Iorio et al. 2005, 2006). This was mainly due to problems with the specificity of the p63 antibody used.

Although these negative and positive markers for limbal stem cells do exist, the problem is not really separating limbal stem cells from differentiated corneal epithelial cells but from the transient amplifying cells that form the intermediate differentiated cell type. This is much more difficult and is mainly where the challenge now lies. CCAAT/enhancer binding protein delta or C/EBP-delta and BMI1 have also been described as positive markers for limbal stem cells (Barbaro et al. 2007). C/EBP-delta is a transcription factor that has been implicated in controlling limbal stem cell self-renewal by alteration of the cell cycle. BMI1 is responsible for hematopoietic stem cell self-renewal and plays a similar role in limbal stem cells. Delta-N-p63-alpha, on the other hand, is present in limbal stem cells; but its role is putatively for cell proliferation. It has therefore been hypothesised that C/EBP-delta, BMI1, and delta-N-p63-alpha are all expressed in resting limbal stem cells. Delta-N-p63-alpha, but not C/EBP-delta or BMI1, are expressed by activated limbal stem cells and early transient amplifying cells. Delta-N-p63-alpha expression is then down-regulated; and delta-N-p63-beta and gamma are expressed in the more differentiated late transient amplifying cells. A greater understanding of mechanisms such as these is crucial for being able to identify markers that are more specific to limbal stem cells than the early transient amplifying cells of the corneal epithelium. Indeed, the results of gene microarray studies on limbal stem cells are currently being investigated and used to identify new markers (Bian et al. 2010). It is hoped that in the near future there will be a cohort of limbal stem cell enriching markers, preferably on the cell surface, that can be used to purify for limbal stem cells as is done for hematopoietic stem cells.

9.4.2 Where Are Corneal Epithelial Stem Cells Located?

It has long been believed that the stem cells for the corneal epithelium reside in the limbus. There are many reasons for this and various forms of evidence. Early studies on guinea pig eyes with pigmentation at the limbus showed migration of this pigment when corneal epithelial wounds healed (Davanger and Evensen 1971). The slow cycling nature of stem cells is exhibited by cells located in limbal epithelium, as shown by radiolabeling studies such as those with tritiated thymidine (Cotsarelis et al. 1989). Cultures established from limbal epithelial cells display a greater proliferative potential than those from the peripheral or central cornea

(Kruse and Tseng 1991, 1992). Proliferative potential is an important property of stem cells. Clinically, it has been shown that limbal stem cell deficiency can be treated successfully using limbal tissue. A vast amount of evidence has accumulated over the past two or three decades indicating that the basal layer of the limbal epithelium is the site for the corneal epithelial stem cell or limbal stem cell as it is more commonly known owing to its anatomical location. It has recently been proposed that the serial transplantation of corneal epithelium, in contrast to limbal epithelium, in the mouse shows that corneal epithelial stem cells are indeed located in the cornea itself as well (Majo et al. 2008). These studies, however, are unable to show whether the transient amplifying cells or the stem cells are enabling survival of the transplanted corneal epithelium or whether the corneal epithelial transplants can maintain the corneal epithelium long term. These studies go against the vast amount of evidence indicating that corneal epithelial stem cells are located at the limbus.

The limbal epithelium is stratified epithelium with undulations of its basal layer, which increases the surface area of the basal layer exposed to the stroma and its underlying vascular supply. The radiolabeling studies previously mentioned indicate that slow cycling cells, consistent with stem cells, are found in the basal layers of the limbal epithelium. Immunohistochemical studies using the cytokeratin 3/12 dimer, connexion 43, p63, and ABCG2 all indicate that limbal stem cells are found in the basal layer of the limbal epithelium. All this evidence indicates that the basal layer of the limbal epithelium is the site of the corneal epithelial stem cell. As already noted, the basal layer of the limbal epithelium undulates and has downgrowth into the limbal stroma. Imaging and histological analysis have revealed that it is likely that the limbal stem cells are found in the bottom of these downgrowths, or crypts, as they have become to be known (Dua et al. 2005; Shortt et al. 2007b). It is not unexpected that limbal stem cells should be found in the base of these limbal crypts. Close proximity to a blood supply is an important prerequisite for the survival of stem cells, bringing them nutrition and access to vital growth factors and cytokines.

9.4.3 Are Limbal Stem Cells Unipotent?

The potential of stem cells gives an indication of their ability to differentiate into various cell types. Embryonic stem cells are pluripotent in that they can theoretically give rise to all cell types of the body (Stojkovic et al. 2004). Hematopoietic stem cells, on the other hand, have more limited differentiation potential; and they are termed multipotent in that they can give rise to many but not all cell types. Most other adult stem cells have a much more limited potential and are, on the whole, either unipotent or bipotent, giving rise to one or two cell types, respectively. We know that limbal stem cells are able to generate epithelial cells both in vitro and in vivo. There is, however, some evidence that limbal stem cells also have a neurogenic potential both in vitro and in vivo (Zhao et al. 2002, 2008; Chacko et al. 2003). This is not unexpected in that we know the corneal epithelium and its nerve supply are almost intermingled and that the health of the corneal epithelium is very much reliant upon its nerve supply, as shown by the significantly impaired corneal epithelial stability in neurotrophic keratopathies. In addition, the recent importance of nerve growth factors and their receptors in limbal stem cells would also not be surprising (Qi et al. 2007, 2008). It would therefore not be surprising if limbal stem cells were indeed bipotent, having both epithelial and neurogenic potential.

9.4.4 *What Is the Best Way to Culture Human Limbal Epithelium?*

The main means of culturing limbal stem cells is by culturing limbal stem cells containing human limbal epithelium. There are various debates regarding the best method for culturing human limbal epithelium to maintain and amplify the limbal stem cells (Shortt et al. 2007a; Grueterich et al. 2003a; Osei-Bempong et al. 2009). The first is whether mitotically inactivated mouse 3T3 fibroblasts alone or human amniotic membrane, or a combination of the two, are better for culturing human limbal stem cells. This question remains to be fully addressed. Human amniotic membrane is relatively immune-privileged and does not require recipient immune suppression.

The second question occurs in the case where human amniotic membrane is used as a substrate to culture the limbal epithelium. Human amniotic membrane has its own single-layered epithelium; and some say that the limbal stem cells culture best with an intact amniotic membrane epithelium, whereas others suggest it should be removed or denuded (Koizumi et al. 2000; Grueterich et al. 2003b). The data are conflicting data as to whether intact or denuded amniotic membrane is best.

The third question is whether limbal epithelial cells should be cultured directly from pieces of limbal tissue (i.e., explants) (Fig. 9.3) or the epithelial cells should be removed from the limbal explant and then cultured as a cell suspension (Koizumi et al. 2002; Zhang et al. 2005). Study groups have used one or the other method, but there has been no direct comparison of the two techniques. The answer to which system is best for culturing human limbal epithelium therefore remains to be addressed.

9.4.5 *What Is the Long-term Outcome of Cultured Limbal Epithelial Transplants?*

The technique of culturing human limbal epithelial cells and then transplanting them in patients with limbal stem cell deficiency was first described in 1997 (Pellegrini et al. 1997). Since then, there have been multiple cases and case series of such transplants. There have also been various modifications of both the culture

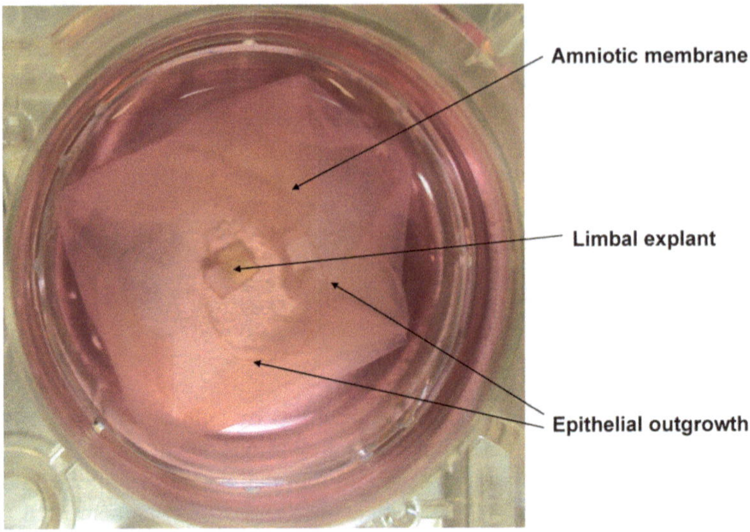

Fig. 9.3 Human limbal explant culture on a human amniotic membrane shows epithelial outgrowth

and transplantation techniques, the most notable being the use of human amniotic membrane as a substrate to culture the limbal epithelium rather than using 3T3 fibroblasts as a feeder layer (Tsai et al. 2000). Many causes of limbal stem cell deficiency have been treated using this technique, from chemical burns to inflammatory eye disease. The main problem with most of the case series to date, and in some cases these are large case series, is that in these studies there were different types of limbal stem cell deficiency treated with different causes; and in some cases, varying culture or transplantation techniques were used (Ahmad et al. 2010b; Shortt et al. 2007a). In many cases, the outcome data from the patients were either not provided or inadequate. For these reasons it is often difficult to make a valid scientific assessment as to whether cultured limbal epithelial transplantation is successful, even if the clinical results suggest it is.

A review of the case series in 2007—a decade after the first cultured limbal epithelial transplants were performed—showed that the overall success of the transplants was in the region of 80%: approximately 85% for autologous cultured limbal epithelial transplants and 75% for allogeneic transplants (Shortt et al. 2007a). This was based on improved visual acuity as the main outcome measure. More recent longer-term results in 112 patients, reported by the group that originally proposed and developed the technique of culturing and transplanting autologous limbal epithelium for limbal stem cell deficiency, showed a success rate of 76.6% (Rama et al. 2010). They suggested that if failure occurs it tends to happen during the first year after transplantation. They also suggested that success of the transplant very much depends on the proportion of cultured limbal epithelial cells expressing p63. A review of the data to date suggested that cultured limbal epithelial transplantation is effective in treating limbal stem cell deficiency, but it remains difficult to obtain a more accurate estimate of the success owing to the variability within each study and between studies.

9.4.6 How Do Cultured Limbal Epithelial Transplants Work?

If someone has limbal stem cell deficiency and he or she is treated with a transplant that contains limbal stem cells, it would reasonable to suggest that it works because it reverses the loss of limbal stem cells. The answer, however, is not that simple. Data from both allogeneic whole tissue and cultured limbal epithelial transplants show that the donor cells or their progeny cannot be identified in the recipient corneal epithelium even as early as a few months after the procedure (Daya et al. 2005; Henderson et al. 1997, 2001a, b). This is despite continued reversal of the patient's limbal stem cell deficiency. It has been hypothesized that the transplanted limbal epithelial cells somehow stimulate the few remaining recipient limbal stem cells to self-renew and proliferate (Daya et al. 2005). This theory would indeed account for why the limbal stem cell deficiency is reversed despite the absence of any donor cells. It remains to be determined, though, how transplanted limbal epithelial cells, either as whole tissue or cultured transplants, reverse limbal stem cell deficiency. It does not seem to be a simple compensation for the limbal stem cell loss by transplanting more limbal stem cells. Modern clinical imaging techniques may give some evidence to the fate of the transplanted cells.

9.4.7 How Can the Technique of Cultured Limbal Epithelial Transplantation Be Used More Widely?

The technique of culturing human limbal epithelial cells and then transplanting them has various limitations. The first is the requirement of expertise: being able to culture the limbal epithelial cells successfully and then to evaluate and transplant them into patients. The second is the requirement of specialized clean laboratories that utilize good manufacturing practice procedures to process transplant grade cells and tissues. The third is regulatory approval for culturing human limbal epithelium for transplantation purposes. The fourth is the expense of regulating and culturing the tissue. For these main reasons, among others, the technique cannot be used in ophthalmology units on a more widespread basis. Also, it is not viable for all ophthalmology units to be able to perform this technique. One way of bypassing some of these problems is to have a centralized facility for culturing human limbal tissue and then transporting the cultured tissue to surgeons for transplantation in patients with limbal stem cell deficiency. This process would require optimization of transportation strategies for cultured human limbal epithelium and assessment of their efficacy following transportation.

9.4.8 *What Regulatory Issues Are There Regarding Culturing Limbal Epithelium for Transplantation?*

The technique of culturing human limbal tissue for transplantation has to be a regulated process (Daniels et al. 2006) to ensure both the surgeon and the patient that the material is safe to use and of adequate quality. There are various requirements for this regulation process. First, the tissue must be cultured using good manufacturing practice in a clean laboratory. Second, the end product (i.e., the cultured limbal epithelial cells) must be microbiologically sterile. Third, the culture must be of a sufficient standard. In the case of cultured human limbal epithelium, the various methods used to assess the culture could be simply morphological, histological, or on a more molecular level. To determine the standard, a series of trial or validation cultures must be established so each culture can be compared to this standard. This whole process is regulated using good manufacturing practice and must be approved on a national or more regional basis. At present, one problem is that in some countries there is no regulatory authority for this relatively new area of translational research. The other problem is what to do when tissue being processed through regulated means in one country needs to be transported to another country where either there is a different regulatory authority. These are all gray areas that will need to be tackled in the future as the technique becomes more widely used.

9.4.9 *How Can Bilateral Limbal Stem Cell Deficiency Be Treated?*

A more effective treatment of bilateral limbal stem cell deficiency remains a major challenge in limbal stem cell biology (Ahmad et al. 2006, 2010a, b). The technique of transplanting cultured human limbal epithelium is particularly effective in the treatment of unilateral limbal stem cell deficiency, where the patient's other, healthy eye can be used as the donor source for the culture. Of course allogeneic limbal epithelial cells, either as whole tissue or cultured, can be used to treat bilateral limbal stem cell deficiency. However, immune rejection after the transplant remains a problem and therefore requires potent immune suppression. Patients with bilateral disease are therefore often reluctant to have allogeneic transplants owing to the risk of life-threatening infections or neoplasia arising from immunosuppressant agents.

Cultured autologous oral mucosal epithelium has been proposed as an effective treatment modality for bilateral limbal stem cell deficiency (Inatomi et al. 2006). Oral mucosal epithelium, once cultured, has been shown to exhibit properties of corneal epithelium. However, the results from these transplants have shown them to be less effective than using cultured allogeneic limbal epithelial transplants. These patients also commonly develop peripheral corneal vascularization, which is obviously a characteristic feature of the original limbal stem cell deficiency. The long-term results of cultured autologous oral mucosal epithelial transplants remain to be determined. Other autologous cells being suggested as possibilities for treating limbal stem cell deficiency include mesenchymal stem cells isolated from bone marrow (Ma et al. 2006) or cells from the skin (Blazejewska et al. 2009) or dental pulp (Gomes et al. 2010).

9.5 Conclusions

Limbal stem cell biology encompasses areas from basic stem cell science to cellular therapy. All the aspects of limbal stem cell biology, from its science to its clinical application, have numerous questions that remain to be answered. In this chapter we addressed some of these controversies and dealt with how these questions might be answered. Research into limbal stem cell biology has a hopeful future, not least because its greater understanding has already resulted in the treatment of a painful and blinding disease, namely limbal stem cell deficiency.

References

Ahmad S, Figueiredo F, Lako M (2006) Corneal epithelial stem cells: characterization, culture and transplantation. Regen Med 1(1):29–44

Ahmad S et al (2010a) Stem cell therapies for ocular surface disease. Drug Discov Today 15(7–8):306–313

Ahmad S et al (2010b) The culture and transplantation of human limbal stem cells. J Cell Physiol 225(1):15–19

Armstrong L et al (2004) Phenotypic characterization of murine primitive hematopoietic progenitor cells isolated on basis of aldehyde dehydrogenase activity. Stem Cells 22:1142–1151

Barbaro V et al (2007) C/EBPdelta regulates cell cycle and self-renewal of human limbal stem cells. J Cell Biol 177(6):1037–1049

Bian F et al (2010) Molecular signatures and biological pathway profiles of human corneal epithelial progenitor cells. Int J Biochem Cell Biol 42(7):1142–1153

Blazejewska EA et al (2009) Corneal limbal microenvironment can induce transdifferentiation of hair follicle stem cells into corneal epithelial-like cells. Stem Cells 27(3):642–652

Budak MT et al (2005) Ocular surface epithelia contain ABCG2-dependent side population cells exhibiting features associated with stem cells. J Cell Sci 118(Pt 8):1715–1724

Chacko DM et al (2003) Transplantation of ocular stem cells: the role of injury in incorporation and differentiation of grafted cells in the retina. Vision Res 43(8):937–946

Cotsarelis G et al (1989) Existence of slow-cycling limbal epithelial basal cells that can be preferentially stimulated to proliferate: implications on epithelial stem cells. Cell 57(2):201–209

Daniels JT et al (2006) Stem cell therapy delivery: treading the regulatory tightrope. Regen Med 1(5):715–719

Davanger M, Evensen A (1971) Role of the pericorneal papillary structure in renewal of corneal epithelium. Nature 229(5286):560–561

Daya SM et al (2005) Outcomes and DNA analysis of ex vivo expanded stem cell allograft for ocular surface reconstruction. Ophthalmology 112(3):470–477

de Paiva CS et al (2005) ABCG2 transporter identifies a population of clonogenic human limbal epithelial cells. Stem Cells 23(1):63–73

DeSousa JL, Daya S, Malhotra R (2009) Adnexal surgery in patients undergoing ocular surface stem cell transplantation. Ophthalmology 116(2):235–242

Di Iorio E et al (2005) Isoforms of deltaNp63 and the migration of ocular limbal cells in human corneal regeneration. Proc Natl Acad Sci USA 102(27):9523–9528

Di Iorio E et al (2006) Q-FIHC: quantification of fluorescence immunohistochemistry to analyse p63 isoforms and cell cycle phases in human limbal stem cells. Microsc Res Tech 69:983–991

Dua HS et al (2005) Limbal epithelial crypts: a novel anatomical structure and a putative limbal stem cell niche. Br J Ophthalmol 89(5):529–532

Espana EM et al (2002) Idiopathic limbal stem cell deficiency. Ophthalmology 109(11): 2004–2010

Geerling G, Maclennan S, Hartwig D (2004) Autologous serum eye drops for ocular surface disorders. Br J Ophthalmol 88(11):1467–1474

Gomes JA et al (2010) Corneal reconstruction with tissue-engineered cell sheets composed of human immature dental pulp stem cells. Invest Ophthalmol Vis Sci 51(3):1408–1414

Grueterich M, Espana E, Tseng SC (2002) Connexin 43 expression and proliferation of human limbal epithelium on intact and denuded amniotic membrane. Invest Ophthalmol Vis Sci 43(1):63–71

Grueterich M, Espana EM, Tseng SC (2003a) Ex vivo expansion of limbal epithelial stem cells: amniotic membrane serving as a stem cell niche. Surv Ophthalmol 48(6):631–646

Grueterich M, Espana EM, Tseng SC (2003b) Modulation of keratin and connexin expression in limbal epithelium expanded on denuded amniotic membrane with and without a 3T3 fibroblast feeder layer. Invest Ophthalmol Vis Sci 44(10):4230–4236

Henderson TR et al (1997) Do transplanted corneal limbal stem cells survive in vivo long-term? Possible techniques to detect donor cell survival by polymerase chain reaction with the amelogenin gene and Y-specific probes. Eye (Lond) 11(Pt 6):779–785

Henderson TR et al (2001a) Identifying the origin of single corneal cells by DNA fingerprinting. Part I. Implications for corneal limbal allografting. Cornea 20(4):400–403

Henderson TR et al (2001b) Identifying the origin of single corneal cells by DNA fingerprinting. Part II. Application to limbal allografting. Cornea 20(4):404–407

Inatomi T et al (2006) Midterm results on ocular surface reconstruction using cultivated autologous oral mucosal epithelial transplantation. Am J Ophthalmol 141(2):267–275

Kenyon KR, Rapoza PA (1995) Limbal allograft transplantation for ocular surface disorders. Ophthalmology 102(Suppl):101–102

Kenyon KR, Tseng SC (1989) Limbal autograft transplantation for ocular surface disorders. Ophthalmology 96(5):709–722, discussion 722–3

Koizumi N et al (2000) Cultivation of corneal epithelial cells on intact and denuded human amniotic membrane. Invest Ophthalmol Vis Sci 41(9):2506–2513

Koizumi N et al (2002) An evaluation of cultivated corneal limbal epithelial cells, using cell-suspension culture. Invest Ophthalmol Vis Sci 43(7):2114–2121

Kolli S et al (2010) Successful clinical implementation of corneal epithelial stem cell therapy for treatment of unilateral limbal stem cell deficiency. Stem Cells 28(3):597–610

Kruse FE, Tseng SC (1991) A serum-free clonal growth assay for limbal, peripheral, and central corneal epithelium. Invest Ophthalmol Vis Sci 32(7):2086–2095

Kruse FE, Tseng SC (1992) Proliferative and differentiative response of corneal and limbal epithelium to extracellular calcium in serum-free clonal cultures. J Cell Physiol 151(2):347–360

Kurpakus MA, Stock EL, Jones JC (1990) Expression of the 55-kD/64-kD corneal keratins in ocular surface epithelium. Invest Ophthalmol Vis Sci 31(3):448–456

Ma Y et al (2006) Reconstruction of chemically burned rat corneal surface by bone marrow-derived human mesenchymal stem cells. Stem Cells 24(2):315–321

Majo F et al (2008) Oligopotent stem cells are distributed throughout the mammalian ocular surface. Nature 456(7219):250–254

Osei-Bempong C, Henein C, Ahmad S (2009) Culture conditions for primary human limbal epithelial cells. Regen Med 4(3):461–470

Pellegrini G et al (1997) Long-term restoration of damaged corneal surfaces with autologous cultivated corneal epithelium. Lancet 349(9057):990–993

Pellegrini G et al (2001) p63 identifies keratinocyte stem cells. Proc Natl Acad Sci USA 98(6): 3156–3161

Puangsricharern V, Tseng SC (1995) Cytologic evidence of corneal diseases with limbal stem cell deficiency. Ophthalmology 102(10):1476–1485

Qi H et al (2007) Patterned expression of neurotrophic factors and receptors in human limbal and corneal regions. Mol Vis 13:1934–1941

Qi H et al (2008) Nerve growth factor and its receptor TrkA serve as potential markers for human corneal epithelial progenitor cells. Exp Eye Res 86(1):34–40

Rama P et al (2010) Limbal stem-cell therapy and long-term corneal regeneration. N Engl J Med 363(2):147–155

Ramaesh T et al (2003) Corneal abnormalities in Pax6+/– small eye mice mimic human aniridia-related keratopathy. Invest Ophthalmol Vis Sci 44(5):1871–1878

Ramaesh K et al (2005) Evolving concepts on the pathogenic mechanisms of aniridia related keratopathy. Int J Biochem Cell Biol 37(3):547–557

Shortt AJ et al (2007a) Transplantation of ex vivo cultured limbal epithelial stem cells: a review of techniques and clinical results. Surv Ophthalmol 52(5):483–502

Shortt AJ et al (2007b) Characterization of the limbal epithelial stem cell niche: novel imaging techniques permit in vivo observation and targeted biopsy of limbal epithelial stem cells. Stem Cells 25(6):1402–1409

Stojkovic M et al (2004) Derivation of human embryonic stem cells from day-8 blastocysts recovered after three-step in vitro culture. Stem Cells 22(5):790–797

Tsai RJ, Li LM, Chen JK (2000) Reconstruction of damaged corneas by transplantation of autologous limbal epithelial cells. N Engl J Med 343(2):86–93

Wolosin JM et al (2002) Changes in connexin43 in early ocular surface development. Curr Eye Res 24(6):430–438

Zhang X et al (2005) Comparison of cell-suspension and explant culture of rabbit limbal epithelial cells. Exp Eye Res 80(2):227–233

Zhao X et al (2002) Adult corneal limbal epithelium: a model for studying neural potential of non-neural stem cells/progenitors. Dev Biol 250(2):317–331

Zhao X et al (2008) Derivation of neurons with functional properties from adult limbal epithelium: implications in autologous cell therapy for photoreceptor degeneration. Stem Cells 26(4): 939–949

Chapter 10
Type 1 Diabetes and Stem Cells: A New Approach

Erdal Karaöz

Abstract The development of diabetes could be reasonably considered the result of disruption of the balance of the production and destruction of beta cells. Therefore, improving the regeneration potential of beta cells to the recovery level of damage caused by the autoimmune system was proposed as an alternative model of therapy for type 1 diabetes. Some in vivo and in vitro studies have noted that the presence of pancreatic progenitor/stem cells in islets may be a potential source for transplantable insulin-producing cells and that they had phenotypic markers identical to those in mesenchymal stem cells (MSCs) from bone marrow. The immunosuppressive activities of MSCs on the T cells, natural killer cells, B cells, and dendritic cells have been demonstrated. It is critical to determine whether these known functions of the MSCs can be fulfilled by pancreatic islet-like stem cells. Potentially, the immunological interactions between beta cells and pancreatic islet stem cells could have a role in the development of type 1 diabetes.

10.1 Introduction

From the point of view of the beta cell mass being a critical component of glucose homeostasis, studies on humans and rodents have demonstrate their plasticity and ability to expand in response to increasing insulin demands. In humans, a 50% higher volume of beta cells were seen in nondiabetic obese subjects than in lean nondiabetic controls (Butler et al. 2003). Furthermore, the autopsy series by Meier et al. showed that the pancreas of patients with longstanding type 1 diabetes had retained some of their regenerative capacity, as indicated by the presence of insulin-producing beta cells even after decades of autoimmune attacks (Brüning et al. 1997).

E. Karaöz (✉)
Department of Stem Cell, Center for Stem Cell and Gene Therapies Research and Practice, Institue of Health Sciences, Kocaeli University, Kocaeli, 41380 Turkey
e-mail: ekaraoz@hotmail.com

K. Turksen (ed.), *Adult and Embryonic Stem Cells*, Stem Cell Biology and Regenerative Medicine, DOI 10.1007/978-1-61779-630-2_10,

The development of diabetes could be reasonably considered the result of disruption of the balance between production and destruction of beta cells. Therefore, improving the regeneration potential of beta cells to a level of recovery after the damage caused by the autoimmune system was proposed as an alternative therapeutic model for type 1 diabetes (Guz et al. 2001; Claiborn and Stoffers 2008). Accordingly, many studies have been performed using experimental animal models. Although different strategies were applied in these studies, the endpoints focused on three main points: (1) to demonstrate the presence of intrapancreatic and islet-derived stem or progenitor cells; (2) to test whether these stem or progenitor cells contribute to the regeneration of pancreatic beta cells in the experimentally damaged pancreas or islets; and (3) to determine the effects of regeneration of beta cells on diabetes.

Much evidence has been accumulated regarding the formation of mature islet cells from stem or progenitor cells by differentiation in experimentally damaged pancreas models (Bonner-Weir et al. 1993; Sandgren et al. 1990). The development of new islets from pancreatic duct epithelium, for example, was found to be induced by partial pancreatectomy in rodents (Bonner-Weir et al. 1993). The induction of endocrine cell regeneration from intra-islet progenitors was illustrated in another model, in which alloxan and streptozotosin (STZ) were used to decrease the viability of beta cells (Korcakova 1971; Cantenys et al. 1981). Numerous other experimental models were introduced to confirm the role of pancreatic progenitor or stem cells in the regeneration of islets, such as partial duct obstruction (Rosenberg and Vinik 1992), application of steroids (Kem and Logothetopoulos 1970), insulin antibodies (Logothetopoulos and Bell 1966), trypsin inhibitor (Weaver et al. 1985), specific growth factors (Otonkoski et al. 1994), creating beta-cell damage by caerulein (Elsasser et al. 1986), and overexpression of Reg1 (Yamaoka et al. 2000), interferon γ I (Gu and Sarvetnick 1993), or transforming growth factor α (Sandgren et al. 1990).

10.2 Pancreas-Derived Progenitor or Stem Cells

There are several attempts to identify stem/progenitors cells in pancreatic tissue as a potential source for transplantable insulin-producing tissue. Unfortunately, the origin of new beta cells in adult pancreas is still unknown. Some in vivo and in vitro studies thought that pancreatic progenitor/stem cells were in the islets (Guz et al. 2001; Lechner et al. 2002; Banerjee and Bhonde 2003; Cornelius et al. 1997; Gershengorn et al. 2004, 2005; Atouf et al. 2007; Davani et al. 2007; Morton et al. 2007), whereas others reported that new adult beta cells might originate from preexisting beta cells (Treutelaar et al. 2003; Parnaud et al. 2008). Additional studies supported the idea that human and rodent epithelium of pancreatic duct cells could differentiate into a pancreatic endocrine phenotype. This location might be the place for stem cells in the pancreas that function naturally to renew islets for the life span of the host (Guz et al. 2001; Cornelius et al. 1997; Treutelaar et al. 2003; Peck and Cornelius 1995; Bonner-Weir et al. 1993, 2000; Taguchi and Otsuki 2004). The transdifferentiation of acinar cells to islets has also been proposed (Teng et al. 2007; Gu et al. 1994; Baertschiger et al. 2008; Minami et al. 2005; Sphyris et al. 2005).

Gershengorn et al. demonstrated that fibroblast-like cells from adult human islets donated postmortem could proliferate readily in vitro (Gershengorn et al. 2004, 2005). These mesenchymal-type cells, which exhibit no hormone expression, could then be induced to differentiate into hormone-expressing islet-like cell aggregates that restore the typical epithelial character of islet cells. It was thought that precursor cells were obtained from the insulin-expressing cells by epithelial-to-mesenchymal transition; and after the expansion, they were more easily differentiated into insulin-expressing cells by mesenchymal-to-epithelial transition compared with other precursor cells or by transdifferentiation. However, Gershengorn et al. more recently reported that human islet-derived precursor cells (hIPCs) were a type of mesenchymal stem cell (MSC) (Gershengorn et al. 2005; Davani et al. 2007). Moreover, they provided evidence that mouse IPCs in long-term culture were not derived from beta cells (Morton et al. 2007). Experiments carried out using mouse islets have demonstrated that murine beta cells did not undergo epithelial-to-mesenchymal-to-epithelial transition (EMET) (Meier et al. 2006; Atouf et al. 2007; Morton et al. 2007). A few recent studies demonstrated the dedifferentiation of pancreatic cells from islets into precursor cells, which later expanded and redifferentiated into the pancreatic lineage progenies (Lechner et al. 2005; Zhao et al. 2007).

In many research reports, different names were used to describe the pancreatic progenitor cells in islets, including intra-islet progenitor cells (Guz et al. 2001; Banerjee and Bhonde 2003), pancreatic stem cells (Schmied et al. 2001; Suzuki et al. 2004), small cells (Zhao et al. 2007), islet-derived progenitor cells (Linning et al. 2004; von Mach et al. 2004; Wang et al. 2004), multipotent stem cells (Choi et al. 2004), nestin-positive islet-derived progenitor (NIP) cells (Cornelius et al. 1997; Lechner et al. 2002; Abraham et al. 2004; Zhang et al. 2005; Eberhardt et al. 2006), monoclonal pancreatic stem cells (Xiao et al. 2008), and beta-stem cells (Duvillie et al. 2003).

Recently, we and some other research groups showed that nestin-positive progenitor/stem cells from islets of human and murine pancreas had phenotypical markers identical to those of MSCs from bone marrow and were able to proliferate and differentiate into insulin-producing cells in vitro (Lechner et al. 2002; Gershengorn et al. 2005; Zhang et al. 2005; Atouf et al. 2007; Davani et al. 2007; Karaoz et al. 2010a) (Table 10.1, Figs. 10.1, 10.2, 10.3, 10.4). Additionally, the expression patterns of Oct4, Rex1, and Sox2, which are generally recognized as the master regulators of stem cell renewal and differentiation, were analyzed by our group and were found to be expressed by rat pancreatic islet-derived progenitor/stem cells (Karaoz et al. 2010a) (Table 10.2). We also showed by reverse transcription-polymerase chain reaction (RT-PCR) that the nestin-positive cells in the pancreatic islets express neither the hormones insulin, glucagon, somatostatin, or pancreatic polypeptide nor the markers of embryonic development of endocrine pancreas (Fig. 10.3). Therefore, based on our observations and scientific evidence, we called them "pancreatic islet-derived stem cells" (PI-SCs) (Karaoz et al. 2010a).

Table 10.1 Immunocytochemical properties of rPI-SCs

Antibody/marker	rPI-SCs (passage 3)[a]
CD 31/PECAM-1	∅
CD 34	∅
CD 45	∅
CD 71	∅
CD105/Endoglin	+
c-Fos	+
Collagen II Ab-2	+
Collagen Ia1	−/+
β-Tubulin	+
Nestin	+
Vimentin	+
Desmin	+
Fibronectin	+
Smooth muscle α-actin	+
Actin	+
MyoD	∅
Myosin IIa	+
S100	+
Myogenin Ab-1	+
Tropomyosin	+
MAP 2a,b, Ab-2	+
GFAP Ab1	+
β3 Tubulin	+
NSE/γ enolase	+
HNK-1ST	∅
Osteocalcin	+
Osteonectin (SPARC)	+
Osteopontin	+[b]
BMP-2	+
BMP-4	+
Glut2	∅
Insulin	∅
Glucagon	∅
Somatostatin	∅
C-Peptide	∅
PDX-1	∅
VWF	∅
CD146	+
Cytokeratin 19	∅
Cytokeratin 18	∅
PCNA	+
Ki67	+
BrdU	+
STAT3	+
Connexin43	+

rPI-SCs pancreatic islet stem cell receptors
[a]+ positive, ∅ lack, −/+ weak
[b]Immunoreactivity was positive in 10–20% of the cells

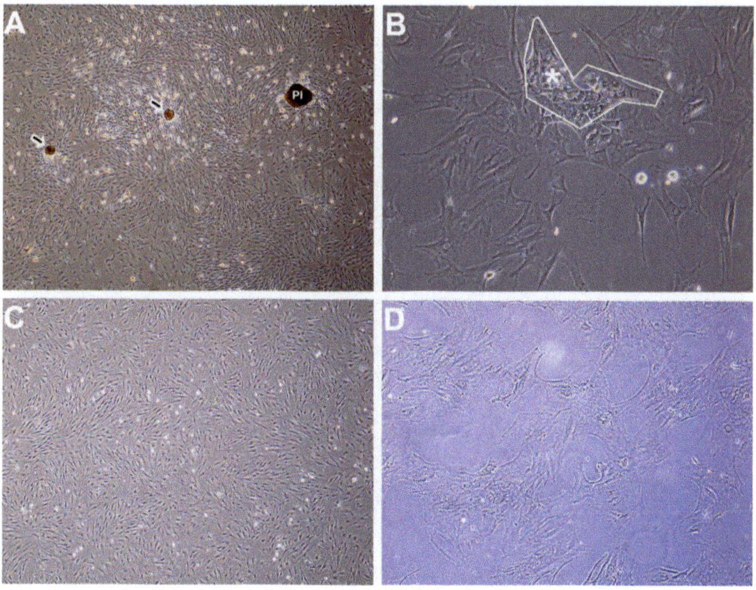

Fig. 10.1 Pancreatic islet stem cell receptors (rPI-SCs) in culture. Fibroblast-like cells is observed growing out and away from pancreatic islets (*PI*) (*arrows*) after 9 days (**a**). Three-dimensional architecture of the pancreatic islets was initiated to disrupt (indicated by the *asterisk*) during the following days (day 11). Degenerated islets are observed around the PI-SCs (**b**). After the next passages, most of these stem cells exhibited large, flattened, or fibroblast-like morphology. (**c**) P_1, 4th day; (**d**) P_{15}, 5th day. (**a** =×40; **b–d** = ×200; **c** = ×100)

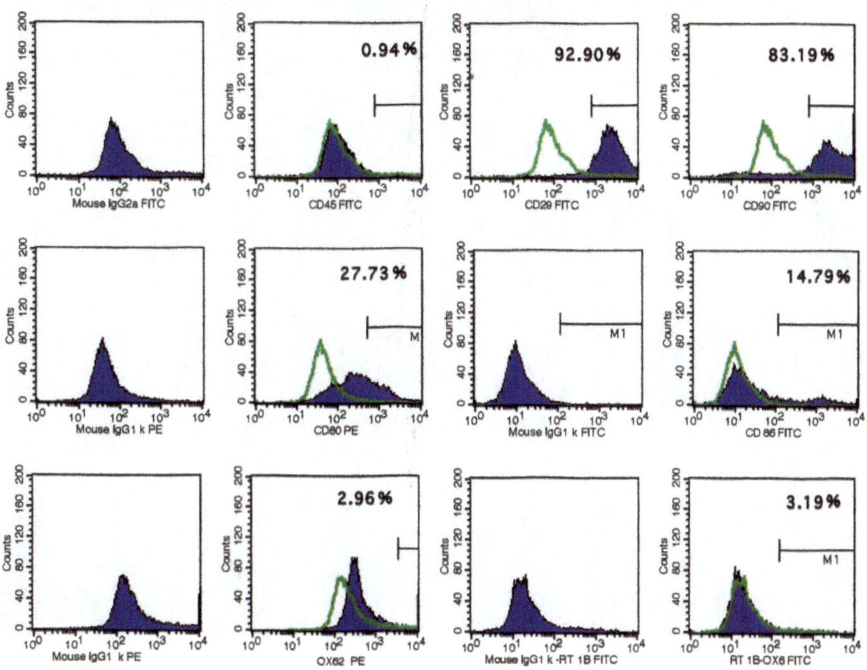

Fig. 10.2 Representative flow cytometry analysis of cell-surface markers in rPI-SCs at the P_3

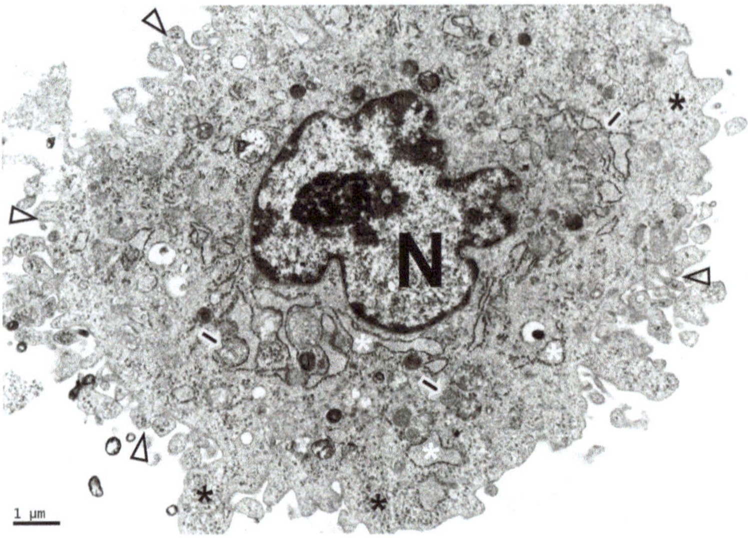

Fig. 10.3 Electron microscopic analysis of rPI-SC. The cells show pale, eccentric, irregularly shaped, large nuclei (*N*), usually with multiple nucleoli. Chromatin forms a thin and dense layer inside the perinuclear cisternae. The cytoplasm had many rough endoplasmic reticulum cisternae with dilated and contained electron-dense material (*white asterisks*) and mitochondria (*arrows*). Free ribosomes (*black asterisks*) and lots of empty vacuoles were present in the cytoplasm. The plasma membrane of rPI-SC has several thick pseudopodia-like structures (*arrowheads*). *Scale bar* 1 μ

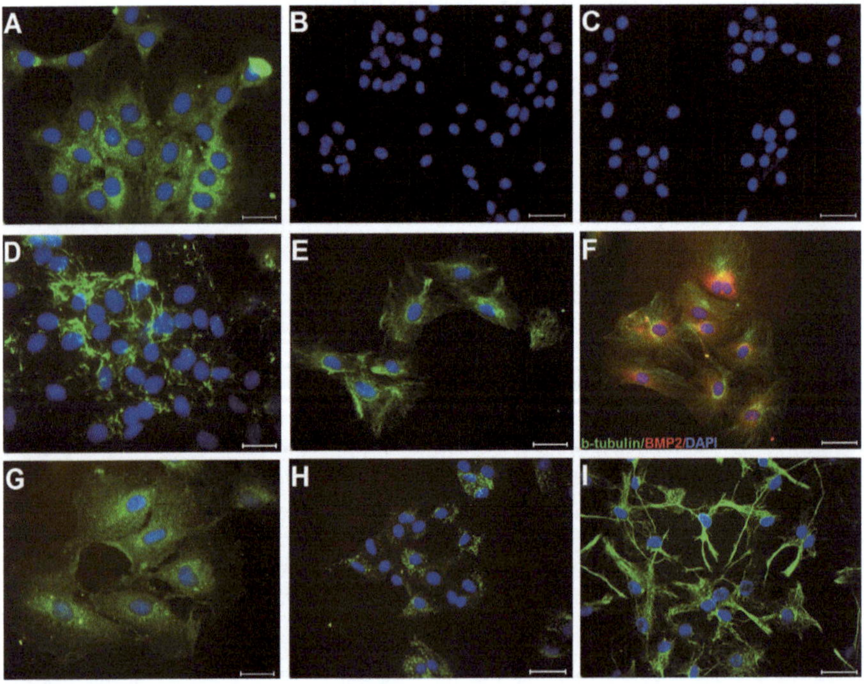

Fig. 10.4 Representative panels of immunofluorescence detection of some markers on rPI-SCs. Almost all of the rPI-SCs expressed connexin43 (**a**), fibronectin (**d**), vimentin (**e**), β-tubulin (**f**), BMP-2 (**f**), myogenin (**g**), β3-tubulin (**h**), and MAP2a,b (**i**). There was no immunoreaction for cytokeratin 18 (**b**) or cytokeratin-19 (**c**). Nuclei were labeled with DAPI (*blue*). *Scale bars* = 50 μm

Table 10.2 Expression of markers of different lineages by rPI-SCs assessed by RT-PCR

Gene	Gen bank (primer bank ID)	Primer sequences (forward; reverse)	Annealing temperature (°C)	Expression in rPI-SCs
Insulin I	NM_019129	CCGTCGTGAAGTGGAG CAGTTGGTAGAGGGAGCAG	57	Ø[a]
Insulin II	NM_019130	ATGGCCCTGTGGATCCGCTT CTAGTTGCAGTAGTTCTCCA	53	Ø
Glucagon	NM_012707	ATCATTCCCAGCTTCCCAGA CGGTTCCTCTTGGTGTTCAT	54	Ø[a]
Somatostatin	NM_012659	CAGGAACTGGCCAAGTAC AGTTCTTGCAGCCAGCTTTG	54	Ø[a]
IAPP	NM_012586	AGTCCTCCCACCAACCAATGT AGCACAGGCACGTTGTTGTA	54	Ø
GLUT-2	J03145	TTACTCTCCATTTCAGTCCTTTGT TAGAGCAGCTCTTTATTCCAGATTT	53	Ø[a]
GK	NM_012565	ACCAGAAAGGGGAGGCCT ATTAAAAACTCCCCCACAGTCC	51	Ø
GLP-1R	NM_012728	TCCTGTTAAAGCTGCAAGGC TTGTCCGAGAGGAAGGCTG	54	Ø
PDX-1	NM_022852	GGTGCCAGAGTTCAGTGCTAA CCAGTCTCGGTTCCATTCG	53	Ø[a]
Ngn3	NM_021700	CTTCACAAGAAGTCTGAGAACACCAG CTGCGCATAGCGGACCACAGCTTC	57	Ø
NeuroD1	NM_019218	TGTCGTTACTGCCTTTGGAA CGATCTGAATACAGCTACACGAA	53	Ø
PAX-6	NM_013001	CGACAAGATTTGCCATGGAT CAACCTTTGGAAAAACCAACA	54	Ø
Nkx2.2	X81408	CACGCAGGTCAAGATCTG TGCCCGCCTGGAAGGTGGCG	55	Ø
GAPDH	NM_017008	CACCCTGTTGCTGTAGCCATATTC GACATCAAGAAGGTGGTGAAGCAG	57	+
SOX9	XM_001081628.1	AGGAGAACACGTTCCCCAAG GTTGTGCAGATGCGGGTACT	54	+
COMP	NM_012834	ACCTAGCCCCACAGATGCTT CTCATTAACGTCGGTGCAGT	54	+/–
OPN	NM_012881	CCGATGAATCTGATGAGTCCTT TCCAGCTGACTTGACTCATGG	54	+[a]
Runx2	XM_001066762	GCCGGGAATGATGAGAACTA TTGGGGAGGATTTGTGAAGA	53	+
MEPE	NM_024142	GGCACCAAAGCTGAATGAAG TTTCTGTCGCTCTGTACTTCAGTC	52	+/–
Osteonectin	NM_012656	GGAAGCTGCAGAAGAGATGG TGCACACCTTTTCAAACTCG	53	+[a]
BMP2	NM_017178	ACCCGCTGTCTTCTAGTGTTG TTCTTCGTGATGGAAGCTGAG	54	+[a]
BMP4	NM_012827	ACTTCGAGGCGACACTTCTG GTCCACCTGCTCCCGAAATA	54	+
PPARg	NM_013124	GAAGACATCCCGTTCACAAGA GTGGATCCGACAGTTAAGATCA	51	+
Adiponectin	NM_144744	ATGCTACTGTTGCAAGCGCT GTGATACATGTAAGCGGCTTCTC	53	Ø
ADFP	NM_001007144	ATGGCATCAGTAGCAGTGGA AATCTGTGGCTCCAGCTTCT	56	+

(continued)

Table 10.2 (continued)

Gene	Gen bank (primer bank ID)	Primer sequences (forward; reverse)	Annealing temperature (°C)	Expression in rPI-SCs
Monoglyceride lipase (Mgll),	NM_138502	CCTCTTTTGTAGGTACTGGAAGC GAAGTCCGATACCACCATCCT	55	+
NF-H	NM_012607	AGTGGTTCCGAGTGAGGTTG CTGCTGAATAGCGTCCTGGT	54	+/–
NF-L	NM_031783	CCATGCAGGACACAATCAAC CGCCTTCCAAGAGTTTTCTG	53	+
GFAP	NM_017009	TCCTGGAACAGCAAAACAAG CAGCCTCAGGTTGGTTTCAT	53	+[a]
TUBB 3	NM_139254	TGGATGTCGTGAGGAAAGAAT TCATCCGTGTTCTCCACTAGC	53	+[a]
Enolase gamma, neuronal (Eno2)	NM_139325	ATGTGATCAACGGTGGCTCT TATTTGCCATCGCGGTAAAA	53	+[a]
Rex-1	NM_001012114	CACAAGCATGGATGATGATGA TGATGGCTTTGAGCTATCCAC	53	+
Sox-2	NM_001109181	ATGATGGAGACGGAGCTGAA GCTTGCTGATCTCCGAGTTG	54	+
Oct-4	EU419996	GCCGTGAAGTTGGAGAAGGT TCACACGGTTCTCAATGCTAGTC	55	+
ACTA2	NM_031004	GTGTGAAGAGGAAGACAGCACA TCGTCCCAGTTGGTGATGAT	59	+[a]
ACTb	NM_031144	ACCCGCGAGTACAACCTTCT CTTCTGACCCATACCCACCA	54	+[a]
Myogenin	NM_017115	GGCCAGTGGAAGGAACAAGT GACACAGACTTCCTCTTACACACCT	56	+[a]
Desmin	NM_022531	GACGCAGTGAACCAGGAGTT TTGGTGAGGACCTCCACTTG	53	+[a]
MyoD1	NM_176079	GACTTCTATGATGATCCGTGTTTC ATGCCATCAGAGCAGTTGGA	53	∅[a]
Nestin	NM_012987	GTGGCTCACATGGAAAGCTC CCACAGCCAGCTGGAACTTA	54	+[a]
c-fos	X06769	GGAGTGGTGAAGACCATGTCA CAACGCAGACTTCTCGTCTTC	57	+[a]
Vimentin	NM_031140	CTCCAACCGGAGCTATGTGA CTCCTGCAATTCCACCTTCTC	54	+[a]
Mapkapk2	NM_178102	GTTCCCTCAGTTCCACGTCA CCATGACAATCAGCAAGCAC	54	+
MHC II	NM_001101017	GCTCGTGACCAGACACATCTAC TCCAGTCCCCGTTCCTAATA	54	∅

CD40	NM_134360	GCCGGGAAACCGACTAGTTA CATCTGCACGACTCCAAAGC	54	+
CD80	NM_012926	CTAACAACTACTCCTTTAGCCTCCT TTGAAGTCTAGTTGGCTACTAATGG	55	+
CD86	NM_020081	TGTCGTCAAGACATGTGTAACCT GCGCCCAAATAGTGTTCGTA	53	∅
Interleukin 6 receptor, alpha	NM_017020	TGAGTCCTGGGACCCAAGTT TGCACCACATGCTTTACTCC	55	+
Tnip1	NM_001108826	ACAAGGGATAAAGATGTTAGGAGAG AGTTTCCAATGGTGGTGGTG	59	+
Bcl3	NM_001109422	AGGCCGGAGGCTCTTTACTA GGCTGAGAATTCGGTAGACG	54	+
Ncf1	NM_053734	GACACCTTCATTCGCCACAT GTGAGGGATGACTCTGTTTTCTG	52	+
Cx3cr1	NM_133534	GTCCAAGAGCATCACTGACATC GTAATCACCCAAACATTCGTTG	52	+

IAPP Islet amyloid polypeptide, *GK* Glucokinase, *COMP* rat2 cartilage oligomeric matrix protein, *OPN* osteopontin rat secreted phosphoprotein 1 (Spp1), *OCN* osteocalsin Bone gamma-carboxyglutamate(gla) protein, *Runx2* runt related transcription factor 2, *MEPE* matrix extracellular phosphoglycoprotein, *BMP2* bone morphogenetic protein 2, *BMP4* bone morphogenetic protein 4, *Mapkapk2* mitogen-activated protein kinase-activated protein kinase 2, *PPARg* peroxisome proliferator-activated receptor gamma, *ADFP* Adipophilin, adipose differentiation related protein, *NF-H* neurofilament heavy polypeptide, *NF-L* neurofilament light polypeptide, *GFAP* rat glial fibrillary acidic protein, *TUBB 3* tubulin beta 3, *Rex-1* RNA exonuclease 1, *Sox-2* SRY (sex determining region Y)-box 2, *ACTA2* smooth muscle alpha-actin, *ACTb* actin beta, *MyoD1* myogenic differentiation 1, *IFNg* interferon gamma, *Tgfb1* transforming growth factor beta 1, *Tnip1* TNFAIP3 interacting protein 1, *Bcl3* B-cell CLL/lymphoma 3, *Pf4* platelet factor 4, *Ncf1* neutrophil cytosolic factor 1, *Ccr2* chemokine (C-C motif) receptor 2, *Cx3cr1* chemokine (C-X3-C motif) receptor 1

[a]Results cross-confirmed at the protein level by immunolocalization analysis

10.3 Immunosuppressive and Immunomodulatory Properties of Adult Stromal MSCs

Up to now in studies related to islet-like MSCs, the main idea was to use those cells as an endogenous source for cell-mediated treatment of type 1 diabetes. However, the immunosuppressive activity of MSCs from various sources, and especially from bone marrow, on the T cells of both animals and humans, natural killer (NK) cells (Guo et al. 2006; Spaggiari et al. 2008), and B cells (Comoli et al. 2008) has been demonstrated.

In studies aiming to explain the mechanisms of immunosuppressive activity of the MSCs, it has been declared that when the MSCs are co-cultured with T cells, dendritic cells (DCs), effector T cells, or NK cells, they change the cytokine secretion profiles of the related cells to encourage an antiinflammatory effect or a tolerant phenotype (Chang et al. 2006). It was noted that the MSCs were activated by interferon-γ(IFNγ) secreted by some cells (e.g., T cells, NK cells) and by interleukin-1 (IL-1)-like cytokines from monocytes. Whereas MSCs had an inhibitory effect on immune cell activities by releasing cytokines such as IL-10 (Krampera et al. 2003) and IL-6 (Noel et al. 2007) and soluble factors including the transforming growth factor β (TGFβ) (Di Nicola et al. 2002; Aggarwal and Pittenger 2005), hepatocyte growth factor β (HGFβ) (Rasmusson 2006), indoleamine-2,3-deoxygenase (IDO) (Meisel et al. 2004), prostaglandin E_2 (PGE_2) (Noel et al. 2007), human leukocyte antigen-G (HLA-G), and nitric oxide (NO) (Nasef et al. 2007).

Recently, it has been determined that the immunosuppressive effect of human MSCs observed when they were indirectly co-cultured with lymphocytes (with paracrine mechanisms-soluble factors) and with cell-to-cell contact in the case of rodent MSCs (Di Nicola et al. 2002; Krampera et al. 2003). Suva et al. (2008) demonstrated withy time-lapse imaging that active T lymphocytes were tightly attached to allogeneic MSCs in vitro; they migrated underneath them within 4 h (transmigration) and remained there for 60 h. This transmigration demonstrated that MSC–T lymphocyte contact is necessary for the efficiency of the inhibitor or stimulator signal molecules. Synthesis of some adhesion molecules by MSCs—e.g., intracellular adhesion molecule-2 (ICAM-2: CD102), lymphocyte function antigen-3 (LFA-3), vascular cell adhesion molecule-1 (VCAM-1: CD106)—which provide the T-lymphocyte interactions, explained how the MSC–T lymphocyte contact is established (Majumdar et al. 1998; Conget and Minguell 1999).

In vivo immunosuppressive effects of MSCs were first observed in monkeys, in whom rejection of tissue-inharmonious skin grafts was delayed by infusion of donor-derived MSCs (Bartholomew et al. 2002). Guo et al. reported suppression of lymphocyte proliferation by mouse MSCs that was induced by allogeneic splenocytes and prolonged the skin grafts' life span with in vivo immunosuppressive effects (Guo et al. 2006). Observations have indicated a decrease in acute and chronic graft-versus-host disease (GVHD) when the MSCs are transported together with HLA-identical hematopoietic stem cells (HSCs) (Tyndall et al. 2007).

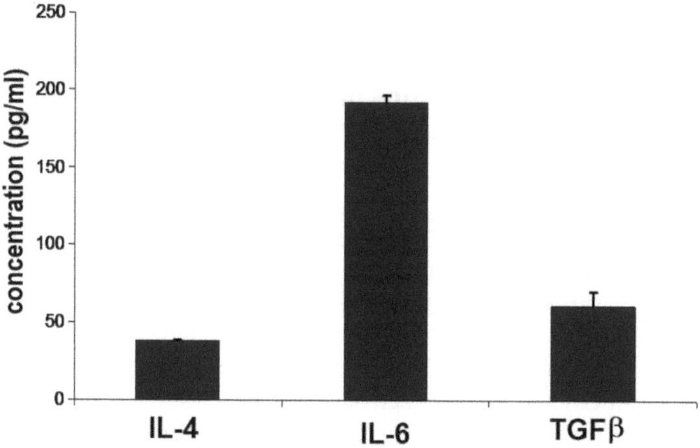

Fig. 10.5 Expression of cytokines—interleukin-4 (IL-4), interleukin-6 (IL-6), transforming growth factor β (TGFβ)—by rPI-SC

In the report by Lee et al. (2002), rapid engraftment was accomplished in an acute leukemia patient who was given MSCs and peripheral blood stem cells by his father who was HLA-haplo-identical. With standard immunosuppressive treatment, the patient survived without acute or chronic GVHD until 31 months after the transplantation. Ultimately, in two other reports, it was declared that the MSCs have considerably optimistic results in treating and/or preventing GVHD (Ball et al. 2007; Fang et al. 2007). In the study by Fang et al., severe acute GVHD developed after a blood transfusion to 15- and 12-year-old children who suffered from acute myeloid leukemia (AML) and acute lymphoblastic leukemia (ALL), respectively, and in whom conventional pharmaceutical therapies had failed (Fang et al. 2007). The researchers implemented MSCs obtained from subcutaneous abdominal adipose tissue by means of plastic surgery, to the children as a second treatment and attained a successful result in that both had revived within 15 days. Ball et al. performed donor-derived MSC infusions in 14 children simultaneously with HLA-inharmonious peripheral blood HSC transplantation. Graft failure was not experienced in this group, although such failure was previously observed in 15% of 47 transplantations for which MSC infusion was not employed (Ball et al. 2007).

With respect to all these data, it is critical to know whether these known functions of the MSCs can be fulfilled by pancreatic islet-like MSCs. In studies performed by our group, rat bone marrow and islet-derived MSCs showed substantially similar phenotypical, structural, and molecular characteristics (Karaoz et al. 2010a) (Fig. 10.5, Table 10.2). In general, this similarity led us to suggest that the known immunosuppressive effects of bone marrow MSCs (BM-MSCs) are likely to be shown by the islet-like MSCs.

10.4 Presence of Transcripts Coding for Antigen-Presenting Surface Proteins in Addition to the Genes with Known Antiapoptotic Functions in PI-SCs

It was reported that MSCs were able to have regulating effects on the immune responses by multiple mechanisms through their suppressive effects on the functions and proliferation of both CD34+-derived and monocyte-derived DCs (Jiang et al. 2005; Beyth et al. 2005; Nauta et al. 2006; Djouad et al. 2007). Furthermore, the studies demonstrated a bidirectional effect; that is, MSCs might be suppressive and/or activating according to the type of stimulation. Studies by at least three independent groups have shown that IFNγ-stimulated MSCs could act as antigen-presenting cells (APCs) and that this derivation would have the activating effects of MSCs on the immune responses (Stagg et al. 2006; Chan et al. 2006; Morandi et al. 2008). These experimental results were in good agreement with the idea that the presence of APCs in islets are considered to be responsible for the development of type 1 diabetes by presenting beta cell-derived peptides to immune cells (Lacy et al. 1979; Nauta et al. 2006). The significant question is whether the MSCs isolated from the pancreatic islets behaved like APCs.

We determined that rPI-SCs expressed CD80 and CD40 without any stimulation (Karaoz et al. 2010c) (Fig. 10.2, Table 10.2). This finding was verified by the results of Klein et al. (2005), who used total RNA isolated from human and nonhuman primate islets to show the expression of CD40 (Klein et al. 2005). Our study however, took this a step further and determined that the source of CD40 mRNA was pancreatic islet stem cells. We failed to demonstrate the presence of the CD86 transcript although the CD80 transcript was clearly present. Lei et al. (2005) observed a similar case, in which murine-derived keratinocyte stem cells (KSCs) expressed CD80 but not CD86, indicating that KSCs could act as APCs (Lei et al. 2005). The absence of major histocompatibility complex (MHC) class II expression in rPI-SCs, whose constitutive expression was observed only in professional APCs of the immune system (Karaoz et al. 2010c), supports that these cells are nonprofessional APCs.

CX3CR1, identified as the human receptor for fractalkine (Imai et al. 1997), was found to be expressed in monocytes, subsets of NK cells, DCs, and brain microglial cells in a knockout mouse via replacement of the CX3CR1 gene with green fluorescent protein gene (Jung et al. 2000). More interestingly, the expression of CX3CR1 was observed in PI-SCs, which might be previously defined as DCs of pancreatic islets also expressing CX3CR1 (Lacy et al. 1979; Adorini 2001; Calderon et al. 2008).

Genes having known antiapoptotic functions that are expressed under stress, including *MAPKAP2*, *TNIP1*, and *BCL3*, were studied in PI-SCs. MAPKAPK2 is a direct substrate of p38 MAP kinase in response to cellular stress, such as mechanical stress, heat shock, osmotic stress, ultraviolet (UV) irradiation, bacterial lipopolysaccharide (LPS), several inflammatory cytokines, and growth factors (Ono and Han 2000). TNIP1 was originally described as an anti-apoptotic tumor necrosis factor α (TNFα)-induced gene in endothelial cells (Opipari et al. 1990). *BCL3*, in complex with *p52*, can promote transcription of the genes encoding the cell cycle regulator cyclin D1 and the antiapoptotic BCL2 protein (Kashatus et al. 2006).

We showed the expression of all three genes in rPI-SCs (Karaoz et al. 2010c) (Table 10.2), implying a role played by rPI-SCs in the regulation of immunity and tissue homeostasis in islets and highlighting the antiapoptotic influence mechanisms that might have a protective role on pancreatic islet cells.

Only some types of cell, mainly hepatocytes and several leukocytes, express the IL-6 receptor (IL-6R). Surprisingly, we determined in PI-SCs that IL-6R interacts with IL-6, an inflammatory cytokine with a well-documented role in inflammation and cancer (unpublished data). In addition, high levels of IL-6 and IL-6R have been reported in patients with several chronic inflammatory and autoimmune diseases (Rose-John et al. 2007). Because type 1 diabetes is an autoimmune disease, IL-6R-positive rPI-SCs might have a pivotal role in its pathogenesis.

It has been reported that cultured MSCs spontaneously secrete TGFβ1 and IL-6, but not IFNγ, IL-5, or IL-10 (Boumaza et al. 2009; Karaoz et al. 2010b). By direct inhibition of lymphocyte apoptosis, IL-6 secretion might play a major role in the antiapoptotic function of MSCs (Xu et al. 2007) The suppression of various immune functions through the release of immunosuppressive factors (e.g., TGFβ1) by MSCs demonstrated their antiapoptotic effects (Puissant et al. 2005) In our experimental studies (unpublished data), not only IL-6, but also the proinflammatory cytokine TGFβ1 secretion was observed in the culture medium of rPI-SCs (Fig. 10.5). TGFβ1 is believed to be the inducer of the regulatory T cells or CD8+ T cells, which inactivate T cells to an anergic state and prevent autoimmune diseases such as diabetes type 1 (Bettini and Vignali 2009; Kishi et al. 2010). Kishi et al. showed that TGFβ1 induces CD8+Foxp3+ T cells ex vivo, which suppresses diabetogenic T cells in vitro and in vivo.

10.5 Do Pancreatic Islet-Derived Stem Cells Have a Key Role in Type 1 Diabetes Pathogenesis?

In view of all these data, it is important to know to what degree the known characteristics of the MSCs are also found in pancreatic islet-derived MSCs. It was determined previously that rat bone marrow and islet-derived MSCs substantially had phenotypical, structural, and molecular characteristics in common (Karaoz et al. 2010a). Generally, such evidence has led us to consider that known immunosuppressive effects of BM-MSCs are likely to be displayed also by islet-like MSCs. As a result, the evidence that has been demonstrated with regard to immunological interactions may indicate that the beta cells and MSCs may play a role in the development of type 1 diabetes.

The APCs, which have an important role in the pathogenesis of type 1 diabetes and are responsible for presenting beta cell-derived peptides to immune-responsible cells, have been known to be present in islets of Langerhans since the 1970s (Calderon et al. 2008). It is crucial to know whether the MSCs that we and other researchers have isolated from the pancreatic islets, and which resemble BM-MSCs, are able to respond like the APCs.

Although expressing MHC class I antigens and the ability to function as APCs, as already mentioned, it is thought that these cells have an effect in the pathogenesis of autoimmune type 1 diabetes. It has been suggested that these cells could have the potential for treating autoimmune diseases. It was especially noted that MSCs could be useful in treating autoimmune diseases by increasing the production of T_{reg} cells (Jones and McTaggart 2008; Di Ianni et al. 2008). Chang et al. (2006) declared that the suppression of stimulated lymphocyte reactivation by placenta-derived MSCs was due to decreased cell proliferation and increased numbers of regulatory T cells but not by cell death. Finally, it has been indicated that the BM-MSCs, obtained from both healthy and autoimmune-disease persons, suppressed proliferation of the stimulated autologous and allogeneic peripheral blood mononuclear cells as much as 90% (Bocelli-Tyndall et al. 2007).

Apart from the knowledge related to the pathogenesis of type 1 diabetes, the role of MSCs in the islets and their regeneration and immunosuppressive characteristics should be investigated. We need to investigate the connection of malfunctioning MSCs in the islets and immune issues. Also the ability to produce endocrine cells, including beta cells, remain to be investigated. In a recent report it was argued that autoreactive (self-reactive) immune cells comprise 2.5% of the cell population of the circulating blood in healthy living individuals. The question remains why these cells do not become involved in addressing autoimmune diseases (Duty et al. 2009). Considering this information, the question of whether the self-reactive cells easily destroy the beta cells because of the incapability of MSCs in the islets comes to mind. We wonder if transduced cells in a viral infection, which have been frequently cited for being responsible for the occurrence of type 1 diabetes, are not MSCs, rather than beta cells. These cells naturally possess MHC class 1 receptors, which play a role in presenting the viral peptide parts to the cell surface. Answers to these questions could define the new role or roles of the pancreas (islets) in the radical treatment for the type 1 diabetes. In our opinion, even if we produce beta cells or islets using exogenous stem cells, we cannot be successful with such replacement treatment so long as the autoimmune attacks continue. To develop strategies for preventing the autoimmune attacks, we must completely and clearly determine the causal mechanisms.

References

Abraham EJ, Kodama S, Lin JC, Ubeda M, Faustman DL, Habener JF (2004) Human pancreatic islet-derived progenitor cell engraftment in immunocompetent mice. Am J Pathol 164:817–830

Adorini L (2001) Interleukin 12 and autoimmune diabetes. Nat Genet 27:131–132

Aggarwal S, Pittenger MF (2005) Human mesenchymal stem cells modulate allogeneic immune cell responses. Blood 105:1815–1822

Atouf F, Park CH, Pechhold K, Ta M, Choi Y, Lumelsky NL (2007) No evidence for mouse pancreatic beta-cell epithelial-mesenchymal transition in vitro. Diabetes 56:699–702

Baertschiger RM, Bosco D, Morel P et al (2008) Mesenchymal stem cells derived from human exocrine pancreas express transcription factors implicated in beta-cell development. Pancreas 37:75–84

Ball LM, Bernardo ME, Roelofs H et al (2007) Cotransplantation of ex vivo expanded mesenchymal stem cells accelerates lymphocyte recovery and may reduce the risk of graft failure in haploidentical hematopoietic stem-cell transplantation. Blood 10:2764–2767

Banerjee M, Bhonde RR (2003) Islet generation from intra islet precursor cells of diabetic pancreas: in vitro studies depicting in vivo differentiation. JOP 4:137–145

Bartholomew A, Sturgeon C, Siatskas M et al (2002) Mesenchymal stem cells suppress lymphocyte proliferation in vitro and prolong skin graft survival in vivo. Exp Hematol 30:42–48

Bettini M, Vignali DA (2009) Regulatory T cells and inhibitory cytokines in autoimmunity. Curr Opin Immunol 21(6):612–618

Beyth S, Borovsky Z, Mevorach D et al (2005) Human mesenchymal stem cells alter antigen-presenting cell maturation and induce T-cell unresponsiveness. Blood 105:2214–2219

Bocelli-Tyndall C, Bracci L, Spagnoli G et al (2007) Bone marrow mesenchymal stromal cells (BM-MSCs) from healthy donors and auto-immune disease patients reduce the proliferation of autologous- and allogeneic-stimulated lymphocytes in vitro. Rheumatology (Oxford) 46:403–408

Bonner-Weir S, Baxter LA, Schuppin GT, Smith FE (1993) A second pathway for regeneration of adult exocrine and endocrine pancreas: a possible recapitulation of embryonic development. Diabetes 42:1715–1720

Bonner-Weir S, Taneja M, Weir GC et al (2000) In vitro cultivation of human islets from expanded ductal tissue. Proc Natl Acad Sci USA 97:7999–8004

Boumaza I, Srinivasan S, Witt WT, Feghali-Bostwick C, Dai Y, Garcia-Ocana A, Feili-Hariri M (2009) Autologous bone marrow-derived rat mesenchymal stem cells promote PDX-1 and insulin expression in the islets, alter T cell cytokine pattern and preserve regulatory T cells in the periphery and induce sustained normoglycemia. J Autoimmun 32:33–42

Brüning JC, Winnay J, Bonner-Weir S, Taylor SI, Accili D, Kahn CR (1997) Development of a novel polygenic model of NIDDM in mice heterozygous for IR and IRS-1 null alleles. Cell 88:561–572

Butler AE, Janson J, Bonner-Weir S, Ritzel R, Rizza RA, Butler PC (2003) Beta-cell deficit and increased beta-cell apoptosis in humans with type 2 diabetes. Diabetes 52:102–110

Calderon B, Suri A, Miller MJ, Unanue ER (2008) Dendritic cells in islets of Langerhans constitutively present beta cell-derived peptides bound to their class II MHC molecules. Proc Natl Acad Sci USA 105:6121–6126

Cantenys D, Portha B, Dutrillaux MC, Hollande E, Roze C, Picon L (1981) Histogenesis of the endocrine pancreas in newborn rats after destruction by streptozotosin: an immunocytochemical study. Virchows Arch B Cell Pathol Incl Mol Pathol 35:109–122

Chan JL, Tang KC, Patel AP et al (2006) Antigen-presenting property of mesenchymal stem cells occurs during a narrow window at low levels of interferon-gamma. Blood 107:4817–4824

Chang CJ, Yen ML, Chen YC et al (2006) Placenta-derived multipotent cells exhibit immunosuppressive properties that are enhanced in the presence of interferon-γ. Stem Cells 24:2466–2477

Choi Y, Ta M, Atouf F, Lumelsky N (2004) Adult pancreas generates multipotent stem cells and pancreatic and nonpancreatic progeny. Stem Cells 22:1070–1084

Claiborn KC, Stoffers DA (2008) Toward a cell-based cure for diabetes: advances in production and transplant of beta cells. Mt Sinai J Med 75:362–371

Comoli P, Ginevri F, Maccario R et al (2008) Human mesenchymal stem cells inhibit antibody production induced in vitro by allostimulation. Nephrol Dial Transplant 23:1196–1202

Conget PA, Minguell JJ (1999) Phenotypical and functional properties of human bone marrow mesenchymal progenitor cells. J Cell Physiol 181:67–73

Cornelius JG, Tchemev V, Kao KJ, Peck AB (1997) In vitro generation of islets in long-term cultures of pluripotent stem cells from adult mouse pancreas. Horm Metab Res 29:271–277

Davani B, Ikonomou L, Raaka BM et al (2007) Human islet-derived precursor cells are mesenchymal stromal cells that differentiate and mature to hormone-expressing cells in vivo. Stem Cells 25:3215–3222

Di Ianni M, Del Papa B, De Ioanni M et al (2008) Mesenchymal cells recruit and regulate T regulatory cells. Exp Hematol 36:309–318

Di Nicola M, Carlo-Stella C, Magni M et al (2002) Human bone marrow stromal cells suppress T-lymphocyte proliferation induced by cellular or nonspecific mitogenic stimuli. Blood 99:3838–3843

Djouad F, Charbonnier LM, Bouffi C et al (2007) Mesenchymal stem cells inhibit the differentiation of dendritic cells through an interleukin-6-dependent mechanism. Stem Cells 25:2025–2032

Duty JA, Szodoray P, Zheng NY et al (2009) Functional anergy in a subpopulation of naive B cells from healthy humans that express autoreactive immunoglobulin receptors. J Exp Med 206:139–151

Duvillie B, Attali M, Aiello V, Quemeneur E, Scharfmann R (2003) Label-retaining cells in the rat pancreas: location and differentiation potential in vitro. Diabetes 52:2035–2042

Eberhardt M, Salmon P, von Mach MA et al (2006) Multipotential nestin and Isl-1 positive mesenchymal stem cells isolated from human pancreatic islets. Biochem Biophys Res Commun 345:1167–1176

Elsasser HP, Adler G, Kern HF (1986) Time course and cellular source of pancreatic regeneration following acute pancreatitis in the rat. Pancreas 1:421–429

Fang B, Song Y, Lin Q et al (2007) Human adipose tissue-derived mesenchymal stromal cells as salvage therapy for treatment of severe refractory acute graft-vs.-host disease in two children. Pediatr Transplant 11:814–817

Gershengorn MC, Hardikar AA, Wei C, Geras-Raaka E, Marcus-Samuels B, Raaka BM (2004) Epithelial-to-mesenchymal transition generates proliferative human islet precursor cells. Science 306:2261–2264

Gershengorn MC, Geras-Raaka E, Hardikar AA, Raaka BM (2005) Are better islet cell precursors generated by epithelial-to-mesenchymal transition? Cell Cycle 4:380–382

Gu D, Sarvetnick N (1993) Epithelial cell proliferation and islet neogenesis in IFN-γ transgenic mice. Development 118:33–46

Gu D, Lee MS, Krahl T, Sarvetnick N (1994) Transitional cells in the regenerating pancreas. Development 120:1873–1881

Guo Z, Li H, Li X et al (2006) In vitro characteristics and in vivo immunosuppressive activity of compact bone-derived murine mesenchymal progenitor cells. Stem Cells 24:992–1000

Guz Y, Nasir I, Teitelman G (2001) Regeneration of pancreatic beta cells from intra-islet precursor cells in an experimental model of diabetes. Endocrinology 142:4956–4968

Imai T, Hieshima K, Haskell C, Baba M, Nagira M, Nishimura M, Kakizaki M, Takagi S, Nomiyama H, Schall TJ, Yoshie O (1997) Identification and molecular characterization of fractalkine receptor CX3CR1, which mediates both leukocyte migration and adhesion. Cell 91:521–530

Jiang XX, Zhang Y, Liu B et al (2005) Human mesenchymal stem cells inhibit differentiation and function of monocyte-derived dendritic cells. Blood 105:4120–4126

Jones BJ, McTaggart SJ (2008) Immunosuppression by mesenchymal stromal cells: from culture to clinic. Exp Hematol 36:733–741

Jung S, Aliberti J, Graemmel P, Sunshine MJ, Kreutzberg GW, Sher A, Littman DR (2000) Analysis of fractalkine receptor CX(3)CR1 function by targeted deletion and green fluorescent protein reporter gene insertion. Mol Cell Biol 20:4106–4114

Karaoz E, Ayhan S, Gacar G, Aksoy A, Duruksu G, Okçu A, Demircan PC, Sariboyaci AE, Kaymaz F, Kasap M (2010a) Isolation and characterization of stem cells from pancreatic islet: pluripotency, differentiation potential and ultrastructural characteristics. Cytotherapy 12:288–302

Karaoz E, Genc ZS, Demircan PC, Aksoy A, Duruksu G (2010b) Protection of rat pancreatic islet function and viability by co-culture with rat bone marrow-derived mesenchymal stem cells. Cell Death Dis. doi:10.1038/cddis.2010.14

Karaoz E, Okçu A, Saglam Ö, Genç ZS, Ayhan S, Kasap M (2010c) Pancreatic islet derived stem cells can express co-stimulatory molecules of antigen-presenting cells. Transplant Proc 42:3663–3670

Kashatus D, Cogswell P, Baldwin AS (2006) Expression of the Bcl-3 proto-oncogene suppresses p53 activation. Genes Dev 20:225–235

Kem H, Logothetopoulos J (1970) Steroid diabetes in the guinea pig: studies on islet-cell ultrastucture and regeneration. Diabetes 19:145–154

Kishi M, Yasuda H, Abe Y, Sasaki H, Shimizu M, Arai T, Okumachi Y, Moriyama H, Hara K, Yokono K, Nagata M (2010) Regulatory CD8+ T cells induced by exposure to all-trans retinoic acid and TGF-beta suppress autoimmune diabetes. Biochem Biophys Res Commun 394(1):228–232

Klein D, Barbé-Tuana F, Pugliese A, Ichii H, Garza D, Gonzalez M, Molano RD, Ricordi C, Pastori RL (2005) A functional CD40 receptor is expressed in pancreatic beta cells. Diabetologia 48:268–276

Korcakova L (1971) Mitotic division and its significance for regeneration of granulated B-cells in the islets of Langerhans in allozan-diabetic rats. Folia Morphol (Praha) 19:24–30

Krampera M, Glennie S, Dyson J et al (2003) Bone marrow mesenchymal stem cells inhibit the response of naive and memory antigen-specific T cells to their cognate peptide. Blood 101:3722–3729

Lacy PE, Davie JM, Finke EH (1979) Prolongation of islet allograft survival following in vitro culture (24 degrees C) and a single injection of ALS. Science 204:312–313

Lechner A, Leech CA, Abraham EJ, Nolan AL, Habener JF (2002) Nestin-positive progenitor cells derived from adult human pancreatic islets of Langerhans contain side population (SP) cells defined by expression of the ABCG2 (BCRP1) ATP-binding cassette transporter. Biochem Biophys Res Commun 293:670–674

Lechner A, Nolan AL, Blacken RA, Habener JF (2005) Redifferentiation of insulin-secreting cells after in vitro expansion of adult human pancreatic islet tissue. Biochem Biophys Res Commun 327:581–588

Lee ST, Jang JH, Cheong JW et al (2002) Treatment of high-risk acute myelogenous leukaemia by myeloablative chemoradiotherapy followed by co-infusion of T cell-depleted haematopoietic stem cells and culture-expanded marrow mesenchymal stem cells from a related donor with one fully mismatched human leucocyte antigen haplotype. Br J Haematol 118(4):1128–1131

Lei J, Cheng J, Li Y, Li S, Zhang L (2005) CD80, but not CD86, express on cultured murine keratinocyte stem cells. Transplant Proc 37:289–291

Linning KD, Tai MH, Madhukar BV et al (2004) Redox-mediated enrichment of self-renewing adult human pancreatic cells that possess endocrine differentiation potential. Pancreas 29:64–76

Logothetopoulos J, Bell EG (1966) Histological and autoradiographic studies of the islets of mice injected with insulin antibody. Diabetes 15:205–211

Majumdar MK, Thiede MA, Mosca JD, Moorman M, Gerson SL (1998) Phenotypic and functional comparison of cultures of marrow-derived mesenchymal stem cells (MSCs) and stromal cells. J Cell Physiol 176:57–66

Meier JJ, Ritzel RA, Maedler K, Gurlo T, Butler PC (2006) Increased vulnerability of newly forming beta cells to cytokine-induced cell death. Diabetologia 49:83–89

Meisel R, Zibert A, Laryea M, Gobel U, Daubener W, Dilloo D (2004) Human bone marrow stromal cells inhibit allogeneic T-cell responses by indoleamine 2,3-dioxygenase-mediated tryptophan degradation. Blood 103:4619–4621

Minami K, Okuno M, Miyawaki K et al (2005) Lineage tracing and characterization of insulin-secreting cells generated from adult pancreatic acinar cells. Proc Natl Acad Sci USA 102:15116–15121

Morandi F, Raffaghello L, Bianchi G et al (2008) Immunogenicity of human mesenchymal stem cells in HLA-class I-restricted T-cell responses against viral or tumor-associated antigens. Stem Cells 26:1275–1287

Morton RA, Geras-Raaka E, Wilson LM, Raaka BM, Gershengorn MC (2007) Endocrine precursor cells from mouse islets are not generated by epithelial-to-mesenchymal transition of mature beta cells. Mol Cell Endocrinol 270:87–93

Nasef A, Mathieu N, Chapel A et al (2007) Immunosuppressive effects of mesenchymal stem cells: involvement of HLA-G. Transplantation 84:231–237

Nauta AJ, Kruisselbrink AB, Lurvink E, Willemze R, Fibbe WE (2006) Mesenchymal stem cells inhibit generation and function of both CD34+-derived and monocyte-derived dendritic cells. J Immunol 177:2080–2087
Noel D, Djouad F, Bouffı C, Mrugala D, Jorgensen C (2007) Multipotent mesenchymal stromal cells and immune tolerance. Leuk Lymphoma 48:1283–1289
Ono K, Han J (2000) The p38 signal transduction pathway: activation and function. Cell Signal 12:1–13
Opipari AW Jr, Boguski MS, Dixit VM (1990) The A20 cDNA induced by tumor necrosis factor alpha encodes a novel type of zinc finger protein. J Biol Chem 265:14705–14708
Otonkoski T, Beattie GM, Rubin JS, Lopez AD, Baird A, Hayek A (1994) Hepatocyte growth factor/scatter factor has insulinotropic activity in human fetal pancreatic cells. Diabetes 43:947–953
Parnaud G, Bosco D, Berney T et al (2008) Proliferation of sorted human and rat beta cells. Diabetologia 51:91–100
Peck AB, Cornelius JG (1995) In vitro growth of mature pancreatic islets of Langerhans from single, pluripotent stem cells isolated from pre-diabetic adult pancreas. Diabetes 44:10A
Puissant B, Barreau C, Bourin P, Clavel C, Corre J, Bousquet C, Taureau C, Cousin B, Abbal M, Laharrague P, Penicaud L, Casteilla L, Blancher A (2005) Immunomodulatory effect of human adipose tissue-derived adult stem cells: comparison with bone marrow mesenchymal stem cells. Br J Haematol 129:118–129
Rasmusson I (2006) Immune modulation by mesenchymal stem cells. Exp Cell Res 312: 2169–2179
Rose-John S, Waetzig GH, Scheller J, Grötzinger J, Seegert D (2007) The IL-6/sIL-6R complex as a novel target for therapeutic approaches. Expert Opin Ther Targets 11:613–624
Rosenberg L, Vinik AL (1992) Trophic stimulation of the ductular-islet cell axis: a new approach to the treatment of diabetes. Adv Exp Med Biol 321:95–104
Sandgren EP, Luetteke NC, Palmiter RD, Brinster RL, Lee DC (1990) Overexpression of TGF alpha in transgenic mice: induction of epithelial hyperplasia, pancreatic metaplasia, and carcinoma of the breast. Cell 61:1121–1135
Schmied BM, Ulrich A, Matsuzaki H et al (2001) Transdifferentiation of human islet cells in a long-term culture. Pancreas 23:157–171
Spaggiari GM, Capobianco A, Abdelrazik H, Becchetti F, Mingari MC, Moretta L (2008) Mesenchymal stem cells inhibit natural killer-cell proliferation, cytotoxicity, and cytokine production: role of indoleamine 2,3-dioxygenase and prostaglandin E2. Blood 111:1327–1333
Sphyris N, Logsdon CD, Harrison DJ (2005) Improved retention of zymogen granules in cultured murine pancreatic acinar cells and induction of acinar-ductal transdifferentiation in vitro. Pancreas 30:148–157
Stagg J, Pommey S, Eliopoulos N, Galipeau J (2006) Interferon-gamma-stimulated marrow stromal cells: a new type of nonhematopoietic antigen-presenting cell. Blood 107:2570–2577
Suva D, Passweg J, Arnaudeau S, Hoffmeyer P, Kindler V (2008) In vitro activated human T lymphocytes very efficiently attach to allogenic multipotent mesenchymal stromal cells and transmigrate under them. J Cell Physiol 214:588–594
Suzuki A, Nakauchi H, Taniguchi H (2004) Prospective isolation of multipotent pancreatic progenitors using flow-cytometric cell sorting. Diabetes 53:2143–2152
Taguchi M, Otsuki M (2004) Co-localization of nestin and PDX-1 in small evaginations of the main pancreatic duct in adult rats. J Mol Histol 35:785–789
Teng C, Guo Y, Zhang H, Zhang H, Ding M, Deng H (2007) Identification and characterization of label-retaining cells in mouse pancreas. Differentiation 75:702–712
Treutelaar MK, Skidmore JM, Dias-Leme CL, Hara M, Zhang L, Simeone D et al (2003) Nestin-lineage cells contribute to the microvasculature but not endocrine cells of the islet. Diabetes 52:2503–2512
Tyndall A, Walker UA, Cope A et al (2007) Immunomodulatory properties of mesenchymal stem cells: a review based on an interdisciplinary meeting held at the Kennedy Institute of Rheumatology Division, London, UK, 31 October 2005. Arthritis Res Ther 9:301

von Mach MA, Hengstler JG, Brulport M et al (2004) In vitro cultured islet-derived progenitor cells of human origin express human albumin in severe combined immunodeficiency mouse liver in vivo. Stem Cells 22:1134–1141

Wang J, Song LJ, Gerber DA et al (2004) A model utilizing adult murine stem cells for creation of personalized islets for transplantation. Transplant Proc 36:1188–1190

Weaver CV, Sorenson RL, Kaung HC (1985) Immunocytochemical localization of insulin-immunoreactive cells in the pancreatic ducts of rats treated with trypsin inhibitor. Diabetologia 28:781–785

Xiao M, An L, Yang X et al (2008) Establishing a human pancreatic stem cell line and transplanting induced pancreatic islets to reverse experimental diabetes in rats. Sci China C Life Sci 51:779–788

Xu G, Zhang Y, Zhang L, Ren G, Shi Y (2007) The role of IL-6 in inhibition of lymphocyte apoptosis by mesenchymal stem cells. Biochem Biophys Res Commun 361:745–750

Yamaoka T, Yoshino K, Yamada T et al (2000) Diabetes and tumor formation in transgenic mice expressing Reg I. Biochem Biophys Res Commun 278:368–376

Zhang L, Hong TP, Hu J, Liu YN, Wu YH, Li LS (2005) Nestin-positive progenitor cells isolated from human fetal pancreas have phenotypic markers identical to mesenchymal stem cells. World J Gastroenterol 11:2906–2911

Zhao W, Hirose T, Ishikawa M et al (2007) Neonatal pancreatic cells redifferentiate into both neural and pancreatic lineages. Biochem Biophys Res Commun 352:84–90

Chapter 11
Successful Scale-Up and Quality Assessments of Human Embryonic Stem Cells for Cell Therapy: Challenges and Overview

Mohan C. Vemuri, Geetha M. Swamilingiah, Shruthi Pal, Jasmeet Kaur, and Udaykumar Kolkundkar

Abstract Large-scale production of human embryonic stem cells (hESCs) is the most inevitable choice to fulfill biomedical research needs to produce mature, functional, and pure derivatives of cell types that can be utilized for transplantation purposes or in drug discovery. Currently, the status of technology suggests that large-scale culturing of hESCs is complex, and several challenges must be addressed. There is a great need for convenient, inexpensive culture systems that can facilitate the propagation of hESCs in serum-free and feeder-free culture conditions, so the cells produced in large scale can meet the demands of cell therapy applications and in screening purposes for toxicology, pharmacology, and drug discovery. As a parallel effort it is equally important to assess the quality of hESCs obtained in this scaled-up procedure and develop a reliable cost-effective method to qualify the cells that can be used for various downstream purposes. In this chapter, we describe various methods used to culture hESCs in large scale and other advances in eliminating xenogenic components from the culture systems.

11.1 Introduction

The current model for therapy using embryonic stem cells or other embryonic stem cell (ESC)-like cells is based on the premise that large numbers of "normal" cells are available and that these cells migrate, integrate, and survive for a sufficient

M.C. Vemuri (✉) • J. Kaur
Life Technologies, 7335, Executive Way, Frederick, MD 21704, USA
e-mail: mohan.vemuri@lifetech.com

G.M. Swamilingiah • S. Pal • U. Kolkundkar
Primary and Stem Cells Systems, Invitrogren BioServices India Pvt. Ltd.,
First Technology Place, 3EPIP, Whitefield, Bangalore 560 066, India

Research & Development, Primary and Stem Cell Systems, 7335 Executive Way,
Frederick, MD 21704, USA

K. Turksen (ed.), *Adult and Embryonic Stem Cells*, Stem Cell Biology and Regenerative Medicine, DOI 10.1007/978-1-61779-630-2_11,

length of time to be clinically useful. Alternatively, cells are used in vitro to generate artificial organs or engineered in some fashion to either rescue a genetic defect or deliver a therapeutic protein or growth factor. It has become clear that ESCs cannot be used directly for cell therapy because signals to direct their differentiation invariably do not exist. Furthermore, in many cases similar signals are used to direct differentiation into multiple phenotypes and, as such, at least limited propagation in culture and directed differentiation in vitro is required. Thus, an important issue facing investigators is growing cells in sufficient numbers for treating patients in an aseptic environment with adequate safeguards, sterility, and traceability. Guidelines for the manufacture of biologicals have been developed and modified under Chemistry and Manufacturing Controls (CMC) and Good Manufacturing Practices (GMP) guidelines. This expertise on CMC and GMP practices and associated regulations is widespread given the use of antibodies, viral gene delivery, and peptide and growth factor treatments. There are nevertheless several additional issues with the use of cells in general and stem cells in particular (Terstegge et al. 2007; Zeng and Rao 2006).

The current gold standard for growing ESCs is using mouse feeders or, alternatively, in feeder-free conditions but using a substrate such as Matrigel or Geltrex in human ESC (hESC) culture medium. The current status of various substrates and serum-free media together with select growth factors used by researchers are summarized in Table 11.1. Translating this to a scalable process and toward GMP requires several changes (Fig. 11.1, Table 11.2). We review the advances made in each of these areas and end by describing a possible process for scalable culture, quality assessment, and the limitations of current state of the art projects.

11.2 Would Good Manufacturing Practice Aid in Cell Therapy?

Good manufacturing practice (GMP) is a quality assurance system in the pharmaceutical industry that follows regulatory guidelines issued by the Food and Drug Administration (FDA) in the United States or the European Medical Agency (EMA) in Europe to ensure that the end-product meets the preset specifications throughout the process of manufacturing and testing the final product. It requires traceability of raw materials and that production follows validated standard operating procedures (SOPs). If hESCs have to be produced in large scale for cell therapy purposes, the defined quality characteristics of hESCs must be met to ensure safety for the patient. These cells are better suited if they are manufactured in GMP settings and are cultured in cell culture media and matrices that are produced under GMP conditions. The GMP process ensures the history and traceability of each ESC line, requiring precise growth and storage conditions, isolation and differentiation protocols, detailed records of the product's life cycle for cells and media with foolproof

Table 11.1 Composition of various hESC culture media from the published literature

Media name	Patent	Substrate	Base media	Special additives	Passaging conditions	GFS	Reference
Geron®	Yes	Matrigel® (1:30) or hLaminin (2 mg/cm2)	CM-DMEM/ X-VIVO 10	KO-SR (20%) b-FGF SCF hFlt-3 h-LIF	Collagenase IV	hLaminin (2 mg/cm^2)	Li et al. (2005)
Bresagen	Yes	Low dose Matrigel® (1:200)	DMEM/F12	Transferrin Ascorbicacid LR-IGF Activin Heregulin bFGF	Accutase®	Fibronectin (IVGN)	Wang et al. (2007)
Peter Donovan	Not known	Matrigel® (1:30)	KO-DMEM	Wnt3a Some other factors CM BMP4 RA	Trypsin	F/N	Dravid et al. (2006)
Martin Pera	Not known	Matrigel®	DMEM	KO-SR (15%) bFGF PDGF/SIP	Collagenase IV	ECM	Pera MF and Alan OT (2004, p. 291)
Sheng HZ	No		DMEM/F12	KO-SR bFGF	Trypsin for dissociation	MEFs	Fang et al. (2006, p. 250)
Snyder Yale	No	Matrigel®	DME/F12	KO-SR(20%) b-FGF ITS Wnt3a April or BAFF Cholesterol lipid-supplement	Collagenase IV	Fibronectin (25mg/mL)	Lu et al. (2006, p. 249)

(continued)

Table 11.1 (continued)

Media name	Patent	Substrate	Base media	Special additives	Passaging conditions	GFS	Reference
Scripps	No	Matrigel®	DME/F12	KO-SR or B27 N2 bFGF2	Collagenase IV	HAS (0.5 mg/mL)	Yao et al. (2006, p. 246)
Vemuri et al.	No	FN/HSA	DME/F12	B27 N2 bFGF2	Collagenase IV	HSA (0.5 mg/mL)	Modification of Yao et al. (2006, p. 246)
Technion	Yes	Fibronectin	DMEM	KO-SR (15%) bFGF TGFb1	Collagenase IV	Fibronectin	Amit et al. (2004, p. 190)
Vallier	No	Gelatin (porcine)	KO-DMEM/ IMDM/	KO-SR/ F12/NUTMIX Insulin Transferrin Monothioglycerol	Collagenase IV	Gelatin	Vallier et al. (2005, p. 293)
Liu et al.	NO	Matrigel® CM	DMEM/F12	B27 N2 bFGF2 Noggin	Collagenase IV	Matrigel® only	Liu et al. (2006, p. 237)
TeSR1 Ludwig et al.	Yes	Matrigel®	DMEM/F12	bFGF TGF beta1 ITS Glutathione Pipeolic Acid GABA LiCl	Collagenase IV	Combinations of: Collagen IV Fibronectin Laminin Vitronectin	Ludwig et al. (2006, p. 229)
Inzunza et al.	Not known	HFF	KO-DMEM	KO-SR(20%) ITS bFGF	Collagnease IV	Mostly on HFF	Inzunza et al. (2005, p. 26)

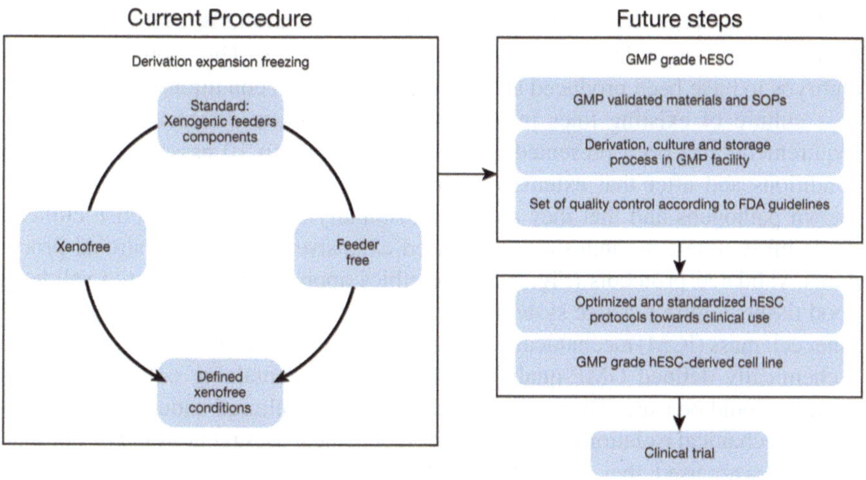

Fig. 11.1 Overview of current human embryonic stem cell (hESC) status with optimization possibilities and future requirement to enter clinical trials

Table 11.2 Good manufacturing practice (GMP) requirements for scalable clinical-grade cells

Need	Comment
Feeder-free derivation	Immunosurgery involves animal-derived substances—mouse antibodies and guinea pig complement—that are not desirable when considering xeno-free culture and cell transplantation
	Different media such as KOSR, xenofree media, GFs
Culture media	Need GMP formulations of AOF conditions
	Need clinical grade, transplant-ready cells in GMP conditions
Substrate	Most current culture conditions use a substrate (Matrigel, Geltrex, CellStart) and this will be a hassle for large-scale cultures
Large-scale expansion	Bioreactor-based expansion in GMP conditions currently suboptimal; process in evaluation
Passaging	Mechanical, enzymatic, AOF-dissociating enzymes have limitations
Freezing and shipping cells	GMP-compatible freezing, shipping, and thawing systems for clinical-grade cells
Culture quality	Cells expanded for clinical use need to be qualified for safety and stability in GMP following FDA guidelines
Transplant-ready, clinical-grade cells	Optimally, a chemically defined GMP-quality culture medium containing only human substances should be used to enable clinical transition of hESCs into regenerative therapies

regulatory compliance. The rigorous implementation of quality control steps for use of stem cells in clinical cell therapy or in nonclinical applications will pave the way for the generation of transplant-ready stem cells cultured under regulatory compliant culture conditions.

The safest option for obtaining clinical-grade cells would be to derive the lines from the very beginning in GMP-compatible conditions. This would require the embryos to have been produced and cultured under such conditions. As an alternative, culture of existing lines in laboratories and culture systems meeting GMP requirements has been presented. It would require up to 10 passages under such conditions and after that extensive testing to show that the cells do not contain known pathogens and that they are chromosomally normal. To establish clinical-grade hESC lines, the industry has devised extensive standard operational procedures, validated protocols (Fig. 11.1) for ethics approval and consent, a validated good tissue culture practice system for embryo cultures, a method for isolating the inner cell mass (ICM) mechanically, and a way to plate whole blastocysts. Optimally, a chemically defined GMP-quality culture medium containing only human substances should be used. The feeders would have been cultured under similar conditions. Mechanical isolation of the ICM, and mechanical passaging would be optimal. If enzymes are used, they should be only GMP-quality human proteins. Feeder-free derivation and culture on a GMP-quality human extracellular matrix would be perfect, but such a system does not yet exist.

Generation of clinical-grade hESCs is an important first step toward a wide range of possible future treatments. However, clinical-grade cell lines are not the final product. From a regulatory point of view, at present it is not even clear whether hESC cells should be considered as active pharmaceutical ingredients, raw material, or intermediate products. Once the cell lines have been established, the next phase will be setting up procedures for scale-up and expansion of the cells into a product that may be given to large patient groups. To realize this, GMP adaptation of the subsequent steps, which involves induced differentiation toward other cell types (Fig. 11.2), might be an even greater challenge. So far, possible protocols for hESC culture (Li et al. 2005) and differentiation (Okamura et al. 2007) have been published, but complete FDA approval for a clinical trial is awaited.

11.3 Current hESC Culture Conditions Using Serum-Free, Xeno-Free, Animal Origin-Free, and Chemically Defined Media

Large numbers of hESCs are required as a startup material to produce differentiated cell types with adequate bioprocess control, including safety, sterility, and traceability. Culturing cells in sufficient numbers required by industry or clinicians is not easy. hESCs are sensitive to culture conditions, and maintaining them in an undifferentiated state is labor-intensive, as the maintenance of optimal culture conditions require manual inputs on a daily basis such as multiple media feeding schedules, weaning out differentiated colonies, and manual subpassaging. In addition, options to use different substrates, culture media, use of dissociating enzymes during subpassaging have to be carefully chosen to maintain the reproducibility, quality control,

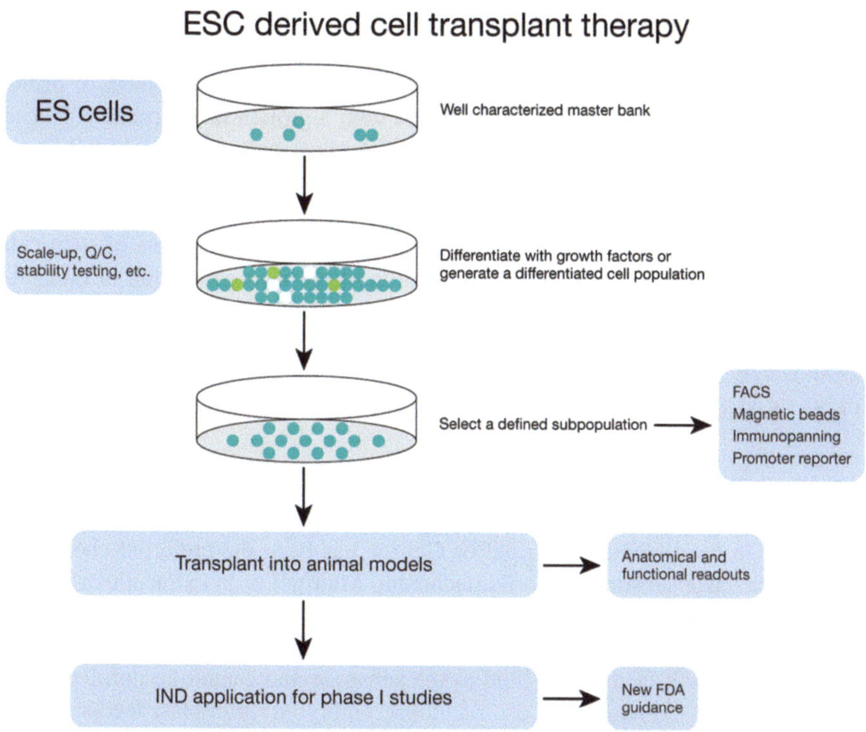

Fig. 11.2 Potential of hESCs for cell therapy

and scalability of the culture propagation process and to keep the cost of large-scale production of hESCs low to keep clinical cell therapy as a viable alternative.

The first medium used in hESC (Thomson et al. 1998) cultures contained fetal bovine serum (FBS). FBS is a complex mixture, undefined, and xenoxenic in nature; and the serum batches vary. As a result, FBS or FBS-containing medium is less than optimal for the production of clinically useful stem cell banks. To overcome serum-related issues, a serum replacement termed KnockOut serum replacement (KSR) was developed. It allows expansion of hESCs when used with basic fibroblast growth factor (bFGF). KSR is widely used in conventional cultures as it provides better defined culture conditions. However, the presence of AlbuMAX, a lipid-rich albumin fraction of bovine serum, an ingredient in KSR, makes it really not free of animal origin components (Garcia-Gonzalo and Izpisua Belmonte 2008). Over the past decade, various culture media have been developed by researchers to meet the needs of a defined hESC culture medium (Table 11.1) involving either replacement of the xeno components in the medium or providing better defined media with supplements. Notable among them are X-VIVO 10 with bFGF (Li et al. 2005); mTeSR1 with bFGF, LiCl, γ-aminobutyric acid (GABA), transforming growth factor β (TGFβ), and pipeolic acid (Ludwig et al. 2006); StemPro hESC SFM with an insulin-like growth factor 1 (IGF1) analog, heregulin-1 (a ligand for ERBB2/ERBB3), FGF2, and activin A that supported long-term growth of hESC lines (Wang et al. 2007); and KSR Xenofree with a growth factor cocktail. These media have been repeatedly evaluated by researchers over multiple hESC lines. Recent advances made toward developing xeno-free media that sustain long-term growth of hESCs with karyotypic stability and the ability to differentiate into multiple lineages should not only facilitate expansion and production of stem cell banks for therapy but should enable derivation of the next generation of clinical-grade hESC lines under GMP conditions.

11.4 Substrate Selection

In a serum-free and feeder cell-free medium formulation, use of a substrate is critical to facilitate cell attachment that supports continued growth of cells in a pluripotent state. In most studies, Matrigel or Geltrex has been the preferred choice for investigators as a substrate for cell attachment. Matrigel is an extracellular matrix extracted from Engelbreth-Holm-Swarm tumor (Kleinman et al. 1982) that contains laminin and collagen and several uncharacterized components; it includes variable amounts of growth factors that bind to the substrate and cannot be purified away from the matrix. Although the use of mouse embryonic fibroblast feeder conditioned medium (MEF) together with Matrigel as substrate resulted in successful propagation of feeder-free hESC cultures. Such cells grown in MEF conditioned media are not compatible for cell therapy applications due to the xenoxenic nature, and it is not easy to make them for large-scale production in bioreactors. Studies carried out using laminin (Beattie et al. 2005) fibronectin (Amit et al. 2004), human serum (Stojkovic et al. 2005), 3D substrates such as alginate matrix (Gerecht-Nir et al. 2004) and a synthetic polyamide matrix (Nur et al. 2006) GelTrex (Kate et al. 2008) suggest that each can support pluripotent growth of hESCs to a differing extent in xeno-free cultures with some advantages and limitations. In contrast, a humanized and xeno-free substrate, CELLstart, has been shown to support the growth in limited large-scale cultures of hESCs when used with defined, as well as standard, media. Cultures grown on CELLstart replicated in a manner similar to that of cultures expanded on Matrigel. Such culture systems that promote large-scale expansion of hESCs retaining their stem cell phenotype are critical for drug discovery and therapeutic applications.

11.5 Potentially Scalable Process

There are several aspects to developing a commercially viable process for making GMP-compatible clinical-grade products. The most critical requirement is developing a process that is compliant with regulations and provides the kind of cells that

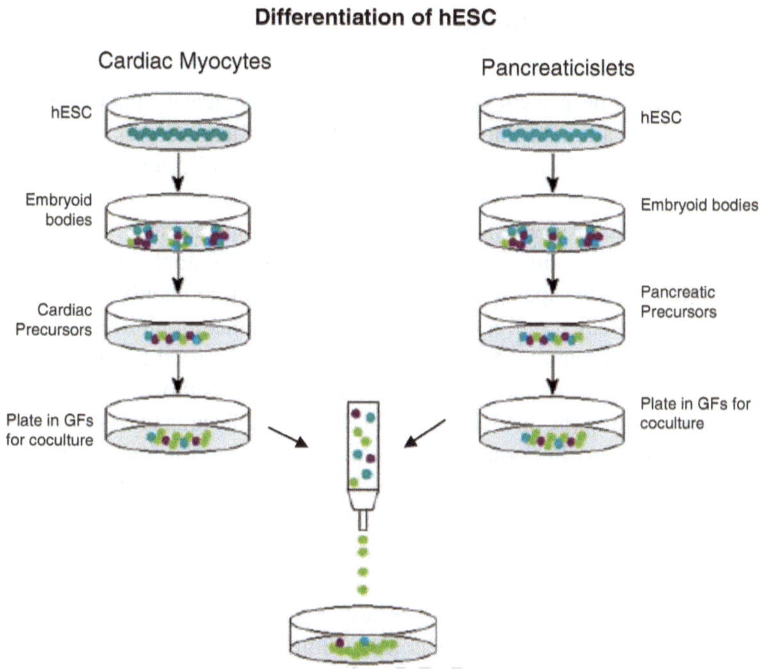

Fig. 11.3 Strategies for differentiation of hESCs into cardiac and pancreatic islet cells

are needed. Less important but nevertheless significant aspects of developing such processes entail determining if they are scalable in a standard manufacturing process, if any aspect of the process can be automated, and if the cost can be reduced to make it commercially viable. It is important to examine a potentially compliant process to determine if there are any unexpected issues in scaling up a product. Such issues include the time of processing, limited automation, number of steps, adherent-culture issues, selection issues, shipping and storage issues. Some of the more critical ones are discussed in detail below.

Perhaps the most critical issue facing large-scale manufacture of cells is growth in adherent culture. A second major roadblock when scaling up ESC cultures has been the realization that most differentiation protocols (Fig. 11.3) do not result in a pure population of cells; therefore, some selection process is required that is robust and can deliver consistently reliable purity without damaging the cells. A third major issue in developing a scalable process is determining long-term storage viability of the product and issues with shipping. These are well-established requirements for cell manufacture. However, the ESC field is relatively young and not enough data have been generated by companies or research groups regarding how long a particular cell type can be stored, its viability, and how well it survives shipment. Automating the transfer process from a holding storage facility to the dry shipper also needs to be developed for ESC-derived products.

11.6 Bioreactor Technology

Bioreactors allow the growth of cells at much higher density than is possible in static cultures and allow in-line monitoring of metabolic processes in a closed system. They thus provide maintenance of sterility and uniformity of the cultures. There are several types of bioreactor, including stirred-tank, airlift, hollow-fiber, and Rotary Cell Culture System (RCCS) designs. The stirred-tank bioreactor is perhaps one of the most commonly used types for industrial applications and laboratory research. Stirred-suspension bioreactors are appealing mainly because of their simple design, scalable configuration, ease of continuous monitoring, and regulating the culture environment. Such bioreactors have been used for the propagation of stem/progenitor cells. The RCCS, invented by the National Aeronautics and Space Administration (NASA), is increasingly used in the area of tissue engineering for medical purposes. Important improvements have been made in the design of traditional bioreactors, and new types of bioreactor are also being developed such as the Couette-Taylor bioreactor, multifunctional-membrane bioreactors, and shaking bioreactors (Wang et al. 2005). By far the most important factor is the ability to use standard technology and the associated reduction of risk. Cost-efficiency is a significant factor as well. Although we believe that a commercially viable scalable process that is clinically compliant is achievable without resorting to suspension culture, there are certainly several advantages to adapting cells to suspension culture. More specialized bioreactors have been developed for specific cell types that incorporate directions strain, provide specialized surface chemistry, or permit multitissue aggregation. We focus on results with the types of suspension technology that can possibly be applied to hESC cultures.

11.6.1 Microcarrier Bead-Based Stirred Suspension Bioreactors

Microcarriers are spherical particles composed of various materials including cellulose, glass, plastic, and polyester, with a typical diameter of 100–250 μm. Because of their high surface area/volume ratio, microcarriers are commonly used to scale culture of anchorage-dependent cells, including human hepatocytes, human retinal pigment epithelial cells, and co-cultures of neurons and astrocytes (Nie et al. 2009). A summary of 12 commercially available microcarrier beads that are likely suitable for large-scale microcarrier-based hESC cultures have been described (Table 11.3). In recent years several investigators have evaluated the use of various types of bioreactors for cultivation of undifferentiated ESCs and proceed toward differentiation with the creation of uniform embryoid bodies as these are critical to the generation of any differentiated phenotype. Some studies also showed that hESCs can be cultured on microbeads under suitable culture conditions that regulate mainly the stirring speed and seeding cell concentration, while the cells retained the ability of stem cell phenotype, and pluripotency (Lock and Tzanakakis 2009). Human ESCs expanded on such microbeads were incubated in media known to induce endoderm differentiation in hESC monolayers. Although it is still a proof of concept,

Table 11.3 Microcarrier material used in large-scale cultures (Adopted from Biotechnol Prog (2009)25: 1,23)

Name	Manufacturer	Material	Visual assessment	
			Attachment	Viability
Cytodex® 1	GE Healthcare	Cross-linked dextran with N, N-diethylamineoethyl groups	Fair	Poor
Cytodex® 3	GE Healthcare	Cross-linked dextran, denatured collagen on surface	Fair	Fair
Cytopore® 1	GE Healthcare	Macroporous cross-linked dextran with N, N-diethylaminoethyl groups, charge density of 1.1 meq/g	Poor[a]	Poor
Cytopore® 2	GE Healthcare	Macroporous cross-linked dextran with N, N-diethylaminoethyl groups, charge density of 1.8 meq/g	Poor[a]	Poor
CultiSphere-S	PerCell	Cross-linked pharmaceutical grade porcine gelatin	Fair	Poor
H11-921	Solo Hill	Cross-linked polystyrene, modified with cationic trimethyl-ammonium	Poor	None
F102	Solo Hill	Cross-linked polystyrene, modified with cationic gelatin	Poor	None
C102	Solo Hill	Cross-linked polystyrene, modified with gelatin	Poor	None
G102	Solo Hill	Cross-linked polystyrene, modified with high silica glass	Poor	None
P102	Solo Hill	Cross-linked polystyrene	Poor	None
PP102	Solo Hill	Cross-linked polystyrene, cationic	Poor	None
PF102	Solo Hill	Cross-linked polystyrene, modified with recombinant fibronectin	Poor	None

these studies support the use of scalable microcarrier suspension cultures for the culture and directed differentiation of hESCs without the intermediary stage of embryoid body (EB) formation.

In a slightly different way, the Itskovitz-Eldor group from Israel developed rotating bioreactors to control the agglomeration of EBs (Gerecht-Nir et al. 2004). The type of rotating vessel had an impact on the process of EB formation and agglomeration. In the slow-turning lateral vessel (STLV), EBs were smaller in size and no large necrotic centers were seen, even after 1 month of cultivation. In the high aspect rotating vessel (HARV), EB agglomeration was massive. These studies served as proof of principle for scalable cell production for clinical and industrial applications but still require commercial adoption and utilization in scale to get the benefit.

11.6.2 Rotary Cell Culture System and Zero Gravity Cultures

Dynamic culture systems such as spinner flasks or, on a larger scale, stirred tanks provide good mass transfer, but these systems use mechanical force which would not only damage the cells but also prevent their aggregation. The RCCS is a new technology for growing anchorage-dependent or suspension cells in scale-up cultures. The RCCS is a horizontally rotated, bubble-free disposable culture vessel with diffusion gas exchange. The system provides a reproducible, complex three-dimensional (3D) in vitro culture system with large cell masses. During cell culture, the rotation speed can be adjusted to compensate for increased sedimentation rates. The unique environment of low shear forces, high mass transfer, and microgravity provide good culture conditions for many cell types, cell aggregates, or tissue particles in a standard tissue culture laboratory. The original purpose of the RCCS bioreactor was to simulate microgravity or zero gravity. During ground experiments using this reactor, it was noted that cells suspended in these reactors tended to form 3D aggregates. Since then, these RCCS bioreactors have been used in several fields of cell and tissue culture. Applications of the RCCS bioreactor range from basic cell biology to space biology, culturing stem cells for regenerative, drug development, and possibly in the future the development of for disease and injury.

11.6.3 Encapsulation Cultures

Controllable, scalable ESC differentiation culture methods have been described (Dang et al. 2004). At present, methods for the differentiation of pluripotent cells such as hESC rely on the generation of EBs in small-scale static culture. These protocols are typically suboptimal owing to issues of scalability, a wide range of EB sizes, and exposing cells in culture to fluctuations in physicochemical parameters. Zandstra group screened various scalable suspension systems for their ability to support the growth and differentiation of hESCs. Homogeneity of initial cell aggregates was improved in two ways: They used the encapsulation method, and employed a micro-printing strategy to generate large numbers of size-specified hESC aggregates. These technologies were integrated into a fully controlled bioreactor system. The utility of such a reactor system in large-scale culture of stem cells in GMP settings and the cost impact of the culture system remains to be evaluated (Niebruegge et al. 2009). Nevertheless, these studies point toward the robust generation of clinically relevant stem cells for therapy.

11.7 Passaging Techniques and Enzymes

Passaging cells is critical for cell expansion in numbers and to allow optimal proliferation to create enough cell doses/banks for therapy. Mechanical passaging has been a preferred choice for most researchers. It is manual and easy to transfer the protocols under GMP settings. Select researchers believe that mechanical passaging allows better karyotype stability, but it is labor-intensive, time-consuming, subject to relativity, and the potential for contamination is particularly high when a large number of banks have to be created. To reduce the labor intensity, automated mechanical passaging (Alexis et al. 2006) was developed, but its efficiency is restricted to small-scale cultures. In contrast, enzymatic passaging by collagenase, dispase, accutase, trypsin, or animal origin-free TrypLE Select is needed in many situations, and it is relatively simple to apply it in large-scale settings such as bioreactor technology under GMP conditions. Each of the dissociating enzymes has distinct advantages for a preferred phenotype of the hESC colony. Use of collagenase or dispase results in gentle dissociation of hESC colony clumps, whereas use of trypsin or AOF TrypLE Select, results in nearly single-cell populations that facilitate increased plating efficiency for single-cell clonality in drug screening and discovery applications. In contrast, accutase dissociation of cells allows growth of hESCs nearly as a monolayer, rather than cell clumps, making cell counting and banking logistics more precise. There are choices with GMP-quality enzymes such as human collagenase or the AOF version of TrypLE Select versus non-GMP-quality trypsin or recombinant trypsin, which have to be carefully scoped and applied in large-scale manufacturing settings.

11.8 Freezing and Shipping

Freezing, cryostorage, and thawing processes should be GMP-complaint, including the quality of the liquid nitrogen in dry shippers used for shipping cells, from the site of manufacture to the site of use. Current cryopreservation protocols for stem cells suspend hESC colonies in a growth medium containing fetal bovine serum (FBS) and dimethylsulfoxide (DMSO), followed by an automated or semiautomated slow freezing protocol overnight with subsequent transfer and storage in liquid nitrogen. Direct contact of frozen stem cell vials with liquid nitrogen should be avoided. Thawing is rapid. This protocol is effective for the preservation of human, murine, and porcine embryonic stem cells. Although this procedure works with all cells, hESCs frozen using this method suffer from low viability, and many cells fail to survive and differentiate upon thawing and expansion. Vitrification and conventional slow cooling have been used to cryostore hESCs, but these protocols need adoption for large-scale cultures in GMP settings, as reviewed in detail by Hunt and Timmons (2007). The zeno-free cryopreservation (Richards et al. 2004) protocol involves vitrification in closed-and-sealed straws using human serum albumin. This protocol is advantageous for small-scale cultures but needs adoption in large-scale settings. Synthe-a-freeze is an animal origin-free freezing medium, a GMP-quality reagent that supports efficient cryopreservation, good revival, and viability of hESCs following cryopreservation. Although few options are available for cryopreservation—use of a combination of conventional cooling protocols (Heng et al. 2005) and vitrification protocols with different freezing media components (Reubinoff et al. 2001; Heng et al. 2005)—most of the protocols and reagents are suboptimal. There is a strong need for GMP-quality freezing reagents and protocols for cryopreserved hESCs for banking, considering the potential of transplant and cell therapy applications of hESCs.

11.9 Develop a Process and Quality Control Procedures

It has been a known fact that hESCs are difficult to cultivate, as they tend to undergo spontaneous differentiation. For large-scale manufacture it is important to have hESCs that are genetically stable; and high-quality input material is a prerequisite. It is therefore important when setting standards to assess the expression of multiple markers (Table 11.4) of the undifferentiated ESC state (Sato et al. 2004) as well as the presence or absence of markers of differentiation. In the interest of public safety, the FDA has published rules and regulations in a detailed form regarding the sources, types, culture procedures, and quality of cells, including developing measures of cell identity and heterogeneity (Dina et al. 2006).

Our rationale for their selection was based on the assumption that cells maintained in culture are under constant selection pressure to divide and self-renew and that changes can occur in a stochastic manner under all culture conditions (Amit et al. 2004). Severely detrimental mutations that do not confer a growth advantage will be lost, whereas those that inhibit death, accelerate growth, or alter differentiation will be selected. In addition, if a cell undergoes sufficient passages, these changes will become fixed in the genome or epigenome, and the cells will be irretrievably changed over time. Measuring the functional ability of cells (Table 11.4), the expression profiles of key genes (e.g., telomerase, cell cycle, key markers of the ESC state), genomic stability, and epigenome (methylation, miRNA, histone acetylation, X chromosome inactivation) (Spivakov and Fisher 2007; Allegrucci et al. 2007) would be a reasonable set of tests to be performed. The manufacturing protocol consists of a master bank of undifferentiated cells that have been carefully tested to ensure their quality from which working lots of cells are withdrawn, amplified to obtain sufficient numbers of cells, differentiated, and an appropriate phenotype selected based on established criteria.

Table 11.4 Measure of stability

Parameter	Marker/assay
Self-renewal markers	Oct 4, Sox2, Nanog
Mitochondrial stability	Mitochondrial activity
Genomic stability	Standard sequencing, microarray, karyotyping, FISH, SNP, SKY, CGH
Epigenetic stability	Methylation changes, histone modifications, X chromosome inactivation

11.10 Tests to Assess Self-renewal Potential

A number of cell surface markers (e.g., SSEA-3, SSEA-4, TRA-1-60, TRA-1-81) are recommended to characterize hESCs. In addition, hESCs also express surface antigens initially described in other stem cell populations, such as AC133, c-kit (CD117), flt3 (CD135), and CD9 (Hoffman and Carpenter 2005). A number of transcription factors play a critical role in maintaining stem cell self-renewal, including OCT4, SOX2, and NANOG (Draper et al. 2004). These three transcription factors share a substantial fraction of their target genes, demonstrate autologus feedback, and control one another's transcription in a large regulatory circuit. Many targets of OCT4, SOX2, and NANOG encode key transcription factors for differentiation and development, but they are transcriptionally inactive in ESC state. OCT4, SOX2, and NANOG also regulate transcriptionally active genes involved in pluripotency maintenance (Babaie et al. 2007; Rodda et al. 2005).

11.11 Tests to Assess Differentiation Capacity

Pluripotency is one of the defining features of hESCs. Perhaps the most common test of pluripotency is the formation of chimeras in mice in which ESCs are injected into the blastocyst. The contribution of ES cells to the resulting chimera determine the differentiation capacity of the injected cells. Although this approach can assess mESCs, the assay is clearly not suitable for hESCs. Therefore, EB formation in vitro and teratoma formation (Bigdeli et al. 2009) after injection into immune compromised mice are currently used to validate pluripotency of hESCs (Chen et al. 2009). This is accomplished by scoring the presence/absence of ectoderm, endoderm, and mesoderm in EBs and teratomas. The important point is that without the demonstration of this key test (i.e., the ability to cause teratoma in mice) stem cells can never be used in regenerative medicine.

Most studies have focused on in vitro differentiation protocols, which assess the differentiated cells via expression analysis of cell-specific markers (Table 11.5). However, few markers are specific for one cell type, and so panels of markers must

Table 11.5 Measure of differentiation ability

Methods	Testing parameter
EBs, differentiation	Mesodermal, ectodermal, and endodermal marker expression
RT-PCR assay	For different cell specific lineages
Array hybridization	Standard sequencing, microarray, karyotyping, FISH, SNP
Massive parallel signature sequencing	SKY, CGH
MicroRNA profiling	Specific microRNAs towards differentiation
Proteomic, functional, and teratoma analysis	Proven differentiation ability

be used in these experiments. Although the generation of ectoderm, mesoderm, endoderm, and trophoblast and germ cells demonstrates the pluripotency of the cells, the full differentiation capacity of hESCs is still a subject of intense study by groups of researchers. The reader is encouraged to find specific reviews on the subject of differentiation (Chin et al. 2010).

11.12 Pathogen Testing and *Mycoplasma*

The most common form of contamination in the hES cell products is bacteria, yeasts, and fungi. The main contamination causing bacteria are gram-positive cocci and *Mycoplasma* species, followed by gram-negative rods and gram-positive rods in a few hESC lines and feeders (Cobo et al. 2007). The risk of viral contamination from products of animal origin (e.g., bovine serum and mouse fibroblasts as a feeder layer for the development of embryonic cell lines) should also be considered.

There are several ways to detect microorganisms, such as by measuring the endotoxin levels or by direct microbial culturing, staining, and biochemical reactions. In the case of *Mycoplasma*, which is fastidious and difficult to grow under in vitro conditions, a visual screening procedure, such as the MycoFluor *Mycoplasma* Detection Kit (Invitrogen, Carlsbad, CA, USA) may be used. It can provide an ultrasensitive, rapid, simple fluorescence microscopic aid for visual identification of *Mycoplasma* infection in laboratory cell cultures (Fig. 11.4). It should be a mandatory standard of current good practice in stem cell banks to carry out routine microbiological controls of the stem cell lines and to work in a controlled environment to reduce the probability of contamination in the final product.

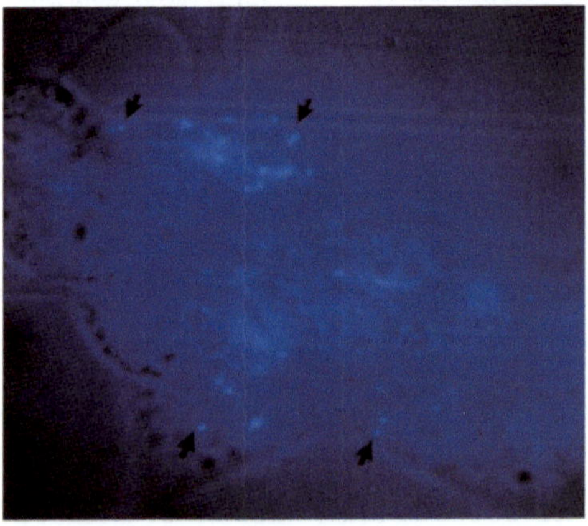

Fig. 11.4 Fluorescence of *Mycoplasma arginini* in live cultured cells stained with the MycoFluor Reagent (Image contributed by Jason A. Kilgore, Invitrogen)

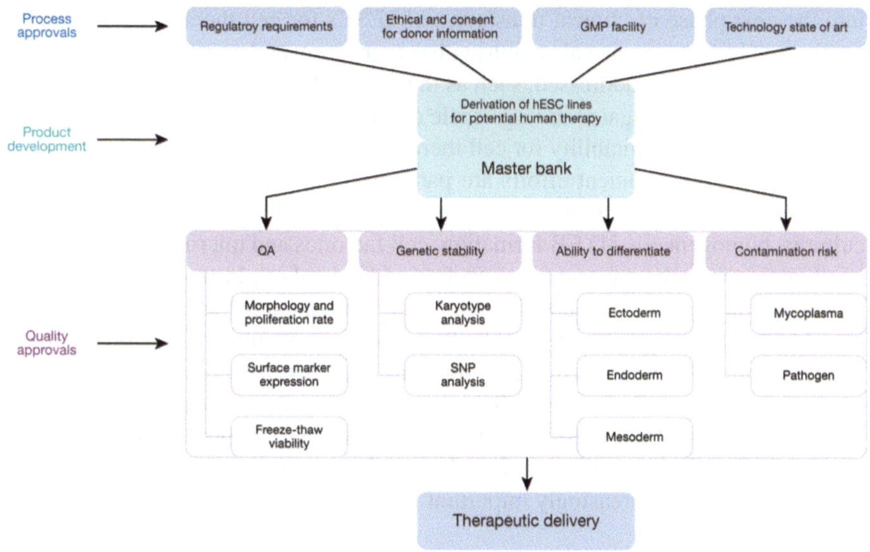

Fig. 11.5 Flow diagram outlining critical points for derivation of clinical-grade hESC lines

11.13 Critical Points to Achieve Clinical-Grade hESCs

We identified several critical control points (Fig. 11.5) with procedures that are known to be essential for successful derivation or end-product quality. To achieve clinical-grade hESCs, one has to consider the ethical approval of both the Human Fertility and Embryology Authority for work involving human embryos and the regulatory requirements of GMP manufacturers. The quality approvals that relate to the physical environment, facility, and equipment maintenance standards are reviewed and recorded. In addition, documented procedures are in place to ensure that any research and development carried out to improve the methodology and efficiency of hESC derivation and characterization is performed in accordance with all GMP regulations but in a manner that segregates it both spatially and temporally to maintain the integrity of the end-products. This allows the outcome of any such research to be of direct clinical relevance and application.

11.14 Conclusion

Recent studies have raised important questions about the ability to amplify stem cell populations in sufficient numbers to be useful for therapy. GMP-compatible bioprocesses for the production of stem cells and their derivatives in sufficient quantity for clinical and biotechnology uses are clearly the need of the hour, allowing us to move into the next phase of clinical translation of hESCs for cell therapy. In the current scenario of hESCs, large-scale culturing has to meet several constraints; and several challenges need to be addressed, such as the criteria for assessing the state of stemness of the cells propagated in large-scale culture conditions, including their traceability, sterility, and suitability for cell therapy.

Bioprocess development efforts are particularly in their infancy, and there is a dire need for substantial improvements in efficiency of bioreactor-based suspension cultures, homogeneous 3D EB formation, cell factories and microcarrier cultures in combination with the use of more xeno-free and animal origin-free reagents such as defined media and defined substrate. For large-scale manufacture, it is important to have hESCs that are genetically stable; high-quality input material is a prerequisite. It is therefore important when setting standards to assess the expression of multiple markers of the undifferentiated ESC state as well as the presence or absence of markers of differentiation as mentioned earlier. Other key areas are the establishment of efficient protocols to manipulate hES cells and derivatives genetically, which will become increasingly important for the application of hESCs in practical research.

Reprogramming technology is another landmark discovery made in stem cell biology, and one needs to emphasize the development of large-scale procedures for induced pluripotent stem cells (iPSC) cells. Currently, most of the reprogramming processes are achieved by genetic modification with viral vectors that fall into the category of gene therapy regulations. As a result, until we have zero footprint technology in reprogramming, the scale-up development for reprogrammed cells is a far-fetched aspect for large-scale production Nevertheless, the critical issues discussed for hESC large scale preparation still remain applicable to iPSCs as well.

Acknowledgments MCV acknowledges the stimulating discussions and the encouragement from Dr. Mahendra Rao currently at NIH, who was responsible for persuading to write this article. MCV is also indebted to Sandra Kuligowski and Maureen Cook at Grand Island, NY for enabling deep insights in to the GMP process and Quality Control systems in cell and media manufacturing.

References

Alexis J, Christelle FR, Kristine W, Maeve C, Alastair C, Nicholas A, Siddharthan C (2006) Automated mechanical passaging: a novel and efficient method for human embryonic stem cells expansion. Stem Cells 24:230–235

Allegrucci C, Wu YZ, Thurston A (2007) Restriction landmark genome scanning identifies culture-induced DNA methylation instability in the human embryonic stem cell epigenome. Hum Mol Genet 16:1253–1268

Amit M, Shariki C, Margulets V, Itskovitz-Eldor J (2004) Feeder layer- and serum-free culture of human embryonic stem cells. Biol Reprod 70:837–845

Babaie Y, Herwig R, Greber B, Brink TC, Wruck W, Groth D, Lehrach H, Burdon T, Adjay J (2007) Analysis of Oct4-dependent transcriptional networks regulating self renewal and pluripotency in human embryonic stem cells. Stem Cells 25:500–510

Beattie GM, Lopez AD, Bucay N, Hinton A, Firpo MT, King CC, Hayek A (2005) Activin A maintains pluripotency of human embryonic stem cells in the absence of feeder layers. Stem Cells 23:489–495

Bigdeli N, Karlsson C, Strehl R, Concaro S, Hyllner J, Lindahl A (2009) Coculture of human embryonic stem cells and human articular chondrocytes results in significantly altered phenotype and improved chondrogenic differentiation. Stem Cells 27:1812–1821

Chen X, Song XH, Yin Z, Zou XH, Wang LL, Hu H, Cao T, Zheng M, Ouyang HW (2009) Stepwise differentiation of human embryonic stem cells promotes tendon regeneration by secreting fetal tendon matrix and differentiation factors. Stem Cells 27:1276–1287

Chin AC, Padmanabhan J, Oh SK, Choo AB (2010) Defined and serum-free media support undifferentiated human embryonic stem cell growth. Stem Cells Dev 19:753–761

Cobo F, Cortes JL, Cabera C, Nieto A, Concha A (2007) Microbiological contamination in stem cell cultures. Cell Biol Int 31(9):991–995

Dang SM, Gerecht-Nir S, Chen J, Itskovitz-Eldor J, Zandstra PW (2004) Controlled, scalable embryonic stem cell differentiation culture. Stem Cells 22:275–282

Dina GH, David A, Kessler (2006) FDA Regulation of Stem-Cell–Based Therapies. The New Eng. J. Med. 355:1730–1735

Draper JS, Moorer HD, Ruban LN, Gokhale PJ, Andrews PW (2004) Culture and characterization of human embryonic stem cells. Stem Cells Dev 13:325–326

Dravid G, Ye Z, Hammond H, Chen G, Pyle A, Donovan P, Yu X, Cheng L (2006) Defining the role of Wnt/beta-catenin signaling in the survival, proliferation and self-renewal of human embryonic stem cells. Stem Cells 10:1489–1501

Fang D, Leishear K, Nquyen TK, Finko R, Cai K, Fukunaga M, Li L, Brafford PA, Kulp AN, Xu X, Smalley KS, Herlyn M (2006) Defining the conditions for the generation of melanocytes from human embryonic stem cells. Stem Cells 24(7):1668–1677

Garcia-Gonzalo FR, Izpisua Belmonte JC (2008) Albumin-associated lipids regulate human embryonic stem cell self-renewal. PLoS One 3:e1384

Gerecht-Nir S, Cohen S, Itskovitz-Eldor J (2004) Bioreactor cultivation enhances the efficiency of human embryoid body (hEB) formation and differentiation. Biotechnol Bioeng 86:493–502

Halme DG, Kessler DA (2006) FDA regulation of stem-cell–based therapies. N Engl J Med 355:1730–1735

Heng BC, Kuleshova LL, Bested SM, Liu H, Cao T (2005) The cryopreservation of human embryonic stem cells. Biotechnol Appl Biochem 41:97–104

Hoffman LM, Carpenter MK (2005) Characterization and culture of human embryonic stem cells. Nat Biotechnol 23:699–708

Hunt CJ, Timmons PM (2007) Cryopreservation of human embryonic stem cell lines. Methods Mol Biol 368:261–270

Inzunza J, Gertow K, Stromberg MA (2005) Derivation of human embryonic stem cell lines in serum replacement medium using postnatal human fibroblasts as feeder cells. Stem Cells 23:544–549

Kate W, MaryLynn T, Sackamone P, Mahendra R, Udaykumar K, Mohan V (2008) Defined cell culture media for pluripotent human embryonic stem cells. Med Sci Dig 33(14):41–48

Kleinman HK, McGarvey M, Liotta LA, Robey PG, Tryggvason K, Martin GR (1982) Isolation and characterization of type IV procollagen, laminin, and heparan sulfate proteoglycan from the EHS sarcoma. Biochemistry 21:6188–6193

Li Y, Powell S, Brunette E, Lebkowski J, Mandalam R (2005) Expansion of human embryonic stem cells in defined serum-free medium devoid of animal-derived products. Biotechnol Bioeng 91:688–698

Liu J, Qian L, Wessells RJ, Bidet Y, Jagla K, Bodmer R (2006) Hedgehog and RAS pathways cooperate in the anterior-posterior specification and positioning of cardiac progenitor cells. Dev. Biol. 290:373–385

Lock LT, Tzanakakis ES (2009) Expansion and differentiation of human embryonic stem cells to endoderm progeny in a microcarrier stirred-suspension culture. Tissue Eng Part A 15:2051–2063

Lu J, Hou R, Booth CJ, Yang SH, Snyder M (2006) Defined culture conditions of human embryonic stem cells. PNAS Proc Natl Acad Sci USA 103(15):5688–5693

Ludwig TE, Bergendahl V, Levenstein ME, Yu J, Probasco MD, Thomson JA (2006) Feeder-independent culture of human embryonic stem cells. Nat Methods 3:637–646

Nie Y, Bergendahl V, Hei DJ, Jones JM, Palecek SP (2009) Scalable culture and cryopreservation of human embryonic stem cells on microcarriers. Biotechnol Prog 25:20–31

Niebruegge S, Bauwens CL, Peerani R, Thavandiran N, Masse S, Sevaptisidis E, Nanthakumar K, Woodhouse K, Husain M, Kumacheva E, Zandstra PW (2009) Generation of human embryonic stem cell-derived mesoderm and cardiac cells using size specified aggregates in an oxygen-controlled bioreactor. Biotechnol Bioeng 102:493–507

Nur EKA, Ahmed I, Kamal J, Schindler M, Meiners S (2006) Three-dimensional nanofibrillar surfaces promote self-renewal in mouse embryonic stem cells. Stem Cells 24:426–433

Okamura RM, Lebkowski J, Au M, Priest CA, Denham J, Majumdar AS (2007) Immunological properties of human embryonic stem cell-derived oligodendrocyte progenitor cells. J Neuroimmunol 192:134–144

Pera MF, Alan OT (2004) Human embryonic stem cells: prospects for development. Development 131:5515–5525

Reubinoff BE, Pera MF, Vajta G, Trounson AO (2001) Effective cryopreservation of human embryonic stem cells by the open pulled straw vitrification method. Hum Reprod 16:2187–2194

Richards M, Fong CY, Tan S, Chan WK, Bongso A (2004) An efficient and safe xeno-free cryopreservation method for the storage of human embryonic stem cells. Stem Cells 22:779–789

Rodda DJ, Chew JL, Lim LH, Loh YH, Wang B, Ng HH, Robson P (2005) Transcriptional regulation of NANOG by OCT4 and SOX2. J Biol Chem 280:24731–24737

Sato N, Meier L, Skalsounis L, Greengard P, Brivanlou AH (2004) Maintenance of pluripotency in human and mouse embryonic stem cells through activation of Wnt signaling by a pharmacological GSK-3-specific inhibitor. Nat Med 10:55–63

Spivakov M, Fisher AG (2007) Epigenetic signatures of stem-cell identity. Nat Rev Genet 8:263–271

Stojkovic P, Lako M, Przyborski S, Stewart R, Armstrong L, Evans J, Zhang X, Stojkovic M (2005) Human-serum matrix supports undifferentiated growth of human embryonic stem cells. Stem Cells 23:895–902

Terstegge S, Laufenberg I, Pochert J (2007) Automated maintenance of embryonic stem cell cultures. Biotechnol Bioeng 96:195–201

Thomson JA, Itskovitz-Eldor J, Shapiro SS, Waknitz MA, Swiergiel JJ, Marshall VS, Jones JM (1998) Embryonic stem cell line derived from human blastocyst. Science 282(5391):1145–1147

Vallier L, Alexander M, Pederson RA (2005) Activin/Nodal and FGF pathways cooperate to maintain pluripotency of human embryonic stem cells. J Cell Sci 118:495–4509

Wang D, Liu W, Han B, Xu R (2005) The bioreactor: a powerful tool for large-scale culture of animal cells. Curr Pharm Biotechnol 6:397–403

Wang L, Schulz TC, Sherrer ES, Dauphin DS, Shin S, Nelson AM, Ware CB, Zhan M, Song CZ, Chen X (2007) Self-renewal of human embryonic stem cells requires insulin-like growth factor-1 receptor and ERBB2 receptor signaling. Blood 110:4111–4119

Yao S, Chen S, Clark J, Hao E, Beattie GM, Hayek A, Ding S (2006) Long term self-renewal and directed differentiation of human embryonic stem cells in chemically defined conditions. PNAS Proc Natl Acad Sci USA 103(18):6907–6912

Zeng X, Rao MS (2006) The therapeutic potential of embryonic stem cells: a focus on stem cell stability. Curr Opin Mol Ther 8:338–344

Chapter 12
Human Embryonic Stem Cells from Laboratory and Clinical Perspectives

Necati Findikli

Abstract Because they can be obtained during the earliest developmental stage and they have the capability for self-renewal and indefinite expansion potential as well as the ability to differentiate in all somatic cell types, human embryonic stem cells (hESCs) have become important research sources in basic science. These scientific areas encompass human embryology, developmental genetics, and disease modeling as well as areas of applied science and medical therapy such as pharmacology, toxicology, and cellular therapy. Their therapeutic potential, on the other hand, still awaits major solution strategies for problems, including immune rejection, possibility tumor formation, and Good Manufacturing Practice (GMP)-grade production among others. In the light of the contemporary literature and laboratory applications, we summarize and discuss the current level of laboratory practice and the problems and alternative solution strategies for using hESCs in clinical practice. We also take note of the novel tools generated from hESC technology.

12.1 Introduction

Human embryonic stem cells (HESCs) are pluripotent stem cells derived from the inner cell mass (ICM) of a growing human blastocyst. hESCs are obtained by isolating ICM cells and culturing them in a special cell culture environment to minimize differentiation and keep the cells undifferentiated for a long period of time in culture. hESCs are generally distinguished by two distinctive properties: their

N. Findikli, M.Sc. (✉)
Department of Bioengineering, Yildiz Technical University, Istanbul, Turkey

Medicana Bahcelievler Hospital IVF Center, Istanbul, Turkey

Istanbul Genetics Group, Istanbul, Turkey
e-mail: necatif@gmail.com

K. Turksen (ed.), *Adult and Embryonic Stem Cells*, Stem Cell Biology and Regenerative Medicine, DOI 10.1007/978-1-61779-630-2_12,

pluripotency (ability to differentiate into the cell types of three germ layers or derivatives) and their ability to replicate indefinitely.

Two independent research groups (Thomson et al. 1998; Reubinoff et al. 2000) introduced the first human embryonic stem cells (ESCs) during the 1990s. Enthusiastic scientists recognized the promise of these cells for research and medicine, and research papers reporting successful derivation of new hESC lines emerged during subsequent years. So far, more than 1000 successful hESC lines have been isolated from human fresh or frozen morula and/or blastocyst-stage human embryos and have been reported by 87 groups from 24 countries worldwide. There is trend for these numbers to steadily increase (Hovatta 2006; Löser et al. 2010).

On the other hand, there have been numerous drawbacks and limitations in the research and development of these cells in regard to their possible therapeutic role. The problems are related to hESC isolation and the cells' characteristics, culture, and differentiation and the use of more advanced manipulation techniques. In many countries, politics and intense ethical debates have taken place, leading to official prohibition of hESC derivation and use, even for research purposes. (In some countries, the debates are ongoing.) From a society's perspective, the main arguments were the destruction of human embryos and possibility of uncontrolled creation and abuse of human gametes specifically for hESC production. From a scientific perspective, there have been several other issues, such as their questionable nature and the lack of standard characterization tools, poorly defined and animal-based culture conditions, unpredicted and tedious differentiation pathways, and most importantly their unknown fate and the immunogenic characteristics in the host when their use in the therapy does become a valid option.

12.2 Issues Related to the Source of Human Embryonic Stem Cells

Starting from the initial source, in nearly all of the reports the possible correlations between embryo development parameters and successful derivation were poorly defined, making a true efficiency for hESC derivation impossible to ascertain from embryological aspects (Stephenson et al. 2006). Embryos that are used for derivation should, by definition, be produced by assisted reproductive technology (ART). Therefore, an embryo to be used for hESC isolation is in fact a product of complex therapeutic regimens including controlled ovarian hyperstimulation, in vitro fertilization, and in vitro culture for several days. Although ART protocols and services have been in clinical use for nearly three decades, the efficiency of the treatment is still limited, and nearly 80% of in vitro-produced gametes/embryos lose their viability before implantation.

Considering that nearly all of the embryos used for hESC derivation are leftover, developmentally inferior embryos (not chosen for fresh embryo transfer), the qualitative and quantitative analysis of the hESCs obtained from such sources can in principle result in diverse differences and variabilities when they are used in further research involving characterization, differentiation, gene expression, and other aspects. Results of human ART have shown that during in vitro culture only 30–40% of the fertilized human oocytes have the capacity to develop successfully to blastocyst stage, and on average only 20% of the fertilized ova can implant in the uterus. That is, approximately 70–80% of these zygotes/embryos exhibit varying degrees of developmental abnormalities including unequal cleavage and excessive fragmentation leading to developmental arrest. Even for the ones that can become expanded blastocysts, some eventually lose viability before or shortly after implantation, possibly due to induced apoptosis and chromosomal/genetic errors in both the inner cell mass (ICM) and trophectoderm lineages. In most hESC isolation studies, fresh or frozen/thawed spare human embryos were used as a source material. Few of the fresh spared embryos can survive and form blastocysts in extended culture, which results in a lower rate of ESC derivation compared to theirfrozen/thawed counterparts (Sjogren et al. 2004).

Although the etiology behind this extensive embryonic loss is largely unknown, several studies have so far pointed out a variety of possible causes, including inadequate oocyte maturation, suboptimal culture conditions, and chromosomal/nuclear abnormalities during early cleavage stages. This situation indicates a need for improved techniques/protocols that can minimize these negative factors in vitro (Moor et al. 1998; Janny and Menezo 1994; Kaye 1997; Jones et al. 1998; Schoolcraft et al. 1999; Munne et al. 1995; Harper et al. 1995; Bielanska et al. 2002). At different stages of development during which hESCs have been derived—morula, late blastocyst, epiblast—there is possible loss; and hence there is scope to refine the methodology.

Current research that is aimed at optimizing human embryo development and minimizing this embryonic loss, thereby increasing the success rate in ART, are expected also to improve the hESC derivation process. Compared to the late 1990s when the first hESCs were reported, today's human embryo culture systems provide more standardized, optimal ingredients and protocols that can be more beneficial for human embryos as well as hESC survival (Summers and Biggers 2003). As an example, it has been reported that a low oxygen concentration increases the viability of preimplantation embryos, assists their normal development, and helps to form healthy blastocysts with well-formed ICMs in greater cell numbers (Dumoulin et al. 1999; Bavister 2004). It has also been indicated that low oxygen can have beneficial effect on hESC cell clonal ability and reduce the risk of chromosomal abnormalities (Forsyth et al. 2006).

In addition to embryo culture conditions, intrinsic (paternal) factors that are inherited from infertile couples can affect the human embryo quality and outcome of hESC derivation. In many cases, the nature of infertility resides in the quality and quantity of oocytes/spermatozoa retrieved. Embryos produced from inferior quality oocytes and/or spermatozoa can carry numerous metabolic and/or genetic problems that may affect their development and implantation. Excess embryos from these couples, if used in hESC derivation, may reach blastocyst stage but still carry abnormal developmental patterns that can affect the derivation and the differentiation profile. An increased rate of imprinting and other epigenetic abnormalities in human in vitro-produced preimplantation embryos (Jacob and Moley 2005) is a clear example of such a consequence. Furthermore, the significantly higher incidence of Beckwith-Wiedemann syndrome (an imprinting disorder caused by loss of imprinting of *IGF2* and other imprinted genes) in babies created by in vitro fertilization (IVF) raises the prospect that the brief in vitro culture of human embryos as part of infertility treatment may cause epigenetic abnormalities (Niemitz and Feinberg 2004). This issue may indicate that initial hESC culture may create relatively diverse epigenetic profiles that can lead to different developmental and differentiation profiles in extended in vitro culture. Whether this finding can explain the differences in cultural behaviors of hESC lines in similar in vitro settings remains to be seen.

As a potential alternative to avoid the above issues, hESC derivation can be realized by using gamete/embryos from fertile donors. hESC isolation from embryos that were specifically created for hESC isolation from donated human sperm and oocytes was first reported by Lanzendorf et al. (2001). In their study, insemination of 142 donated oocytes gave 68% fertilization and 50% blastocyst development rates. As embryos were produced from gametes of healthy donors, it remains to be argued whether better embryo development and hESC derivation rates can be obtained with the current technology.

A possible use of human embryos obtained from fertile couples undergoing preimplantation genetic diagnosis (PGD) for human leukocyte antigen (HLA) typing can in theory be another superior source for hESC derivation (Findikli et al. 2006). The PGD technique, in its simplest terms, involves screening preimplantation embryos for chromosomal abnormalities or for single gene defects. In addition to its diagnostic value and expanding indications—cancer predisposition, dynamic mutations, late-onset disorders—a new feature, preimplantation genetic diagnosis combined with HLA typing, demonstrates its novel therapeutic role in contemporary medicine (Kuliev and Verlinsky 2005; Rechitsky et al. 2006).

In some countries, to obtain a healthy, child who is HLA-compatible with his or her sibling, fertile couples carrying a specific single gene disorder can undergo IVF and PGD-HLA typing procedures. During the course of this treatment, healthy but HLA-incompatible embryos can be donated for hESC derivation purposes (Findikli et al. 2006).

12.3 Studies Involving the Isolation of Human Embryonic Stem Cells by Ethically Acceptable Approaches

To avoid ethical concerns, several recent studies have utilized embryos that have shown developmental arrests and were discarded from routine IVF treatments, embryos produced and discarded after PGD, and embryos created parthenogenetically from donated human oocytes. There have been recent reports of successful generation of hESC lines from biopsied human blastomeres.

Utilization of arrested human embryos for hESC isolation was reported by Zhang et al. (2006). As these embryos are no longer considered viable, this approach had been proposed as an alternative way to derive new hESCs because the disintegration of nonviable human embryos can be considered ethically acceptable. Especially in countries with a nonflexible policy, arrested embryos can provide a more ethical source for hESC derivation and so can resolve some of the political issues surrounding research using human embryos (Landry and Zucker 2004).

Application of PGD has not only helped couples who carry specific genetic problems have disease-free children, it has created a novel source for hESC research. After biopsy and genetic analysis, embryos that are diagnosed as chromosomally abnormal or that carry a specific monogenic disease can be donated for research, resulting in hESC lines that can be the earliest in vitro models for that particular genetic anomaly. On the other hand, Munne and his colleagues proposed that embryos diagnosed as chromosomally abnormal after PGD could in part have a self-correction ability, that is, can revert to a normal karyotype after prolonged culture, resulting in hESC lines with normal, stable karyotypes (Munne et al. 2005). However, this finding must be confirmed by other independent studies.

As the PGD technique involves removal of a cell from a developing embryo without impairing its potential to create a pregnancy (Handyside et al. 1990), it has also been proposed that removed cells could be cultured and expanded for several days, thereby being an ideal and ethically acceptable hESC source for future research and therapy. Retrieval of individual human blastomeres through PGD and expanding them directly in culture mainly for detection of chromosomal aneuploidy have been documented by some groups, and a proof-of-concept study has been performed in mice (Geber et al. 1995; Hlinka et al. 2001; Bielanska et al. 2003; Chung et al. 2006).

It is known that preimplantation-stage embryos and hESC cells largely depend on the presence of adjacent neighboring cells as well as several unknown factors produced by them. Therefore, technical improvements and alternative culture methodologies were sought to clarify the feasibility of this approach. In mice, whether a blastomere is to become a trophoblast or an ICM cell appears to be specified by its position during the first cleavage (Piotrowska et al. 2001). It was also demonstrated that asymmetrical distribution of the *Cdx2* gene product in mouse oocytes and embryos defines the lineage of the trophectoderm (Deb et al. 2006). In correlation with studies in which a signal for ICM or trophectoderm lineage is present in some blastomeres far earlier than when the phenotypical characteristics emerge, the earliest signs of cellular differentiation occur during the compaction stage in human embryos (Piotrowska et al. 2001; Edwards 2005; Hansis et al. 2004). Klimanskaya et al. was first to report the derivation of new hESC cell lines from isolated human blastomeres (Fong et al. 2006; Klimanskaya et al. 2006). Although it was later argued that their technique was not the same as human embryo biopsy for PGD purposes, the same group has more recently announced an extension of their initial reports on embryos obtained from routine PGD cases (Chung et al. 2008). These results indicate that if one or two biopsied blastomeres were enough to expand the line, the same embryo would also be implanted in the uterus, creating a viable offspring and comparable pregnancy rates as well as an unlimited stem cell source for that sibling.

For other alternative sources and potential solutions to immunohistocompatibility issues, several groups have applied somatic-cell nuclear transfer (SCNT) and parthenogenetic activation methods to create patient-specific human and primate embryonic stem cell lines (Byrne et al. 2007; Lin et al. 2003; Vrana et al. 2003; Mai et al. 2007; Revazova et al. 2007). Compared to SCNT, parthenogenetic activation of oocytes is a relatively simple method to create histocompatible stem cells because the technique does not require complex instruments or micromanipulation experience.

Regarding histocompatibility, some researchers argue that a minimum of 40–50 homozygous hESC lines would be necessary to cover 50% of the HLA isotypes in the American population and 150 lines to cover the UK population, minimizing the immune rejection of hESC-derived transplants (Faden et al. 2003; Taylor et al. 2005). Although these numbers may be underestimated due to the ethnically diverse nature of the populations in question, creating master hESC banks for future therapeutic and research applications seems to be a realistic approach. It helps with the standardization of hESC cultures and in reducing the cost and unnecessary derivation of new hESC lines (Rao and Auerbach 2006; Civin and Rao 2006; Stacey and Hunt 2006).

These potential drawbacks and limitations have led to a significant development: isolation of an exciting, promising pluripotent stem cell source, the induced pluripotent stem cell (iPS cell). iPS cells are artificially derived from a terminally differentiated adult somatic cell (e.g., fibroblast, hepatocyte) by introducing a group of four "stemness" genes (*oct3/4*, *sox2* together with *Klf4* and *c-myc* or with *Nanog* and *Lin28*) that are known to be expressed in hESCs. These cells were first produced in 2006 from mouse cells and in 2007 from human cells (Takahashi and Yamanaka 2006; Takahashi et al. 2007; Yu et al. 2007). They give rise to cells derived from all three germ layers in vitro and in vivo. Murine iPSCs injected into murine blastocysts have been shown to contribute to embryonic development (Takahashi and Yamanaka 2006).

This development is considered an important scientific discovery as it allows researchers to obtain ESCs or any other pluripotent stem cell types without the need to destroy a human embryo. Moreover, because the reprogrammed cells theoretically carries the same genome profile of the cell donor, this method can be a valid option for creating patient-specific stem cells. It would further solve the current issues of graft-versus-host disease and other immunity-related problems.

Although this novel technology would potentially diminish the need to destroy human embryos for hESC derivation, owing to the properties of the approach itself reprogramming of adult cells to obtain iPSCs may pose significant risks that could limit its use in humans. Because viral vectors and proto-oncogenes are used to genomically alter the cells, tumor formation may be triggered. More recent studies may have found a solution for this problem by replacing the effect of potentially problematic proto-oncogenes with reprogramming factors or proteins, creating a new terminology such as the protein-induced pluripotent stem cells (piPSs) (Park et al. 2008; Zhou et al. 2009).

12.4 Studies Involved in Improving Isolation and Cell Culture Conditions

Since 1998, derivation methodology of the reported hESC lines included isolation of ICM cells from Trophectoderm cells by immunological, mechanical, chemical, or laser-assisted means or by direct plating of intact blastocysts on feeder cells/dishes coated with extracellular matrix proteins without prior ICM dissection or proper staging of blastocysts for hESC derivation (Thomson et al. 1998; Reubinoff et al. 2000; Heins et al. 2004).

The immunological method, also called immunosurgery, utilizes exposure of embryos to pronase enzyme, animal-derived complement system reagents, and antibodies raised against human cells. Recent experience, however, has shown that this technique is not the optimal derivation method when poor-quality, spare embryos with small or nearly visible ICMs are used. Moreover, hESCs that have been isolated with this method are eventually not suitable for therapeutic use owing to the risk of their carrying xenogenic pathogens (Martin et al. 2005). The whole (or direct) culture method has also been reported and compared to immunosurgery (Findikli et al. 2006; Ellerstrom et al. 2006). This method has been said to be superior to immunosurgery because it not only shows a better success rate but clearly avoids the use of animal-derived antibodies, making the culture system (one step) more suitable for therapeutic use. Alternatively, depending on the quality of the starting material, using a combination of protocols has been proposed to increase the efficiency of the isolation (Suss-Toby et al. 2004).

Chemically removing zona pellucida by means of the acid Tyrode approach is a valid alternative to pronase as in the former method the embryos are not exposed to a bacterial product. However, exposure of embryos to the acidic solution (pH 2.5–3.0) can be hazardous for ICM cells unless the incubation time is carefully optimized and experienced staff perform the procedure.

Finally, application of a noncontact diode laser has become widely used technology in contemporary assisted reproductive technologies for artificially opening the zona pellucida before intracytoplasmic sperm injection (ICSI) (Rienzi et al. 2001), embryo transfer (assisted hatching) (Obruca et al. 1994), or embryo biopsy during preimplantation genetic diagnosis applications (Joris et al. 2003). Use of laser technology for ICM isolation during hES cell derivation has been reported by Turetsky et al. (2008). Their study indicated that this approach is potentially useful in xeno-free hESC derivation because it avoids the use of animal-derived enzymes of immunological substances that have traditionally been used for this purpose. Ström et al. developed another potentially advantageous isolation method wherein a mechanical isolation protocol utilizes two metal needles with sharpened tips that can cut both the zona pellucida and ICM, thereby avoiding exposure of embryos to either acidic environments or animal-derived substances (Strom et al. 2007).

The proper maintenance and expansion of hESCs comprise important issues in hESC biology. Although nearly a decade has passed from the first successful report on hESC isolation and expansion, in vitro culture of these cells still requires direct exposure to one or several undefined culture ingredients of nonhuman origin.

Use of a blood-borne complement system to isolate ICM from an expanded human blastocyst has already been discussed. However, current hESC cultures still need other materials, such as serum, mitotically inactivated mouse embryonic fibroblasts (MEFs), or MEF-derived extracts.

Although the functional significance and mode of action of some products have been delineated in recent studies, the feeder layer or layer-derived products provide certain currently unknown soluble and/or membrane-bound factors that can increase the derivation efficiency and support undifferentiated growth of hESCs (Kim et al. 2005; Aflatoonian et al. 2010; Amit et al. 2004). Although hESC lines were initially cultured on mouse-derived fetal fibroblasts, these feeders have been replaced with human counterparts that are isolated from various tissues (Richards and Bongso 2006). Another recent approach involves the culture of whole blastocysts on defined or purified cell extracts or matrix proteins of human origin, including collagen VI, fibronectin, laminin, and vitronectin (Ludwig et al. 2006).

Previous studies performed to develop feeder-free culture environments to support established hESC lines have indicated that three factors—transforming growth factor β1 (TGFβ1), leukemia inhibitory factor, bone morphogenetic protein (BMP) antagonist Noggin—act together to suppress hESC differentiation and promote self-renewal (Ludwig et al. 2006; Xu et al. 2005). To develop chemically defined media that sustain hESC self-renewal, it is important that signals and mechanisms controlling hESC fate choices (e.g., choosing to differentiate into a particular lineage or continuing to proliferate as undifferentiated progeny) should be understood in detail. Several studies have discovered that during hESC self-renewal a major role is played by members of the Wnt and TGFβ superfamily of signaling molecules (Sato et al. 2004; James et al. 2005). TGFβ family members seem to stimulate hESC self-renewal by inducing phosphorylation of the intracellular mediators Smad2 and/or Smad3. On the other hand, BMPs induce hESC cell differentiation to extraembryonic lineages or to germs cells by phosphorylation of Smad1/5/8 (Xu et al. 2002; Pera et al. 2004; Kee et al. 2006). Fibroblast growth factor β and insulin-like growth factor II also play important roles in hESC self-renewal by inducing expression of TGFβ family molecules such as TGFβ/activin/nodal (Bendall et al. 2007).

In a recent study, albumin-associated lipids, which are essential ingredients for KnockOut Serum Replacement (KSR), have also been found to have strong positive effects on the self-renewal of hESCs, indicating that a deeper understanding of the mechanisms will eventually lead us to produce xeno-free, chemically defined culture media for hESC self-renewal and differentiation (Garcia-Gonzalo and Belmonde 2008). Clearly, a lack of xeno-free reagents during the process of derivation of hESC lines is a huge gap that needs to be addressed. The availability of reagents such as KSR, which is made of completely xeno-free materials, or those with human materials would enable researchers to derive the next generation of hESC lines with greater potential for use in human clinical cell therapy.

The upscaling process involves disaggregation of undifferentiated hESC colony pieces from their original culture, after which they are transferred into a new culture environment that again supports undifferentiated growth of hESCs. The most widely used method to maintain undifferentiated "high quality" hESCs in culture is mechanical passaging. With this method, micropipettes or finely drawn Pasteur pipettes are used to cut the proper colonies in pieces. Optimum numbers of pieces are transferred into new culture dishes every 4–7 days. Although this method seems to be advantageous over enzymatic dissociation in that no animal-derived dissociation enzyme is used, it certainly becomes a limiting technique when large-scale hESC production is required. Passaging and upscaling of hESCs by enzymatic techniques have, on the other hand, recently been questioned in several reports. Compared to those that were mechanically passaged, cells treated with dissociating enzymes have shown accumulated chromosomal abnormalities, indicating a potential technique-induced genetic instability of the hESC lines studied (Brimble et al. 2004; Maitra et al. 2005). It has also been demonstrated that mouse and human ESCs propagated by automated culture maintain their mean specific growth rates, their capacity for multi-germ-layer differentiation, and expression of the pluripotency-associated markers SSEA-1/Oct-4 and Tra-1-60/Tra-1-81/Oct-4, respectively (Terstegge et al. 2007). The feasibility of ESC culture by automation may greatly facilitate the use of this versatile cell source for a variety of biomedical applications.

Delineating the differentiation potential of hESCs and trying to establish acceptable, reproducible differentiation protocols are still challenges from many aspects. From their first isolation, many research groups have been trying to improve the differentiation protocols for modifying hESCs into cardiomyocytes, neurons, and insulin-producing cells among others, taking us one step closer to possible therapeutic options (Murry and Keller 2008; Nizzardo et al. 2010).

Accurate assessment of toxicity and safety is an essential part of drug development. In vitro toxicity assays using human hepatocytes are one way to model this process, but widespread use of this approach is hindered by the scarcity and the quality of donor tissue. The differentiation of hESCs and induced pluripotent stem cells to functional hepatocyte-like cells has been reported (Asgari et al. 2010). Studies show that further development of this technology could lead to the scalable production of hepatocyte-like cells for liver toxicity screening and clinical therapy.

With all these improvements and research on hESCs, the flow and sharing of new information has become essential to provide better research tools and to design efficient research projects with optimal funding. For this reason, the hESC registry (hESCreg) was set up in April 2007 with European Union funding. It was created to offer the research community, legislators, regulators, and the general public at large an in-depth overview on the current status of hESC derivation and research activity in Europe (www.hescreg.eu) (Borstlab et al. 2008). As a first step, a database was developed that is freely accessible and now contains more than 500 hESC lines, including the lines carrying genetic mutations that have been derived in Europe and elsewhere. These tools, especially obtained by isolating embryos carrying monogenic or chromosomal disorders, are no doubt valuable for understanding the early pathogenesis of numerous genetic diseases, which are difficult to analyze by other means.

12.5 Conclusion

Since 1998, the accumulated data on hESC derivation and culture indicate that hESC cell research is expanding exponentially and is moving forward to keep its promise in many scientific and medical disciplines including developmental biology, human embryology, toxicology, pharmacology, genetics, and regenerative and reproductive medicine.

All reported hESC lines up to now have been established using animal-derived material during derivation or cultivation, and so they are not suitable for clinical use. It is anticipated that the recent advances in understanding ESC self-renewal and expansion mechanisms will no doubt bring contemporary stem cell research one step closer to xeno-free hESC lines that can be utilized in medicine.

Finally, to develop a clinically relevant therapeutic cell/tissue replacement product using therapeutic-grade hESCs, contemporary scientists must cope with the following challenges.

- Production of human embryos under Good Manufacturing Practice (GMP) conditions followed by animal- and pathogen-free ICM isolation
- Feeder-free (animal or human origin) hESC culture systems that involve only purified and screened human extracellular matrix proteins and growth factors, cocktails, and so on
- New GMP-compliant products of a xeno-free nature that are clear alternatives to extracellular attachment factors such as Matrigel and completely humanized or xeno-free reagents such as KnockOut Serum Replacement
- Suitable large-scale hESC culture expansion systems that allow mass production and upscaling of hESCs without a negative effect on cellular proliferation, differentiation, or genetic instability
- Well-defined differentiation protocols/systems that allow mass production of certain precursors and ultimate somatic cell types
- Intensive, tedious screening systems that can efficiently remove undifferentiated hESCs from the differentiated cell population

References

Aflatoonian B, Ruban L, Shamsuddin S et al (2010) Generation of Sheffield (Shef) human embryonic stem cell lines using a microdrop culture system. In Vitro Cell Dev Biol Anim 46:236–241

Amit M, Shariki C, Margulets V et al (2004) Feeder layer- and serum-free culture of human embryonic stem cells. Biol Reprod 70:837–845

Asgari S, Pournasr B, Salekdeh GH (2010) Induced pluripotent stem cells: a new era for hepatology. J Hepatol 53:738–751

Bavister B (2004) Oxygen concentration and preimplantation development. Reprod Biomed Online 9:484–486

Bendall SC, Stewart MH, George MD et al (2007) IGF and FGF cooperatively establish the regulatory stem cell niche of pluripotent human cells in vitro. Nature 448:1015–1021

Bielanska M, Tan SL, Ao A (2002) Chromosomal mosaicism throughout human preimplantation development in vitro: incidence, type, and relevance to embryo outcome. Hum Reprod 17:413–419

Bielanska M, Tan SL, Ao A (2003) Chromosomal information derived from single blastomeres isolated from cleavage-stage embryos and cultured in vitro. Fertil Steril 79:1304–1311

Borstlab J, Stacey G, Kurtz A et al (2008) First evaluation of the European hESCreg. Nat Biotechnol 26:859–860

Brimble SN, Zeng X, Weiler DA et al (2004) Karyotypic stability, genotyping, differentiation, feeder-free maintenance, and gene expression sampling in three human embryonic stem cell lines derived prior to August 9, 2001. Stem Cells Dev 13:585–597

Byrne JA, Pedersen DA, Clepper LL et al (2007) Producing primate embryonic stem cells by somatic cell nuclear transfer. Nature 450:497–502

Chung Y, Klimanskaya I, Becker S et al (2006) Embryonic and extraembryonic stem cell lines derived from single mouse blastomeres. Nature 439:216–219

Chung Y, Klimanskaya I, Becker S et al (2008) Human embryonic stem cell lines generated without embryo destruction. Cell Stem Cell 2:113–117

Civin CI, Rao MS (2006) How many human embryonic stem cell lines are sufficient? A U.S. perspective. Stem Cells 24:800–803

Deb K, Sivaguru M, Yong HY et al (2006) Cdx2 gene expression and trophectoderm lineage specification in mouse embryos. Science 311:992–996

Dumoulin JC, Meijers CJ, Bras M et al (1999) Effect of oxygen concentration on human in-vitro fertilization and embryo culture. Hum Reprod 14:465–469

Edwards RG (2005) Genetics of polarity in mammalian embryos. Reprod Biomed Online 11:104–114

Ellerstrom C, Strehl R, Moya K et al (2006) Derivation of a xeno-free human embryonic stem cell line. Stem Cells 24:2170–2176

Faden RR, Dawson L, Bateman-House AS et al (2003) Public stem cell banks: considerations of justice in stem cell research and therapy. Hastings Cent Rep 33:13–27

Findikli N, Candan NZ, Kahraman S (2006) Human embryonic stem cell culture: current limitations and novel strategies. Reprod Biomed Online 13:581–590

Fong CY, Richards M, Bongso A (2006) Unsuccessful derivation of human embryonic stem cell lines from pairs of human blastomeres. Reprod Biomed Online 13:295–300

Forsyth NR, Musio A, Vezzoni P (2006) Physiologic oxygen enhances human embryonic stem cell clonal recovery and reduces chromosomal abnormalities. Cloning Stem Cells 8:16–23

Garcia-Gonzalo F, Belmonde JCI (2008) Albumin-associated lipids regulate human embryonic stem cell self-renewal. PLoS One 3:e1384

Geber S, Winston RM, Handyside AH (1995) Proliferation of blastomeres from biopsied cleavage stage human embryos in vitro: an alternative to blastocyst biopsy for preimplantation diagnosis. Hum Reprod 10:1492–1496

Handyside AH, Kontogianni EH, Hardy K et al (1990) Pregnancies from biopsied human preimplantation embryos sexed by Y-specific DNA amplification. Nature 344:768–770

Hansis C, Grifo JA, Krey LC (2004) Candidate lineage marker genes in human preimplantation embryos. Reprod Biomed Online 8:577–583

Harper JC, Coonen E, Handyside AH et al (1995) Mosaicism of autosomes and sex chromosomes in morphologically normal, monospermic preimplantation human embryos. Prenat Diagn 15:41–49

Heins N, Englund MC, Sjoblom C et al (2004) Derivation, characterization, and differentiation of human embryonic stem cells. Stem Cells 22:367–376

Hlinka D, Dudas M, Herman M et al (2001) Experimental attempts to extend the current preimplantation genetic diagnosis with individual karyotypization of human blastomeres. Reprod Nutr Dev 41:91–106

Hovatta O (2006) Derivation of human embryonic stem cell lines, towards clinical quality. Reprod Fertil Dev 18:823–828

Jacob S, Moley KH (2005) Gametes and embryo epigenetic reprogramming affect developmental outcome: implication for assisted reproductive technologies. Pediatr Res 58:437–446

James D, Levine AJ, Besser D et al (2005) TGFbeta/activin/nodal signaling is necessary for the maintenance of pluripotency in human embryonic stem cells. Development 132:1273–1282

Janny L, Menezo YJ (1994) Evidence for a strong paternal effect on human preimplantation embryo development and blastocyst formation. Mol Reprod Dev 38:36–42

Jones GM, Trounson AO, Lolatgis N et al (1998) Factors affecting the success of human blastocyst development and pregnancy following in vitro fertilization and embryo transfer. Fertil Steril 70:1022–1029

Joris H, De Vos A, Janssens R et al (2003) Comparison of the results of human embryo biopsy and outcome of PGD after zona drilling using acid Tyrode medium or a laser. Hum Reprod 18:1896–1902

Kaye PL (1997) Preimplantation growth factor physiology. Rev Reprod 2:121–127

Kee K, Gonsalves JM, Clark AT et al (2006) Bone morphogenetic proteins induce germ cell differentiation from human embryonic stem cells. Stem Cells Dev 15:831–837

Kim HS, Oh SK, Park YB et al (2005) Methods for derivation of human embryonic stem cells. Stem Cells 23:1228–1233

Klimanskaya I, Chung Y, Becker S et al (2006) Human embryonic stem cell lines derived from single blastomeres. Nature 444:481–485

Kuliev A, Verlinsky Y (2005) Place of preimplantation diagnosis in genetic practice. Am J Med Genet A 134A:105–110

Landry DW, Zucker HA (2004) Embryonic death and the creation of human embryonic stem cells. J Clin Invest 114:1184–1186

Lanzendorf SE, Boyd CA, Wright DL et al (2001) Use of human gametes obtained from anonymous donors for the production of human embryonic stem cell lines. Fertil Steril 76:132–137

Lin H, Lei J, Wininger D et al (2003) Multilineage potential of homozygous stem cells derived from metaphase II oocytes. Stem Cells 21:152–161

Löser P, Schirm J, Guhr A (2010) Human embryonic stem cell lines and their use in international research. Stem Cells 28:240–246

Ludwig TE, Levenstein ME, Jones JM et al (2006) Derivation of human embryonic stem cells in defined conditions. Nat Biotechnol 24:185–187

Mai Q, Yu Y, Li T et al (2007) Derivation of human embryonic stem cell lines from parthenogenetic blastocysts. Cell Res 17:1008–1019

Maitra A, Arking DE, Shivapurkar N et al (2005) Genomic alterations in cultured human embryonic stem cells. Nat Genet 37:1099–1103

Martin MJ, Muotri A, Gage F et al (2005) Human embryonic stem cells express an immunogenic nonhuman sialic acid. Nat Med 11:228–232

Moor RM, Dai Y, Lee C et al (1998) Oocyte maturation and embryonic failure. Hum Reprod Update 4:223–236

Munne S, Alikani M, Tomkin G et al (1995) Embryo morphology, developmental rates, and maternal age are correlated with chromosome abnormalities. Fertil Steril 64:382–391

Munne S, Colls VE, Bermudez MG et al (2005) Self-correction of chromosomally abnormal embryos in culture and implications for stem cell production. Fertil Steril 84:1328–1334

Murry CE, Keller G (2008) Differentiation of embryonic stem cells to clinically relevant populations: lessons from embryonic development. Cell 132:661–668

Niemitz EL, Feinberg A (2004) Epigenetics and assisted reproductive technology: a call for investigation. Am J Hum Genet 74:599–609

Nizzardo M, Simone C, Falcone M (2010) Human motor neuron generation from embryonic stem cells and induced pluripotent stem cells. Cell Mol Life Sci 67:3837–3847

Obruca A, Strohmer H, Sakkas D et al (1994) Use of lasers in assisted fertilization and hatching. Hum Reprod 9:1723–1726

Park IH, Zhao R, West JA et al (2008) Reprogramming of human somatic cells to pluripotency with defined factors. Nature 451:141–146

Pera MF, Andrade J, Houssami S et al (2004) Regulation of human embryonic stem cell differentiation by BMP-2 and its antagonist noggin. J Cell Sci 117:1269–1280

Piotrowska K, Wianny F, Pedersen RA et al (2001) Blastomeres arising from the first cleavage division have distinguishable fates in normal mouse development. Development 128:3739–3748

Rao MS, Auerbach JM (2006) Estimating human embryonic stem-cell numbers. Lancet 367:650
Rechitsky S, Kuliev A, Sharapova T et al (2006) Preimplantation HLA typing with aneuploidy testing. Reprod Biomed Online 12:89–100
Reubinoff BE, Pera MF, Fong CY et al (2000) Embryonic stem cell lines from human blastocysts: somatic differentiation in vitro. Nat Biotechnol 18:399–404
Revazova ES, Turovets NA, Kochetkova OD et al (2007) Patient-specific stem cell lines derived from human parthenogenetic blastocysts. Cloning Stem Cells 9:432–449
Richards M, Bongso A (2006) Propagation of human embryonic stem cells on human feeder cells. Methods Mol Biol 331:23–41
Rienzi L, Greco E, Ubaldi F et al (2001) Laser-assisted intracytoplasmic sperm injection. Fertil Steril 76:1045–1047
Sato N, Meijer L, Skaltsounis L et al (2004) Maintenance of pluripotency in human and mouse embryonic stem cells through activation of Wnt signaling by a pharmacological GSK-3-specific inhibitor. Nat Med 10:55–63
Schoolcraft WB, Gardner DK, Lane M et al (1999) Blastocyst culture and transfer: analysis of results and parameters affecting outcome in two in vitro fertilization programs. Fertil Steril 72:604–609
Sjogren A, Hardarson T, Andersson K et al (2004) Human blastocysts for the development of embryonic stem cells. Reprod Biomed Online 9:326–329
Stacey G, Hunt CJ (2006) The UK Stem Cell Bank: a UK government-funded, international resource center for stem cell research. Regen Med 1:139–142
Stephenson EL, Braude PR, Mason C (2006) Proposal for a universal minimum information convention for the reporting on the derivation of human embryonic stem cell lines. Regen Med 1:739–750
Strom S, Inzunza J, Grinnemo KH et al (2007) Mechanical isolation of the inner cell mass is effective in derivation of new human embryonic stem cell lines. Hum Reprod 22:3051–3058
Summers MC, Biggers JD (2003) Chemically defined media and the culture of mammalian preimplantation embryos: historical perspective and current issues. Hum Reprod Update 9:557–582
Suss-Toby E, Gerecht-Nir S, Amit M et al (2004) Derivation of a diploid human embryonic stem cell line from a mononuclear zygote. Hum Reprod 19:670–675
Takahashi K, Yamanaka S (2006) Induction of pluripotent stem cells from mouse embryonic and adult fibroblast cultures by defined factors. Cell 126:663–676
Takahashi K, Tanabe K, Ohnuki M et al (2007) Induction of pluripotent stem cells from adult human fibroblasts by defined factors. Cell 131:861–872
Taylor CJ, Bolton EM, Pocock S et al (2005) Banking on human embryonic stem cells: estimating the number of donor cell lines needed for HLA matching. Lancet 366:2019–2025
Terstegge S, Laufenberg I, Pochert J et al (2007) Automated maintenance of embryonic stem cell cultures. Biotechnol Bioeng 96:195–201
Thomson JA, Itskovitz-Eldor J, Shapiro SS et al (1998) Embryonic stem cell lines derived from human blastocysts. Science 282:1145–1147
Turetsky T, Aizenman E, Gil Y et al (2008) Laser-assisted derivation of human embryonic stem cell lines from IVF embryos after preimplantation genetic diagnosis. Hum Reprod 23:46–53
Vrana KE, Hipp JD, Goss AM et al (2003) Nonhuman primate parthenogenetic stem cells. Proc Natl Acad Sci USA 100(Suppl 1):11911–11916
Xu RH, Chen X, Li DS et al (2002) BMP4 initiates human embryonic stem cell differentiation to trophoblast. Nat Biotechnol 20:1261–1264
Xu RH, Peck RM, Li DS et al (2005) Basic FGF and suppression of BMP signaling sustain undifferentiated proliferation of human ES cells. Nat Methods 2:185–190
Yu J, Vodyanik MA, Smuga-Otto K et al (2007) Induced pluripotent stem cell lines derived from human somatic cells. Science 318:1917–1920
Zhang X, Przyborski SS, Cooke M et al (2006) Derivation of human embryonic stem cells from developing and arrested embryos. Stem Cells 24:2669–2676
Zhou H, Wu S, Joo JY et al (2009) Generation of induced pluripotent stem cells using recombinant proteins. Cell Stem Cell 4:381–384

Chapter 13
Clinical and Laboratory Aspects of Preimplantation Genetic Diagnosis and Derivation of Affected Human Embryonic Stem Cell Lines

Rıdvan Seçkin Özen

Abstract Preimplantation genetic diagnosis (PGD) consists of diagnostic procedures detecting a genetic condition(s) in the oocyte or embryo produced by in vitro fertilization (IVF) prior to pregnancy. Chromosomal abnormalities and single gene disorders can be tested by PGD, which gives the parents the opportunity to choose unaffected embryos for transfer. Establishment of affected human embryonic stem cell (hESC) lines from affected preembryos with genetic disorders diagnosed by PGD provides a powerful research tool for exploring fundamental biological mechanisms of early stages of development. This, in turn, leads to the development of new approaches for diagnosing, treating, and preventing genetic disorders. Human embryonic stem cell (hESC) lines are pluripotent and can produce all types of cell lineages in the body. Considering the presence of genetic diversities and polymorphisms in populations, there is a need for large hESC line collections to provide the various genetic components for research purposes. There are several advantages and a uniqueness of hESC research. It cannot be done in animal models or with cell culture methods. Also, it has an invaluable place in pharmacogenomic testing and regenerative medicine applications as affected hESCs of some disorders are the only biological tools we have as disease models.

R.S. Özen, M.D., Ph.D. (✉)
Children's Memorial Research Center, Stem Cell Core, Northwestern University`s Feinberg School of Medicine and Istanbul Genetik Grubu, International Reproductive Genetic Diagnosis Center—InteRepGen, Turkali Mah. Nuzhetiye Cad. No:38/A 34357, Besiktas, Istanbul, Turkey
e-mail: ridvanseckinozen@istanbulgenetik.com; ridvanseckinozen@gmail.com; igg@istanbulgenetik.com

K. Turksen (ed.), *Adult and Embryonic Stem Cells*, Stem Cell Biology and Regenerative Medicine, DOI 10.1007/978-1-61779-630-2_13,

13.1 Introduction

Preimplantation genetic diagnosis (PGD) consists of diagnostic procedures detecting a genetic condition(s) in the oocyte or embryo produced by in vitro fertilization (IVF) prior to pregnancy. Based on genetic test results, PGD allows parents to choose which embryos to transfer to the uterus for implantation. Because only unaffected embryos are transferred and initiation of abnormal pregnancies is prevented, PGD eliminates the possibility of having to terminate an unfavorable pregnancy diagnosed by postconceptional prenatal diagnostic procedures—i.e., amniocentesis (AS), chorionic villous sampling (CVS), cordocentesis.

A PGD program involves several steps, requiring clinicians and laboratories to collaborate. Couples should be informed about PGD by a genetic counselor. After ovulation induction and oocyte aspiration performed by an IVF center/clinic, oocyte fertilization and embryo culturing followed by polar body removal and/or blastomere or trophectoderm biopsy steps are done in an IVF laboratory. Biopsied samples are tested, and results are reported separately for each embryo in a genetics laboratory specializing in PGD. Couples are given detailed information about the PGD report and decide which embryos should be transferred under the supervision of an IVF center physician. Surplus embryos that are not transferred but are suitable for transfer can be frozen if so requested by the patients. If pregnancy occurs, prenatal diagnosis (AS, CVS) can still be recommended for confirmation purposes.

The major application of PGD is aneuploidy screening for the most common chromosome abnormalities and diagnosis of unbalanced inheritance of chromosome abnormalities (translocations, inversions, deletions). As it is known that numerical chromosome abnormalities are the most common reason for pregnancy loss, selection of euploid embryos for transfer increases the implantation rate, lowers the spontaneous abortion rate, and reduces the risk of trisomic offspring (e.g., Down syndrome). The main indications for this preimplantation genetics screening (PGS) are advanced maternal age (≥35 years), recurrent pregnancy losses, or repeated IVF failure.

The PGD is also available for a large number of single gene disorders: autosomal recessive (e.g., cystic fibrosis, β-thalassemia, spinal muscular atrophy), autosomal dominant (e.g., Huntington`s disease, myotonic dystrophy), and X-linked (e.g., fragile-X syndrome, hemophilia A, Duchenne muscular dystrophy). Identification is mainly by polymerase chain reaction (PCR)-based techniques using different approaches and detecting normal, carrier, and affected embryos. HLA typing of embryos—together with monogenic disease testing (e.g., β-thalassemia) or with no other single gene testing (e.g., leukemia)—helps couples use their own new PGD baby as a cord blood stem cell donor for a sick sibling. PGD testing for late-onset diseases (some groups consider this unethical because the individuals stay healthy until the onset of the disease, usually in their fourth decade) and cancer predisposition (*BRCA1* and *p53*) are other indications. Nondisclosure PGD (for Huntington disease) indicates the cases when parents do not want to know if they are carriers but still want to have a baby free of the disease.

13.2 Obtaining Cells for Preimplantation Genetic Diagnosis

The PGD can be carried out on polar body 1, polar bodies 1 and 2, blastomere, and trophectoderm cells. In follicle-stimulating hormone (FSH)-stimulated ovarian follicles, the oocyte completes meiosis I while ovulation occurs and divides into two cells, leaving one cell cytoplasmically active and the other compact and sequestrated (polar body 1, or PB1) between the zona pellucida and the active cell (Fig. 13.1) not having any counterpart in the developing embryo. PB1 has "2n"

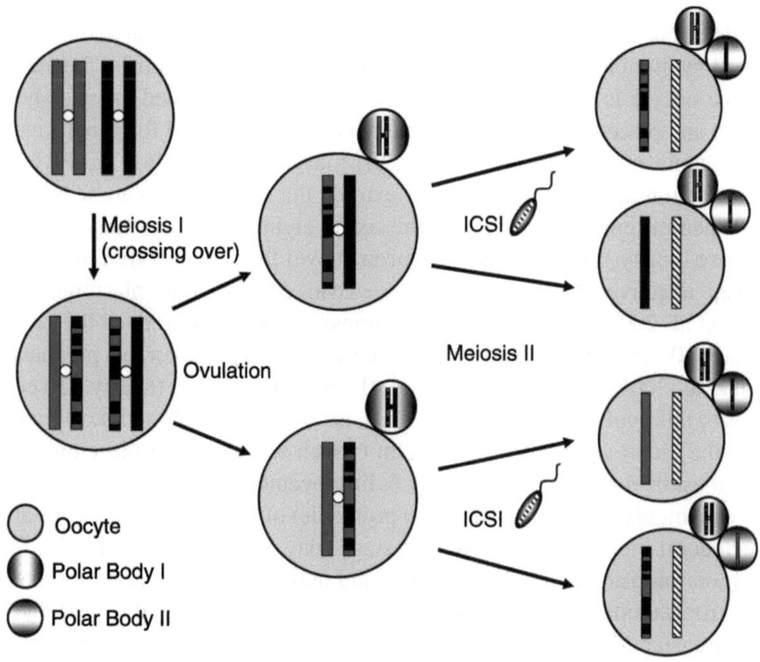

Fig. 13.1 Chromosomal segregation and recombination during meiosis I and II. Crossing-over (recombination) event happens between each homologous chromosome pair in meiosis I. Recombination occurs at several points along the chromosome. One chromatid of a chromosome can recombine with only one chromatid or both chromatids of the other chromosome, depending on where the chiasma occurs. This figure shows the multiple recombination events between only two nonsister chromatids of a homologous chromosome pair. Recombined chromosomes are a mixture of maternal and paternal DNA having heterozygous genetic elements at some loci. When ovulation occurs, the oocyte divides into two cells with chromosomal segregation: one cell with no cytoplasm (polar body 1, or PB1) consisting of half (2n) of the duplicated DNA material. Entrance of sperm into the oocyte triggers meiosis II and the (n) number of chromatids is extruded with PB2. Chromosomal segregation problems occur in both meiosis I and II at rates of, respectively, 41.7% and 35.1% (Verlinsky et al. 2005). PB1 and PB2 analysis enables prediction of the maternal DNA constitution left in the oocyte. PB1 analysis without PB 2 analysis does not provide the exact information about the maternal genomic DNA contribution to the embryo

number of chromosomes, and each has recombined DNA material of maternal and paternal chromosomes of the mother. This recombination event (crossing-over) occurs when the same (homologous) chromosomes are aligned side by side on metaphase stage of meiosis I.

The number and position of recombination events between homolog chromosomes happens in a randomized manner for each oocyte. There are several hot spots for recombination along the chromosome. For a given locus of any chromosome, some oocytes become heterozygous, whereas others keep the homozygous state. IVF protocols stimulate many follicles in the ovaries for maturation compared to one or two in natural cycles. The number and maturation of oocytes decreases, especially after 35 years of age. For IVF, after oocyte retrieval (day 0) and surrounding cumulus cell removal, PB 1 is biopsied by a micropipette through the zona pellucida by creating a hole mechanically or with the aid of laser or chemicals and transferred to another droplet of medium to avoid possible DNA or cell contamination from the medium in which the oocyte is biopsied. As a general rule, every biopsied sample should be picked up and placed into separate medium and then processed for genetic analysis.

Intracytoplasmic sperm injection (ICSI) (day 0) is done after the PB1 biopsy. Sperm injection induces the oocyte the extrude the PB2, which can be biopsied on day 1. When the embryo develops to the six- to eight-blastomere stage (day 3), one blastomere biopsy is done. It has not been shown that polar body removal has an additional negative affect on embryo survival compared to blastomere biopsy (Cieslak et al. 2006). Single blastomere biopsy is thought to reduce the pregnancy rate nearly 10%, and biopsy of two blastomere cells may cause a low pregnancy rate (Cohen et al. 2007). The embryo reaches the blastocyst stage (nearly 120 cells) on day 5 or 6; this evolves into the inner cell mass (ICM), (which will develop into the body of the fetus) and the trophectoderm (which will form the placenta). Embryo transfer must be done no later than day 6. Improvements in cryopreservation procedures (with highly efficient vitrification protocols) of the biopsied embryos allows a trophectoderm biopsy at the blastocyst stage (day 5), which is useful for PGD. It yields more biopsied cells (10–30 cells) and avoids removing an embryonic cell. Frozen trophectoderm-biopsied embryos are thawed during the next cycle for embryo transfer (de Boer et al. 2004; Kokkali et al. 2005; McArthur et al. 2005; Kuwayama 2007; Escribá et al. 2008; Schlenker et al. 2009).

Genetic analysis of blastomere should be finished within 1–2 days to enable day 5 embryo transfer (ET). If PB1 and PB2 analyses (sequential polar body analysis, or SPBA) give successful results for the tested genetic condition, ET can be done on day 3, which is preferred by some IVF centers. Trophectoderm biopsy usually requires freezing of embryos due to a lack of time for genetic analysis unless anintensive effort is spent to give genetic reports on day 5 for day 6 ET.

In addition to each biopsy technique having its own difficulties, there are advantages and requirements associated with its clinical use. The particular genetic condition(s) for which the test is performed, the inheritance mode of the disease, and whether the test yields successful results from the biopsied sample influences the strategy for obtaining a suitable biopsy specimen. SPBA can be done for autosomal recessive disorders, autosomal dominant disorders when the mother is

Table 13.1 Biopsied sample type preference versus mode of inheritance for single gene disorders

	Biopsied sample type		
Mode of inheritance	PB1+PB2	Blastomere	PB1+PB2+blastomere
Autosomal dominant (e.g., myotonic dystrophy, neurofibromatosis 1, tuberous sclerosis)			
Affected mother	+	+[a]	+[b]
Affected father		+	+[c]
Autosomal recessive (e.g., cystic fibrosis, β-thalassemia, spinal muscular atrophy)	+	+[a,d]	+[b]
X-linked (e.g., fragile-X syndrome, DMD, hemophilia A)	+	+[a,d]	+[b]
HLA typing		+	+[c]
Aneuploidy testing		+	

PB: polar body

[a]Blastomere analysis can be done as the first choice depending on the PB biopsy experience at the in vitro fertilizaiton center and analysis capacity of the genetic diagnosis center

[b]Blastomere analysis may follow to strengthen the PB results or when PB testing fails to give interpretable results

[c]PB testing can be done to have prior information about the maternal contribution to the embryo, which helps when analyzing the blastomere results

[d]When PB results reveal a wild-type (normal) genetic contribution to the embryo, blastomere analysis may not follow if parents accept the possibility of having a carrier baby due to transmission of an affected chromosome by sperm

affected, and X-linked disorders if the mother is the carrier. A blastomere sample or trophectoderm biopsy is required if father is affected with an autosomal dominant or X-linked disorder. HLA typing has to be done in blastomere or trophectoderm cells because the paternal genetic information is needed. SPBA or blastomere analysis can be added to the testing strategies mentioned above (Table 13.1).

13.3 Single-Gene Disorder Testing

Preimplantation genetic diagnosis of single gene disorders has become possible for every type of mutation, including single base substitutions (e.g., most commonly *HBB* gene mutations), small deletions (e.g., deltaF508 mutation of the *CFTR* gene), large deletions [e.g., DMD gene exon(s) deletions of the Duchenne muscular atrophy], small insertions (e.g., +TATC1278 insertion of the *HEXA* gene in Tay-Sachs disease), duplications (e.g., duplication of a large region harboring the *PMP22* gene in Charcot-Marie-Tooth neuropathy type 1A), trinucleotide repeat expansion disorders (e.g., CTG trinucleotide repeat expansion in the *DMPK* gene in myotonic dystrophy), and mutation of genes having a pseudogene (e.g., exon or full deletion of the *SMN1* gene having SMN2 pseudogene in spinal muscular atrophy).

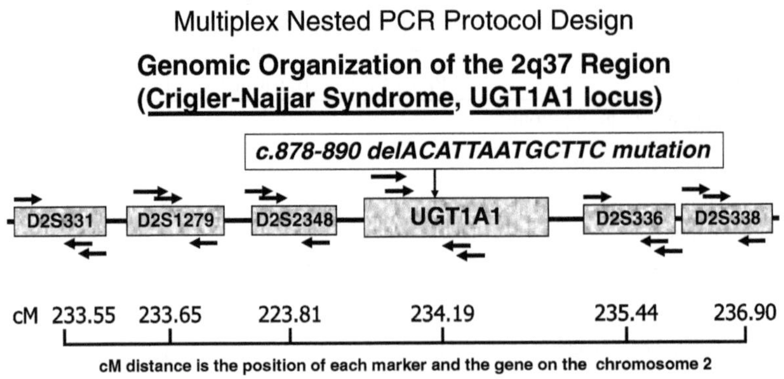

Fig. 13.2 Multiplex nested polymerase chain reaction (PCR) protocol design for single gene disorders. Genomic organization of the 2q37 region between 233.55 and 236.90cM distances on the q arm of chromosome 2 is shown with the *UGT1A1* gene (responsible for Crigller-Najjar syndrome) and flanking STRs linked to the gene. D2S331, D2S1279, and D2S2348 are on the 5′ end of the gene. D2S336 and D2S338 are on 3′ end of the gene (Ozen et al. 2009). For STRs, the position of semi-nested PCR primers are shown by *arrows*; two outside primers are used for the first-round PCR, and one inside primer is used in combination with one of the outside primers for second-round PCR. For *UGT1A1*, the position of full-nested PCR primers are shown by *arrows*; two outside primers are used for the first-round PCR, and two inside primers are used for second-round PCR. Mutation of the *UGT1A1* gene is c.878-890 delACATTAATGCTTC, present in both parents of the family seeking a preimplantation genetic diagnosis (PGD). First-round multiplex PCR includes the outside primers of the *UGT1A* gene and informative (heterozygous) STRs of the parents. Second-round PCR is done separately for the gene and STRs by using nested (semi- or full-) primers

13.3.1 Multiplex Nested Polymerase Chain Reaction

Considering that PGD is done in a single cell, routine genetic analysis methods done in bulky DNA extracted from peripheral blood sampling or other sources (buccal swab, amniocentesis, chorionic villous sampling, cultured specimens) is not applicable. There is a requirement of preamplification of desired DNA elements to reach enough material for following detection steps. Multiplex nested PCR is the most common method used for single gene disorders during the PGD performance and consists of three consecutive steps.

- Step 1. Cell lysis of biopsied sample in a PCR tube containing lysis buffer (commonly used buffers are proteinase K and KOH-DTT). Because the extracted DNA cannot be transferred to a separate PCR tube, inactivation of proteinase K or neutralization of alkaline KOH buffer should be done before following the PCR steps.
- Step 2. First-round PCR: Multiplex PCR mixture is added. It has multiple primer sets designed to amplify the mutation region of the gene and inside gene or closely linked—preferably not more than 2 cM distant from both the 5′ and 3′ sides of the gene to keep the possibility of a recombination event at a very low level—short tandem repeat/single nucleotide polymorphism (STR/SNP) genetic markers (Fig. 13.2) (Ozen et al. 2009). PCR conditions can be strengthened by

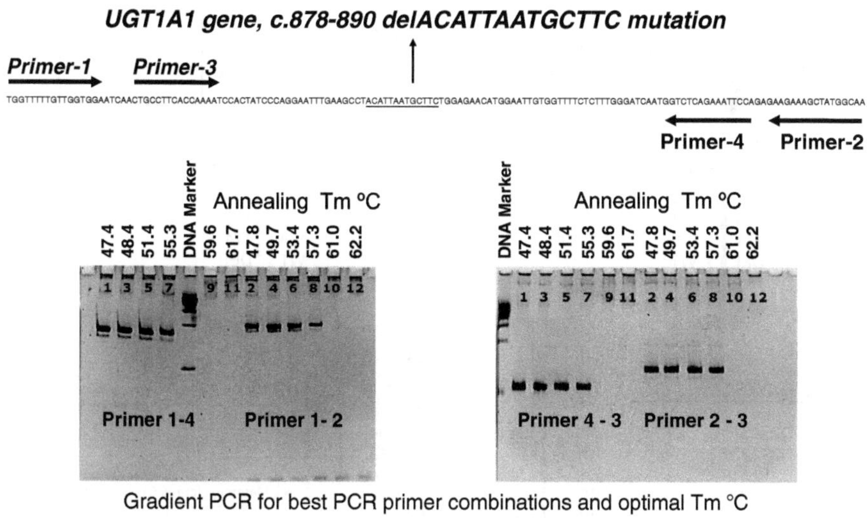

Fig. 13.3 Nested PCR primer optimization for PGD protocol. Primer positions are shown on the genomic DNA sequence. Gradient PCR (between 47°C and 61°C) with primer 1–4, 1–2, 3–4, and 2–3 combinations are done to find the best annealing Tm (°C). Polyacrylamide gel electrophoresis (PAGE) shows the amplification yield for each primer set for different PCR annealing temperatures. All primer sets have amplification products between 47°C and 57°C with no extra bands. In the case of nonspecific annealing of primers to other DNA sequences giving extra products larger or smaller than the desired basepair length, the annealing temperature giving one band pattern is chosen. If no PCR product is obtained or extra bands are present at every annealing temperature, a new primer design is needed

enhancers such as dimethylsulfoxide (DMSO). Nearly 50 loci can be amplified successfully in a single-cell-based optimized multiplex PCR protocol.

- Step 3. Second-round PCR: Separate PCR reactions are done with nested primers (semi-nested or full-nested) for each loci included in the first-round PCR. Each nested PCR primer set should be optimized for the best amplification conditions (Tm °C and $MgCl_2$ concentrations) (Fig. 13.3). A 1.0- to 1.5-ml aliquot of the first-round reaction is used as the DNA template for the second-round PCR.

Detection of the mutation and genetic polymorphisms is dependent on the design of the PCR protocol followed by fluorescence sequencing or fragment analysis, restriction fragment length polymorphism (RFLP) (natural or mismatch primed), amplification refractory mutation system (ARMS), real-time PCR, and minisequencing.

13.3.2 Haplotyping of Family Members

Designing a PGD protocol for a specific single gene disorder requires information about the mutated gene and the exact mutation position. Clinical diagnosis of a proband or carrier status stated by biochemical testing or an affected relative may

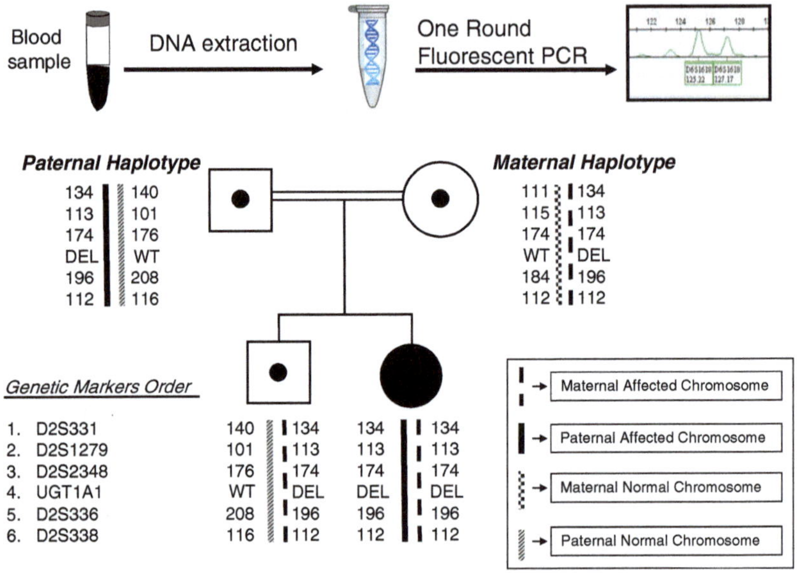

Fig. 13.4 Haplotyping of family members for the genomic region of interest. DNA extracted from peripheral venous blood samples of family members are used for haplotype analysis. *UGT1A1* gene and closely linked STRs are haplotyped by fragment analysis of *UGT1A1* gene mutation (c.878-890 delACATTAATGCTTC deletion, or DEL) and fluorescence PCR of STRs. STR alleles segregating with maternal and paternal chromosomes carrying *UGT1A1* gene mutation (DEL) is based on the genotype of the affected daughter. Allele information is given as basepair lengths of PCR products detected in fragment analysis. Maternal and paternal chromosomes carrying wild-type (normal) and the DEL mutation are shown with differently labeled bars. Paternally inherited chromosomes are placed on the left and maternally inherited chromosomes are placed on the right of the genotype pattern of the children. In the presence of the consanguinity between parents (as in this family), shared alleles can be used to build the haplotype as well. If there were no affected children, allele information of the carrier son would not be helpful for drawing the haplotype because the parental inheritance of mutation was not known

lead a molecular geneticist to test a specific gene, as in β-thalassemia for the *HBB* gene. Once the mutation(s) is found in the parent(s), a PGD set up specifically for the given family is done with the DNA material obtained from the parents and related family members prior to testing the embryo (Fig. 13.4). Haplotype information regarding which STR/SNP alleles are on the same chromosome harboring wild-type and mutated genes is needed to overcome the negative effect of allele dropout of gene alleles on prediction of embryo genotype.

13.3.3 Allele Dropout

Allele dropout (ADO) is the nonamplification and detection failure of one allele of a heterozygous locus. ADO can be encountered at higher percentages for some loci.

STR/SNP	DXS998		DXS297		DXS548		AC1		ATL1		Intron 1		AC2		DXS8091		DXS1193		P23		P26		P39	
PCR	No ADO	ADO	No ADO	ADO	No ADO	ADO	No ADO	ADO	No ADO	ADO	No ADO	ADO	No ADO	ADO	No ADO	ADO	No ADO	ADO	No ADO	ADO	No ADO	ADO	No ADO	ADO
PB1	144	0	50	2	145	1	183	19	48	6	84	4	264	34	133	6	238	8	72	0	203	4	139	3
PB2	134	2	49	0	124	3	157	8	48	1	73	2	242	12	115	6	205	5	65	0	184	3	132	1
Blastomere	51	2	26	0	62	7	72	4	8	0	26	1	85	7	52	1	68	1	13	0	59	3	76	2
TOTAL	329	4	125	2	331	11	412	31	104	7	183	7	591	53	300	13	511	14	150	0	446	10	347	6
ADO %	1.2		1.6		3.32		7.52		6.73		3.83		8.97		4.33		2.74		0		2.24		1.73	

Combined ADO rates for different combinations:

7.52 x 6.73 x 8.97 x 4.33 === 0.002 %

7.52 x 6.73 x 8.97 === 0.045 %

7.52x 8.97 === 0.7 %

3.32 x 8.97 === 0.3 %

3.32 x 4.33 === 0.1 %

Fig. 13.5 Allele dropout (ADO) rates of STR/SNP on the FRAXA region with respect to PBI, PBII, and blastomere analyses. ADO and No ADO results of 12 STR/SNP genetic markers from PGD cases studied with PB1, PB2, and blastomere cells are shown. For the AC1 genetic marker, ADO was less encountered in the blastomere analysis (4/4+72=5.2%) compared to PB1 (19/19+183=9.4%) analysis. For DXS548, ADO was found more in the blastomere analysis (7/7+62=10.1%) compared to the PB1 (1/1+145=0.7%) analysis. P23 marker gave no ADO results with any sample type (Ozen et al. 2005b). Considered with other genetic marker ADO rates, ADO does not seem to be affected by the sample type. When total ADO rates are compared, there is also a big difference between genetic markers: 8.97% for AC2, 4.33% for DX8091, 1.2% for DXS998, and 0% for P23. These numbers show that it would not be appropriate to give a fixed percentage rate for ADO. A combined ADO rate for the four highest genetic markers (AC1, ATL1, AC2, DXS8091) is 0.002%. Even if these four markers are the only informative ones in a given family, misdiagnosis due to ADO while using four genetic markers is infrequent. If AC1 and AC2 are the only ones to be used for a PGD case, the combined ADO rate is 0.7%. Considering that these two STRs are very close to each other, it would be expected to have ADO of one marker together with the other one, but this is not the case most of the time

Reoptimizing the PCR primer sets for a locus giving high ADO rates can help yield better results, but the DNA sequence of the locus may not let this happen. ADO is not specific for the cell type. ADO rates of STRs/SNPs of the FARAXA region for PB1, PB2, and blastomere cells are shown in Fig. 13.5 (Ozen et al. 2005b).

13.3.4 *Place of STR/SNP Polymorphic Genetic Markers in Preimplantation Genetic Diagnosis*

Inclusion of STR/SNP genetic markers in the PGD has several advantages.

13.3.4.1 Detecting Allele Dropout or Preferential Amplification

Autosomal Recessive Single Gene Disorders

When parents share the same mutation of a given gene, for a carrier embryo an ADO of a normal allele may lead one to predict the genotype of the embryo as affected because there is only information regarding an affected allele. Genetic markers closely linked to the gene can give information about the presence of normal chromosome (carrying the wild-type allele) and save these embryos for embryo transfer.

In the cases of parents having different mutations, an affected embryo may give single cell results as normal when ADO occurs on both mutant alleles, leaving only normal allele information in the test results. Genetic markers can reveal the presence of parental chromosomes carrying mutated genes.

Autosomal Dominant Single Gene Disorders

Allele dropout of a mutated allele (with the presence of information indicating a normal allele) in an affected embryo leads to a misdiagnosis of a normal embryo. Genetic markers allele information linked to the mutated and normal chromosomes show the actual genotype of the embryo.

X-Linked Single Gene Disorders

For female embryos, ADO of the mutations may conceal the carriers as being normal. Because X-inactivation occurs randomly, carrier females may present clinical symptoms. For male embryos, ADO of the gene locus may be considered as a failure of amplification. Inclusion of genetic markers can solve both problems for predicting a genotype.

13.3.4.2 Detecting DNA Contamination

Maternal cell or DNA contamination of the biopsied sample can be detected by genetic marker analysis. Normal embryos may give carrier results due to maternal contamination. The same results can be obtained in trisomies of the related chromosome.

Unrelated DNA contamination might be detected if the source of contamination has different alleles. Because unrelated DNA would produce normal gene test results, carrier-affected embryos might be misdiagnosed as carriers especially when the parents share the same mutation.

13.3.4.3 Embryo Identification

Each embryo of a given IVF cycle has its own genetic fingerprint comprised of a combination of its parents. By including highly polymorphic and heterogeneous STR markers in the PGD testing, it would be possible to check retrospectively whether the transferred embryo has the same fingerprint as the biopsied one.

13.3.4.4 Detection of Chromosomal Aneuploidies

Having prior information regarding the parents for a wide range of STR markers of selected chromosomes and choosing two or three of the heterozygotes among them would allow the detection of trisomies or monosomies. To increase the sensitivity of this approach, STRs should be selected from the distant part of the chromosomes to give

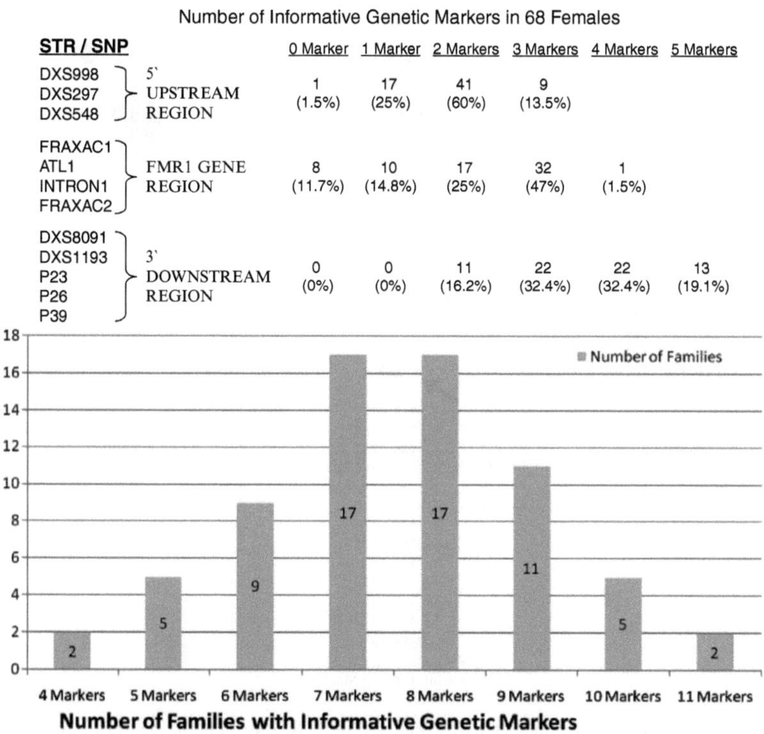

Fig. 13.6 Informative genetic markers in the *FMR1* gene and flanking regions were studied in 68 female subjects. In all, 12 STRs/SNPs were studied, and the number of heterozygous ones in 5´, inside gene, and 3´ regions were determined. All of the subjects were informative for at least four markers. The number of subjects having the same number of informative genetic markers is shown in the chart. Altogether, 65 subjects (96%) had at least two informative markers in both sides of the *FMR1* gene (data not shown). The number of subjects who were informative for at least for two markers in 5′ upstream, inside gene, and 3′ downstream region were 41+9=50, 17+32+1=50, and 11+22+22+13=68, respectively. Percentages in parenthesis, such as 41(60%), indicates the percentage of informative marker numbers in the three regions. Three subjects had only one informative marker on the 5′ side and inside the *FMR1* gene

enough distance between them for recombination to occur because only heterozygous parts of the chromosomes provide both alleles of one parent (Rechitsky et al. 2004).

13.3.4.5 PGD by Linkage Analysis When Mutation Is Not Known

In some families, the mutation type of a known gene responsible for the clinical symptoms cannot be studied by DNA analysis for several reasons. Linkage analysis can be used to perform PGD if the haplotype of parents is known by using the genetic markers' data of the proband or other relatives. PGD with linkage analysis requires a large collection of STRs/SNPs to increase the detection and accuracy level of genetic testing (Fig. 13.6). For example, fragile-X syndrome can be studied for PGD with linkage analysis without *FMR1* gene testing (Fig. 13.7) (Ozen et al. 2005b).

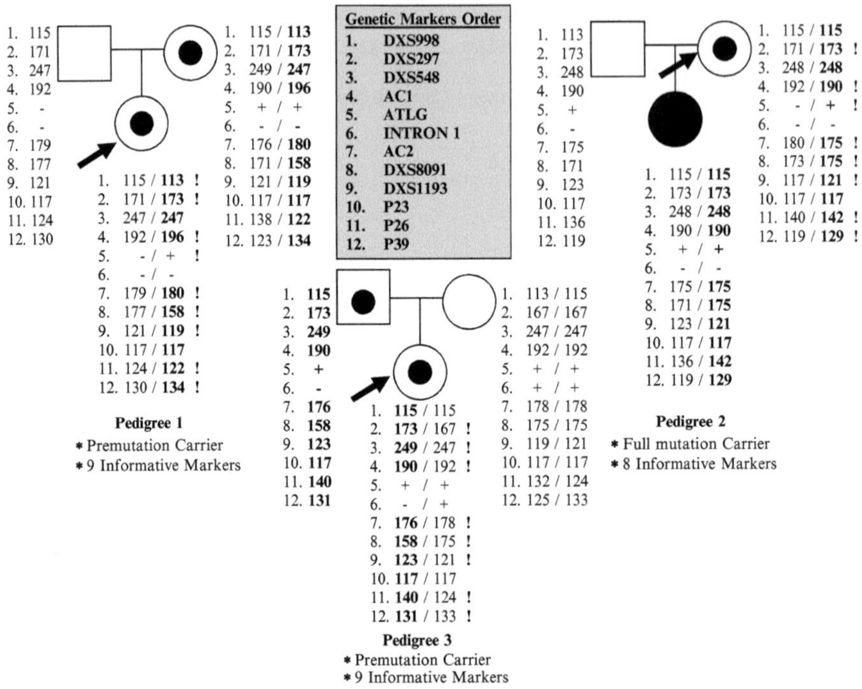

Fig. 13.7 PGD with linkage analysis was done using STR/SNP genetic markers. Three fragile-X family pedigrees with haplotype information of STR/SNP genetic marker alleles are shown. The order of genetic markers is given in the *box*. Allele information is given as basepair lengths of PCR products detected in fragment analysis. *Pedigree 1*: The proband is a permutation carrier (inherited the affected chromosome from a permutation carrier mother) and has nine informative markers. *Pedigree 2*: The proband is a full mutation carrier (inherited the affected chromosome from a permutation carrier mother who amplified the CGG repeats to the full mutation level) and has eight informative markers. *Pedigree 3*: The proband is a permutation carrier (inherited the affected chromosome from a transmitting man) and has nine informative markers

Some genetic disorders are caused by duplication of a chromosome region harboring the gene responsible for the disease. For example, duplication of the *PMP22* gene is responsible for 70–80% of patients with Charcot-Marie-Tooth neuropathy type 1A. A collection of STRs in this repeated region can be used to identify the chromosome that has a duplicated region.

13.4 HLA Typing by STRs

Human leukocyte antigen (HLA) alleles are on the major histocompatibilty complex (MHC) region of chromosome 6. HLA typing of single cells for specific HLA alleles has encountered various difficulties, including PCR setup problems for single

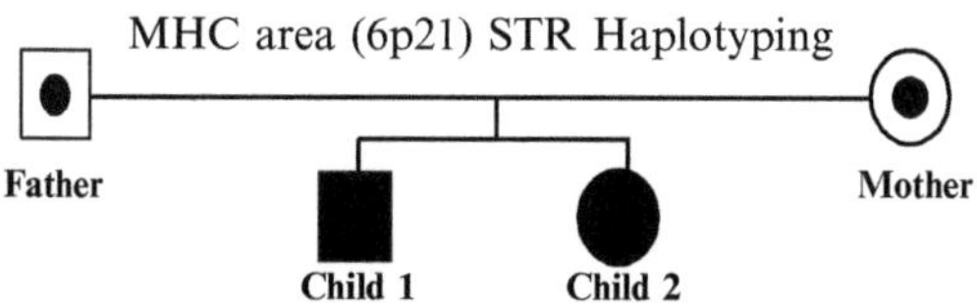

HLA STR MARKERS		Father		Child 1		Child 2		Mother		
1	D6S439	1 **123** 125		1 **123 125**		1 **123 125**		1 127 **125**		
2	D6S1618	2 **127** 125	!	2 **127 123**	!	2 **127 123**	!	2 143 **123**	!	
3	D6S1568	3 **131** 139	!	3 **131 143**	!	3 **131 143**	!	3 135 **143**	!	
4	D6S1629	4 **187** 185	!	4 **187 172**	!	4 **187 172**	!	4 189 **172**	!	
5	D6S1583	5 **155** 153		5 **155 157**		5 **155 157**		5 141 **157**		
6	D6S1560	6 **155** 155		6 **155 164**		6 **155 164**		6 145 **164**		
7	RING	7 **155** 169		7 **155 159**		7 **155 159**		7 159 **159**		
8	TAP-1	8 **207** 220		8 **207 205**		8 **207 205**		8 207 **205**		
9	D6S2444	9 **142** 132		9 **142 140**		9 **142 140**		9 131 **140**		
10	D6S2443	10 **131** 111		10 **131 103**		10 **131 103**		10 131 **103**		
11	DQ CAR	11 **153** 153	!	11 **153 161**	!	11 **153 161**	!	11 140 **161**	!	Class II area
12	D6S1666	12 **103** 103		12 **103 103**		12 **103 103**		12 103 **103**		
13	LH	13 **163** 163		13 **163 175**		13 **163 175**		13 165 **175**		
14	DN	14 **188** 188		14 **188 193**		14 **188 193**		14 203 **193**		
15	D6S1615	15 **131** 137	!	15 **131 133**	!	15 **131 133**	!	15 127 **133**	!	
16	9N2	16 **131** 125	!	16 **131 133**	!	16 **131 133**	!	16 125 **133**	!	
17	D6S273	17 **274** 268		17 **274 274**		17 **274 274**		17 272 **274**		
18	TNF-A	18 **100** 92	!	18 **100 109**	!	18 **100 109**	!	18 92 **109**	!	
19	TNF-B	19 **128** 130	!	19 **128 127**	!	19 **128 127**	!	19 124 **127**	!	Class III area
20	TNF-D	20 **120** 128		20 **120 126**		20 **120 126**		20 128 **126**		
21	62	21 **197** 199	!	21 **197 206**	!	21 **197 206**	!	21 187 **206**	!	
22	MicA	22 **133** 135	!	22 **133 144**	!	22 **133 144**	!	22 135 **144**	!	
23	MIB	23 **124** 124		23 **124 106**		23 **124 106**		23 106 **106**		
24	D6S265	24 **175** 175		24 **175 181**		24 **175 181**		24 192 **181**		
25	D6S510	25 **148** 148		25 **148 163**		25 **148 163**		25 159 **163**		Class I area
26	RF	26 **277** 277		26 **286 286**		26 **286 286**		26 286 **286**		
27	MOG	27 **128** 115	!	27 **128 122**	!	27 **128 122**	!	27 120 **122**	!	
28	MogA	28 **130** 126	!	28 **130 128**	!	28 **130 128**	!	28 135 **128**	!	
29	D6S258	29 **175** 158	!	29 **175 168**	!	29 **175 168**	!	29 158 **168**	!	
30	D6S248	30 **256** 264	!	30 **256 280**	!	30 **256 280**	!	30 264 **280**	!	
31	D6S306	31 **104** 108	!	31 **104 104**	!	31 **104 104**	!	31 110 **104**	!	
32	D6S276	32 **143** 131	!	32 **143 146**	!	32 **143 146**	!	32 117 **146**	!	
33	D6S461	33 **231** 229		33 **231 219**		33 **231 219**		33 231 **219**		

Fig. 13.8 HLA typing with linkage analysis using STRs. A total of 33 STRs collected in a major histocompatibility region (MHC) covering classes I, II, and III region is shown with parental and sibling alleles. Allele information is given as basepair lengths of PCR products detected by fragment analysis. Child 1 and Child 2 are in need of bone marrow transplantation. In this family, STR alleles are the same, indicating that parents gave the same chromosome 6 to each child. In the case of having different chromosomes from their parents, each child would need an HLA-matched embryo, although different ones, even though they are in the same family. Considering 33 markers covering a large genomic segment, recombination may occur in different parts of the region which would be revealed during PGD testing. If the child has one recombinant parental chromosome, STR allele haplotyping of parents using this child will not reflect the real alignment of STR alleles of the related parent. To detect this type of recombination events, other family members (grandparents or healthy children) should be included to the testing

cells and false-positive PGD results. Using a large STR collection covering the MHC area and performing haplotyping analysis during the PGD helps to find embryos having the same chromosome 6 region (and the same HLA alleles) as the baby who is in need of bone marrow transplantation (Fig. 13.8) (Verlinsky et al. 2004). Fragment analysis of STRs for PGD testing to select HLA-matched and HLA-nonmatched embryos is shown in Fig. 13.9 (Umay et al. 2009).

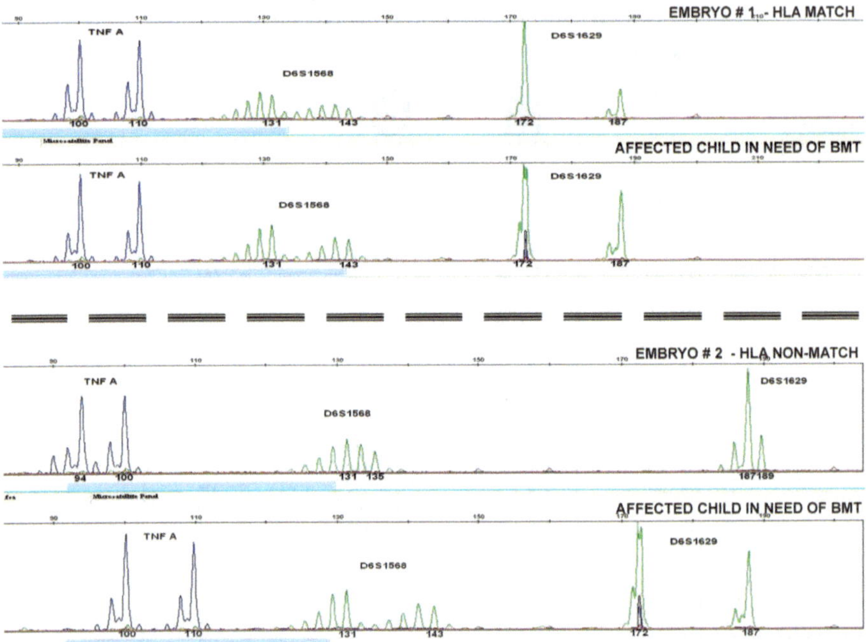

Fig. 13.9 Selecting HLA matched and nonmatch embryos by STR analysis in the MHC region. TNFA, D6S1568, and D6S1629 STR alleles of the affected child in need of bone marrow transplantation are 100/110, 131/143, and 172/187 respectively. Embryo 1 has the same STR alleles as the affected child, which makes this embryo a full HLA match for maternal and paternal chromosomes. Embryo 2 results match in only one allele, showing inheritance of one parental matched chromosome and one mismatched chromosome. Embryo 1 should be used for embryo transfer if it has a carrier or normal PGD results for the disease running in the family. If the affected child has leukemia, this embryo is considered suitable for embryo transfer, expecting other genetic compositions of the embryo to be different from those of the affected child, presumably affecting the clinical picture

13.4.1 Interpretation of Polar Bodies 1 and 2 and Blastomere Analysis

The analysis of PB1 and PB2 is based on prediction of the oocyte genotype using meiosis I and meiosis II segregation information. PB1 has a complete set of chromosomes (2n) recombined in meiosis I. The genomic segment containing the tested gene may be in a heterozygous or homozygous state depending on the position of the recombination events taking part along the chromosome. When the gene is close to telomere, there is more chance for recombination to occur and to be present in a heterozygous state.

An oocyte is deduced to be normal in two combinations found by SPBA (Fig. 13.10).

Embryo #	Cell Type	STR1	STR2	STR3	Gene	STR4	STR5	STR6	Predicted Oocyte/Embryo Genotype	ET
1	PB1	120/**158**	115/**103**	215/**207**	N/**Mutation**	154/**128**	118/**110**	145/**131**	Oocyte: NORMAL	YES
	PB2	**158**	**103**	**207**	**Mutation**	**128**	**110**	**131**		
2	PB1	**158**	**103**	**207**	**Mutation**	**128**	**110**	**131**	Oocyte: NORMAL* Embryo: CARRIER	YES
	PB2	120	115	215	Normal	154	118	145		
	Blast	**168**/120	**117**/115	***207***/215	**Mutation**/N	**146**/154	ADO/118	***145***/*145*		
3	PB1	120/**158**	115/ADO	215/**207**	N/**Mutation**	154/ADO	118/**110**	145/**131**	Oocyte: AFFECTED Embryo: AFFECTED	NO
	PB2	120	115	215	Normal	154	118	145		
	Blast	**168/158**	**117**/***103***	***207***/***207***	**Mutation/ Mutation**	**146**/***128***	**114/110**	***145***/131		
4	PB1	120/**158**	115/**103**	215/**207**	N/**Mutation**	*128*/***128***	118/**110**	145/**131**	Oocyte: Uninterpretable Embryo: TRISOMY	NO
	PB2	FA	FA	FA	FA	FA	FA	FA		
	Blast	140/ 120/**158**	*103*/ 115/***103***	203/ 215/**207**	*N*/ *N*/**Mutation**	*128*/ 154/***128***	*114*/ 118/**110**	139/ *145*/**131**		
FATHER		**168**/140	**117**/103	**207**/203	**Mutation**/ Normal	**146**/128	**114**/114	**145**/139	CARRIER	
MOTHER		120/**158**	115/**103**	215/**207**	Normal/ **Mutation**	154/**128**	118/**110**	145/**131**	CARRIER	
CHILD		**168/158**	**117/103**	**207/207**	**Mutation / Mutation**	**146/128**	**114/110**	**145/131**	AFFECTED	

N = Normal, FA = Failed PCR Amplification, ADO = Allele Drop Out, ET = Embryo Transer *123* = Same allele in both parents
* Homozygous PB 1 results might be followed by Blastomer Biopsy analysis (due to possible ADOs in PB 1)

Fig. 13.10 Sequential PB1/PB2 analysis (SPBA) and blastomere analysis report for PGD. Four embryos are tested for a given gene with three STRs on both sides of the gene. Parental haplotypes are built according to the genotype of the affected child. STR allele information (basepair length of the PCR product in the fragment analysis) on the affected chromosomes are given in *boldface*. For blastomere results, paternal allele information is written on the left and the maternal information on the right. Results of shared alleles between the parents are written in *italics* and *underlined*. *Embryo 1*: Heterozygous PB1 and affected PB2 combination gives a normal oocyte. No blastomere biopsy is requested. and the embryo is found suitable for embryo transfer. Depending on the paternal affected or normal chromosome contribution, the embryo will be carrier or normal, respectively. *Embryo 2*: Homozygously affected PB1 and normal PB2 combination gives a normal oocyte. Because of possible multiple ADOs in PB1 blastomere, a biopsy is requested. Blastomere analysis confirms the SPBA results and indicates the presence of a normal maternal chromosome together with an affected paternal chromosome. Embryo is diagnosed as a carrier and found suitable for embryo transfer. *Embryo 3*: Heterozygous PB1 and normal PB2 combination produces an affected oocyte. Blastomere biopsy is requested. Blastomere analysis confirms the SPBA results and indicates the presence of an affected maternal chromosome together with an affected paternal chromosome. Embryo is diagnosed as affected and found not suitable for embryo transfer. *Embryo 4*: Heterozygous PB1 and no PB2 results combination produces an uninterpretable oocyte. Failed amplification of PB2 may be a PCR failure, or no chromosome is present due to a meiotic segregation error. Blastomere biopsy is requested. Blastomere analysis shows the presence of both maternal chromosomes together with a normal paternal chromosome. Embryo is diagnosed as trisomic and is found not suitable for embryo transfer

- PB1 is heterozygous, and PB2 has a mutated gene: Both mutated genes are discarded to the polar bodies, one by one; and a normal gene is left in the oocyte.
- PB1 is homozygously affected, and PB2 has a normal gene: Both mutated genes are discarded to PB1; and a normal gene is present in PB2 and is the final state of the oocyte.

Homozygous PB1 might be heterozygous with ADOs on the other chromosome, which may lead to misdiagnosis. The chance of this type of misdiagnosis should be considered when predicting the oocyte genotype. Inclusion of more STRs to the PB testing can strengthen the sensitivity of the genetic test and lower the interference of this kind of testing error in the reports. Even if the oocyte is deduced to be normal, some PGD centers prefer to continue with blastomere analysis when PB1 is homozygously affected and PB2 is normal.

Homozygous PB1 results might come from a sister chromatid instead of a chromosome due to a meiosis I error. Depending on the segregation of chromosomes in meiosis II, the resulting embryo may have trisomy of the related chromosome.

Blastomere analysis gives both parental contributions to the embryo. Having prior SPBA information would be helpful for some embryos when the blastomere results make it difficult to define the exact maternal contribution owing to the shared STR alleles between parents.

13.4.2 Advantages of Haploid Genome Analysis

Haploid genome (PB2 and single sperm) analysis (HGA) has the advantages of building the haplotype of parental chromosomes in families having a de novo mutation or no pedigree information. In sequential PB1 and PB2 analyses, PB2 allows defining which maternal chromosome is affected even if no linkage data are known before the IVF cycle. This advantage of SPBA partly depends on the number of the PB sets in a given IVF cycle and the informative value of the results. The paternal haplotype of the couples with no linkage information can be found by single sperm analysis (Figs. 13.11, 13.12) (Tur-Kaspa et al. 2004).

Additional information about the main copy of the gene and the genetic position of the highly homologous second gene copy or pseudogene(s) can be obtained with HGA (Ozen 2005a). The presence of a highly homologous copy of the gene or a pseudogene(s) necessitates differentiation of the active gene status from other copies. Once this problem is overcome, the diagnostic value of PCR analysis may become more reliable (Daniels et al. 2001). Since the presence or absence of a pseudogene copy on chromosomes carrying an active or deleted one is known, pseudogene detection may have a predictive value as it occurs in the SMN1 and SMN2 models. Genomic organization of SMN genes and flanking STRs are shown in Fig. 13.13. SMN 1 and SMN2 differentiation by restriction enzymes is shown in Fig. 13.14 (Ozen et al. 2003).

Sperm #	UGT1A1 877 T/A	UGT1A1 878-890 del ACATTAATGCTTC	D2S331	D2S1279	D2S2348	D2S336	D2S338	GENOTYPE
1	-	-	-	-	-	-	-	No Results
2	A	878-890 del ACATTAATGCTTC	148	ADO	187	189	99	AFFECTED
3	-	-	-	-	-	-	-	No Results
4	ADO	878-890 del ACATTAATGCTTC	148	155	187	185	97	AFFECTED, and Recombinant
5	T	NORMAL	ADO	188	ADO	185	97	NORMAL
6	-	-	-	-	-	-	-	No Results
7	-	-	-	-	-	-	-	No Results
8	A	878-890 del ACATTAATGCTTC	148	155	187	189	99	AFFECTED
9	T	NORMAL	139	188	183	185	97	NORMAL
10	T	NORMAL	139	188	183	ADO	97	NORMAL
11	-	-	-	-	-	-	-	No Results
12	ADO	NORMAL	139	188	183	185	97	NORMAL
13	-	-	-	-	-	-	-	No Results
14	-	-	-	-	-	-	-	No Results
15	T	NORMAL	148	155	183	185	97	NORMAL, and Recombinant
16	-	-	-	-	-	-	-	No Results
17	T	NORMAL	139	188	183	185	97	NORMAL
18	T	NORMAL	ADO	188	183	189	99	NORMAL, and Recombinant
19	-	-	-	-	-	-	-	No Results
20	-	-	-	-	-	-	-	No Results
21	A	878-890 del ACATTAATGCTTC	148	155	187	189	99	AFFECTED
22	A	878-890 del ACATTAATGCTTC	148	ADO	187	ADO	99	AFFECTED
23	-	-	-	-	-	-	-	No Results
24	A	878-890 del ACATTAATGCTTC	148	155	187	189	99	AFFECTED
25	A	878-890 del ACATTAATGCTTC	148	155	187	189	99	AFFECTED

Fig. 13.11 Single sperm analysis of *UGT1A1* gene 877 T/A and 878–890 del ACATTAATGCTTC mutations with linked STRs. Single sperms obtained from the father are tested for *UGT1A1* gene 877 T/A and 878–890 del ACATTAATGCTTC mutations together with D2S331, D2S1279, D2S2348, D2S336, and D2S338 STRs. Sperm numbers 2, 4, 8, 21, 22, 24, and 25 were found to have mutation, whereas 5, 9, 10, 12, 15, 17, and 18 were wild type (normal). STR alleles corresponding to these sperms revealed the haplotype of paternal chromosomes. Sperm numbers 1, 3, 6, 7, 11, 13, 14, 16, 19, 20, and 23 showed failed PCR amplification or ambiguous results and were taken into consideration. Sperm numbers 4, 8, and 18 were recombinant

A pedigree of a spinal muscular atrophy (SMA) family where the haplotyping of parents is possible is shown in Fig. 13.15. Embryo transfer decision for SMA requires detection of at least one normal SMN1 allele in blastomeres or prediction of the existence of maternal SMN1 retained in oocytes by SPBA. Combination of single sperm analysis and SPBA of SMN1 and SMN2 together with linked STR markers reduces the possible consideration of ADO or failed amplification interpretation and provides the opportunity to differentiate carriers from normals (Figs. 13.16, 13.17, 13.18, 13.19). Possible combinations of PB2 results are shown in Fig. 13.20.

13.5 Chromosome Testing

Chromosome numerical and structural abnormalities are diagnosed with fluorescently labeled chromosome-specific probes using fluorescence in situ hybridization (FISH). For chromosome analysis of translocations or other rearrangements, a blood

CASE 1 Single Sperms	STRs			SMN	
	D5S1556	D5S610	D5S351	EXON 7	EXON 8
11 (29%)	? / 130	106	132	- / *SMN2*	- / *SMN2*
18 (47%)	124 / 134	102	126	*SMN1* / *SMN2*	*SMN1* / *SMN2*
9 (24%)	INCONCLUSIVE RESULTS				
Peripheral Blood DNA	124 / 134 // ? / 130	102 / 106	126 / 132	*SMN1* / *SMN2*	*SMN1* / *SMN2*

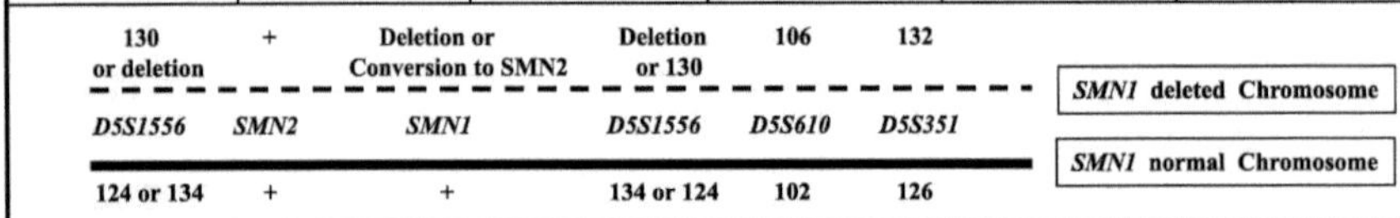

CASE 2 Single Sperms	STRs			SMN	
	D5S1556	D5S610	D5S351	EXON 7	EXON 8
15 (43%)	? / 120	114	146	- / *SMN2*	- / *SMN2*
19 (54%)	132 / del	104	160	*SMN1* / -	*SMN1* / -
1 (3%)	INCONCLUSIVE RESULTS				
Peripheral Blood DNA	132 / del // 120 / ?	104 / 114	160 / 146	*SMN1* / *SMN2*	*SMN1* / *SMN2*

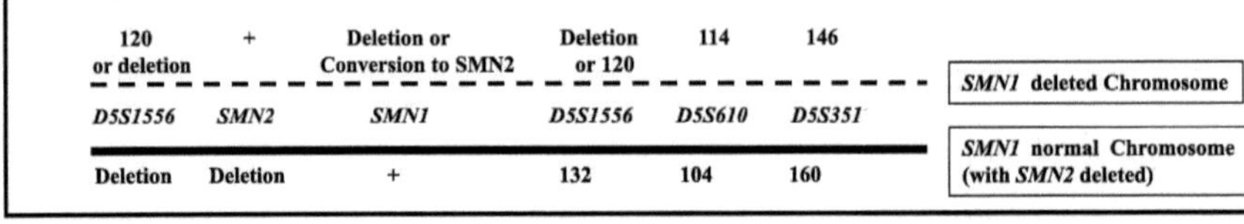

Fig. 13.12 Single sperm analysis of exon 7 and exon 8 of the *SMN* gene and flanking STRs. In single sperm haplotyping, both *SMN* gene and STR results were obtained for each chromosome and then their chromosomal position was located. In case 1, results of 29 (76%) of 38 single sperms were consistent with a haploid genome, and 9 (24%) gave inconclusive results resulting from failed amplification of more than three loci or heterozygote results because of multiple sperms in the same tube. SMN2 was found to be present at least as one copy on each chromosome. D5S1556 alleles on normal chromosomes were 124 and 134, and their location on which *SMN* gene promoter region (SMN1 or SMN2) cannot be determined by single sperm analysis. D5S1556 information on deleted (or SMN2 converted) chromosome is 130, where it may represent two copies located on both promoter regions or one intact copy on SMN2 and a deleted one on SMN1. In case 2, results of 34 (97%) of 35 single sperms were consistent with a haploid genome, and 1 (3%) gave inconclusive results. SMN1 intact chromosome was found to have no copy of SMN2. SMN2 was present on SMN1 deleted chromosome. As the extent of deletion is not known for the SMN2 gene, D5S1556 may represent one or two alleles on the SMN1 intact chromosome

sample from the individual carrying the trait is tested by the cytogenetics laboratory to verify the changes and to test the ability of FISH probes to detect the changes.

In FISH experiments, because of overlapping of excitation and emission wavelengths of fluorophores, DNA probe cocktails can consist of only five differently labeled probes. Using rehybridization cycles (two to three cycles), 10–12 chromosomes (e.g., X, Y, 13, 18, 21, 16, 17, 18, 15, 22) can be tested and nearly 70–90% of aneuploidies can be detected (Griffin et al. 1991; Grifo et al. 1992; Munné et al. 1993; Colls et al. 2007; Munné et al. 2010).

To detect abnormalities for every pair of chromosomes there is a need for more DNA templates to process and standardize techniques providing objective results.

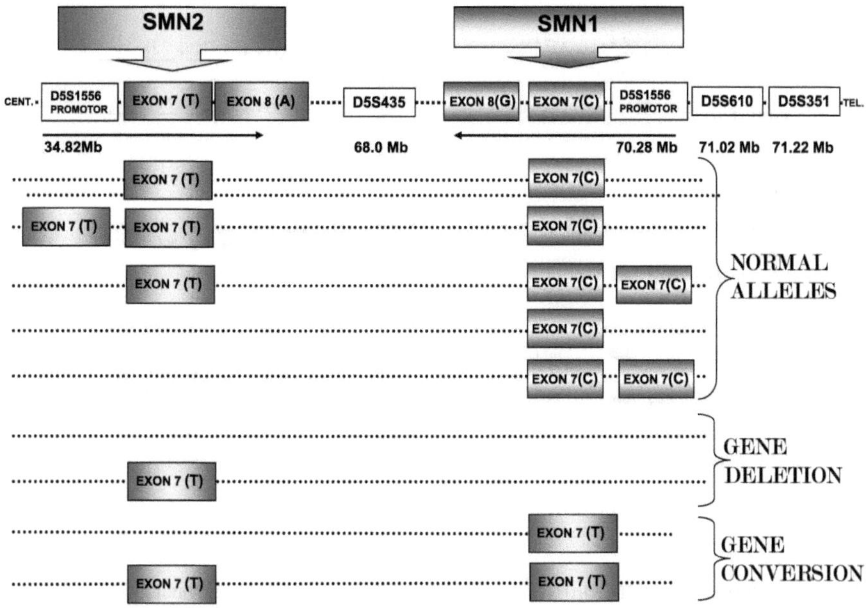

Fig. 13.13 Genomic organization of the *SMN* genes and flanking STRs. *SMN* is duplicated on chromosome 5q11.2-13, giving two inverted copies expressing identical proteins. Both gene copies (SMN1 telomeric copy and SMN2 centromeric copy) are >99% identical and a single nonpolymorphic nucleotide difference (C in SMN1; T in SMN2) is responsible for the alternative splicing patterns. D5S435 (proximal), D5S610 (distal), and D5S351(distal) STRs are single copy markers; and D5S1556 (Ag1-CA) is a multicopy dinucleotide marker located on the promoter region of both SMN1 and SMN2. Normal alleles may have multiple copies of SMN1 with or without SMN2. Affected alleles include a lack of exon 7 of SMN1; the extent of deletion might include only exon 7 or the SMN1 gene together with other neighboring sequences. Gene conversion of SMN2 can cause SMN1 absence as well. Homozygous deletion of SMN1 exon 8 without a homozygous deletion of SMN1 exon 7 does not constitute a disease allele. However, SMN1 exon 7 deletions are usually with exon 8 deletions. For PGD, in addition to its involvement in deletion or conversion mutation, exon 8 stands as the an important marker for SMN1 analysis. When interpreting the results it has to be kept in mind that some hybrid *SMN* genes might have SMN2 exon 7 and SMN1 exon 8 (or vice versa). As this hybrid gene may not be detectable at the DNA (from peripheral blood) level, *SMN* exon 8 should be considered a marker more than a mutation

Whole genome amplification (WGA) protocols can provide enough DNA for hybridization purposes. There are several WGA protocols using different approaches: multiple displacement amplification (MDA) using Phi29-polymerase; the Genomeplex system; degerate oligonucleotide primed (DOP)-PCR. Genome-wide systems testing all chromosomes in a single blastomere to be used for embryo transfer in the same IVF cycle use (1) comparative experiments (comparative genomic hybridization, or CGH) on metaphase plates (Voullaire et al. 2000; Ozen et al. 2002; Wells et al. 2008, 2009) or on BAC or oligonucleotide-based arrays or (2) the data from SNP polymorphism transmission of IVF partners (both numbers and intensity) on SNP arrays (Vanneste et al. 2009; Johnson et al. 2010). CGH is also possible for

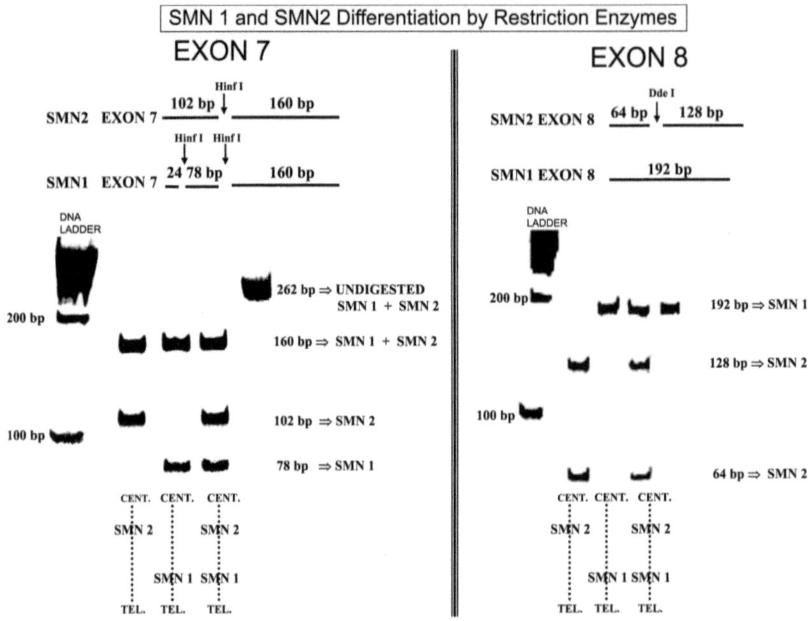

Fig. 13.14 SMN1 and SMN2 differentiation by restriction enzymes. Single nucleotide differences between exon 7 and exon 8 of SMN1 and SMN2 were used to differentiate both copies. PCR amplification involving exon 7 harbors a Hinf I site at the +6 nucleotide position in SMN1, where C→T transition creates SMN2 allele and changes Hinf I site. Exon 8 nucleotide difference in SMN2 creates the DdeI digestion site (Daniels et al. 2001)

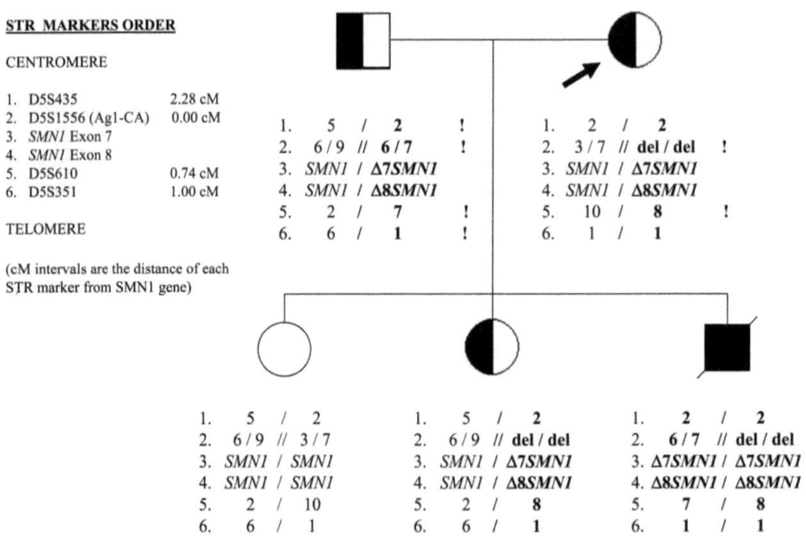

Fig. 13.15 Pedigree of a spinal muscular atrophy family with parental haplotype information obtained from siblings. Δ7SMN1: exon 7 deletion of SMN1; Δ8SMN1: exon 8 deletion of SMN1. Informative STRs, which can be used for PGD, are noted by "!" on the right side of the alleles. In this family, the maternal affected chromosome (harboring deleted SMN1) does not carry a D5S1556 allele, whereas the paternal affected chromosome carries two alleles

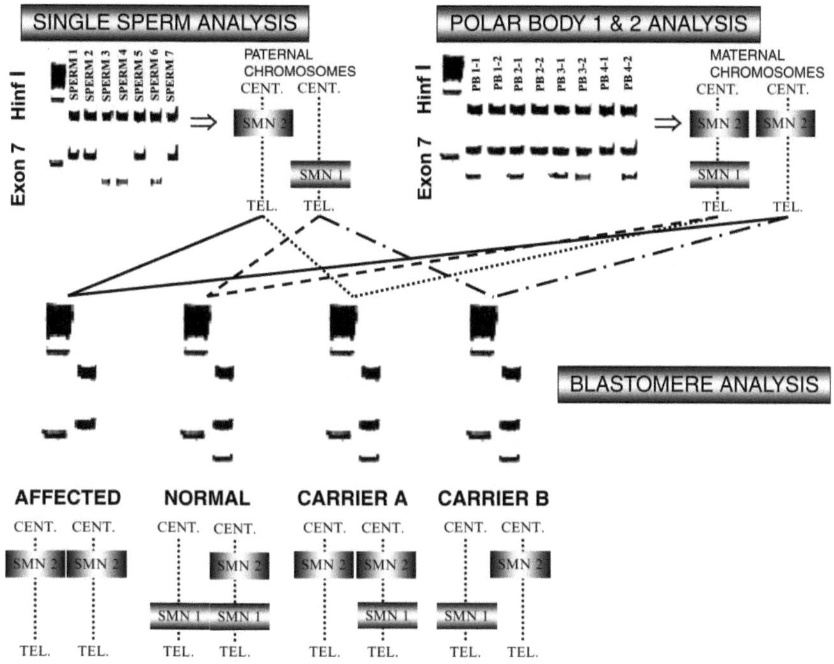

Fig. 13.16 Normals (two copies of SMN1) can be differentiated from carriers (one copy of SMN1). Carrier A has one more SMN2 than carrier B

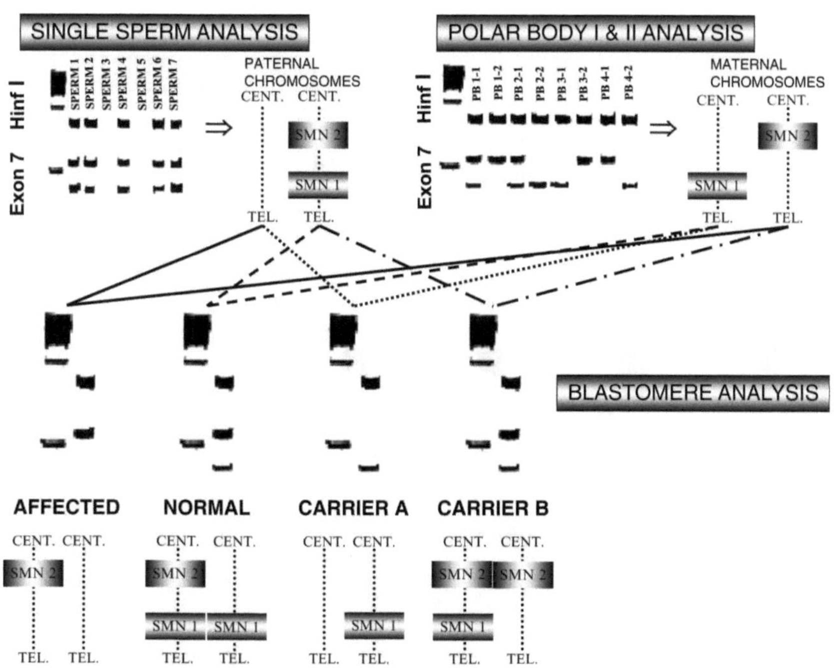

Fig. 13.17 Normals (two copies of SMN1) can be differentiated from carriers (one copy of SMN1). Carrier B has two more SMN2 than carrier A

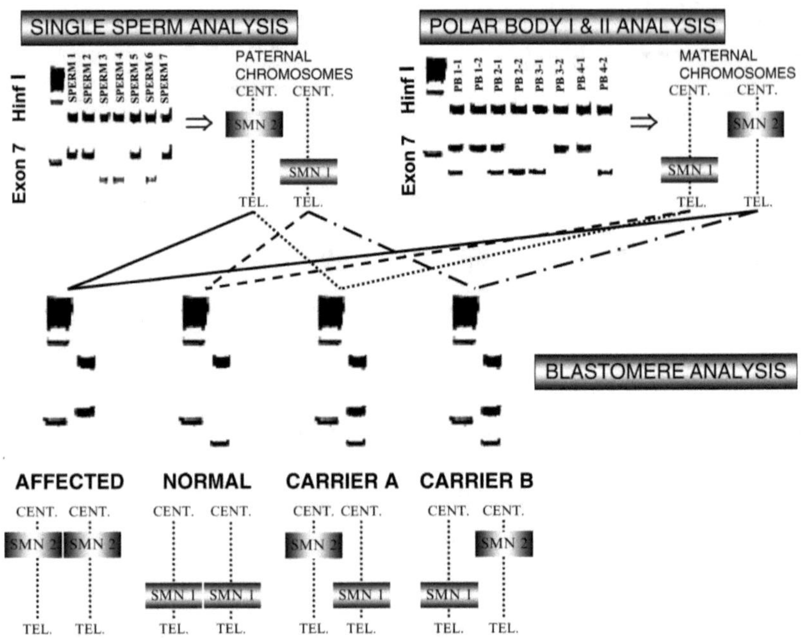

Fig. 13.18 Normals (two copies of SMN1) can be differentiated from carriers (one copy of SMN1) by the existence of SMN2

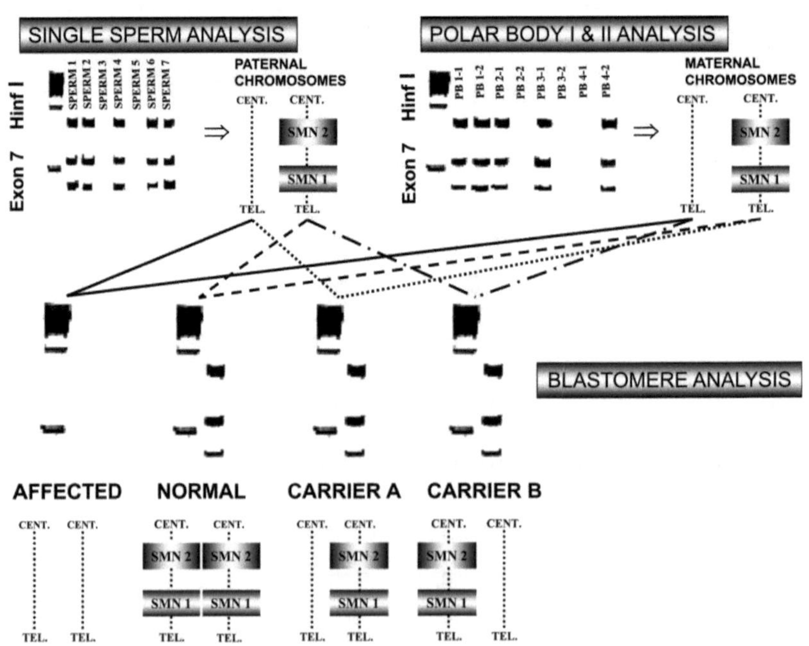

Fig. 13.19 Normals (two copies of SMN1) can be differentiated from carriers

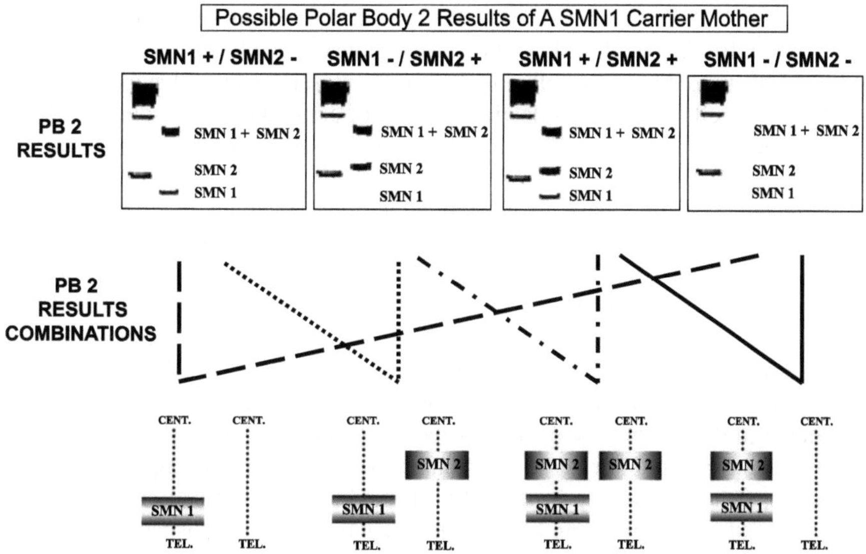

Fig. 13.20 Possible combinations of PB2 results for SMN1 and SMN2. PB2 results for an SMN1 carrier mother can have four possible results. SMN1 positive or negative chromosomes might harbor an SMN2 or not. SMN2 presence on each chromosome can be determined by the combination of PB2 results

blastocyst analysis when vitrification of the embryo is performed for later transfer (Sher et al. 2009). Array technologies require bioinformatics software programs, which run optimized algorithms for single cells developed by analysis of large numbers of samples.

13.6 Affected Embryonic Stem Cell Lines

Human embryonic stem cells (hESCs) are pluripotent cells derived from the inner cell mass of blastocyst-stage preimplantation embryos (Thomson et al. 1998). Embryos of poor quality or at the morula stage and single blastomeres are other sources for hESC establishment (Mitalipova et al. 2003; Strelchenko et al. 2004; Klimanskaya et al. 2007). Human ESCs are undifferentiated cells with the potential to differentiate into three germ layers (endoderm, mesoderm, ectoderm) or trophoblasts (Reubinoff et al. 2000; Gerami-Naini et al. 2004). The hESC lines with different DNA content (normal, carrier, affected, and different polymorphic combinations) will be needed in the future for research and clinical use with regenerative purposes. PGD is a valuable source of embryos to make hESC lines both diseased and normal by using discarded and surplus embryos, respectively (Galat et al. 2004; Kukharenko et al. 2004; Pickering et al. 2005; Verlinsky et al. 2005;

Mateizel et al. 2006; Eiges et al. 2007; Peura et al. 2008; Tropel et al. 2010; Frumkin et al. 2010; Bradley et al. 2011). PGD-tested embryos can be an alternative source for normal euploid lines as well, possibly due to mosaic embryos having aneuploid and euploid cells or in vitro selection in favor of euploid cells (Lavon et al. 2008; Taei et al. 2010). PGD can also provide a selective advantage when choosing the embryos to create hESC lines.

Affected hESC lines have great value in the study of certain disorders compared to animal or cellular models. Animal models do not fully represent the monogenic diseases or chromosomal disorders in humans due to differences at the developmental, anatomical, and gene expression levels; and there are some biochemical differences as well. However, affected hESC lines can serve as in vitro models for the disease phenotype at both the molecular and cellular levels. Cell culture models utilize fibroblasts or endothelial cells obtained from affected individuals and are restricted to the availability of the desired cells to be biopsied. Thus, hESC lines give us another option for nerve cell studies when cell biopsy is impossible. hESC lines propagate indefinitely compared to the short lifetime of cell cultures, which can be transformed for a long life-span but then have a cancerous nature and additional chromosomal anomalies.

The hESC lines might serve as a powerful tool for studying cell differentiation, developmental biology (especially in the early stages of embryogenesis, including X inactivation and gene switching mechanisms), cell replacement therapy of degenerative diseases, pharmacogenomic testing (Stephenson et al. 2009), and gene therapy experiments. The negative effects of chromosome numerical abnormalities on cell viability and cancer multi-step processes can be studied with hESC lines.

Many hESC line banks (e.g., diseased lines with some certain polymorphisms, different HLA haplotypes) will be needed in the future, organizing "hESC working groups," which would bring nationwide and international institutions together, would allow more researchers and clinicians to contribute to expanding the sample types and developing possible future usage.

References

Bradley CK, Scott HA, Chami O et al (2011) Derivation of Huntington's disease-affected human embryonic stem cell lines. Stem Cell Dev 20(3):495–502

Cieslak J, Tur-Kaspa I, Ilkevitch Y et al (2006) Multiple micromanipulations for preimplantation genetic diagnosis do not affect embryo development to the blastocyst stage. Fertil Steril 85:1826–1829

Cohen J, Wells D, Munné S (2007) Removal of 2 cells from cleavage stage embryos is likely to reduce the efficacy of chromosomal tests that are used to enhance implantation rates. Fertil Steril 87:496–503

Colls P, Escudero T, Cekleniak N et al (2007) Increased efficiency of preimplantation genetic diagnosis for infertility using "no result rescue". Fertil Steril 88:53–61

Daniels G, Pettigrew R, Thornhill A et al (2001) Six unaffected live births following preimplantation diagnosis for spinal muscular atrophy. Mol Hum Reprod 7:995–1000

de Boer KA, Catt JW, Jansen RP et al (2004) Moving to blastocyst biopsy for preimplantation genetic diagnosis and single embryo transfer at Sydney IVF. Fertil Steril 82:295–298

Eiges R, Urbach A, Malcov M et al (2007) Developmental study of fragile X syndrome using human embryonic stem cells derived from preimplantation genetically diagnosed embryos. Cell Stem Cell 1:568–577

Escribá MJ, Zulategui JF, Galán A et al (2008) Vitrification of preimplantation genetically diagnosed human blastocysts and its contribution to the cumulative ongoing pregnancy rate per cycle by using a closed device. Fertil Steril 89:840–846

Frumkin T, Malcov M, Telias M et al (2010) Human embryonic stem cells carrying mutations for severe genetic disorders. In Vitro Cell Dev Biol Anim 46:327–336

Galat V, Strelchenko N, Ozen S et al (2004) Human embryonic stem cells from embryos affected by genetic diseases. Presented at the 2nd annual meeting of the international society for stem cell research, Boston, 10–13 June 2004

Gerami-Naini B, Dovzhenko OV, Durning M et al (2004) Trophoblast differentiation in embryoid bodies derived from human embryonic stem cells. Endocrinology 145:1517–1524

Griffin DK, Handyside AH, Penketh RJ et al (1991) Fluorescent in-situ hybridization to interphase nuclei of human preimplantation embryos with X and Y chromosome specific probes. Hum Reprod 6:101–105

Grifo JA, Boyle A, Tang YX et al (1992) Preimplantation genetic diagnosis: in situ hybridization as a tool for analysis. Arch Pathol Lab Med 116:393–397

Johnson AD, Bhimavarapu A, Benjamin EJ et al (2010) CLIA-tested genetic variants on commercial SNP arrays: potential for incidental findings in genome-wide association studies. Genet Med 12(6):355–363

Johnson DS, Genelos G, Baner J et al (2010) Preclinical validation of a microarray method for full molecular karyotyping of blastomeres in a 24-h protocol. Hum Reprod 25:1066–1075

Klimanskaya I, Chung Y, Becker S et al (2007) Derivation of human embryonic stem cells from single blastomeres. Nat Protoc 2:1963–1972

Kokkali G, Vrettou C, Traeger-Synodinos J et al (2005) Birth of a healthy infant following trophectoderm biopsy from blastocysts for PGD of beta-thalassaemia major. Hum Reprod 20:1855–1859

Kukharenko V, Strelchenko N, Ozen S et al (2004) Panel of human embryonic stem cell lines. Presented at the 2nd annual meeting of the international society for stem cell research, Boston, 10–13 June 2004

Kuwayama M (2007) Highly efficient vitrification for cryopreservation of human oocytes and embryos: the cryotop method. Theriogenology 67:73–80

Lavon N, Narwani K, Golan-Lev T et al (2008) Derivation of euploid human embryonic stem cells from aneuploid embryos. Stem Cells 26:1874–1882

Mateizel I, De Temmerman N, Ullmann U et al (2006) Derivation of human embryonic stem cell lines from embryos obtained after IVF and after PGD for monogenic disorders. Hum Reprod 21:503–511

McArthur SJ, Leigh D, Marshall JT et al (2005) Pregnancies and live births after trophectoderm biopsy and preimplantation genetic testing of human blastocysts. Fertil Steril 84:1628–1636

Mitalipova M, Calhoun J, Shin S et al (2003) Human embryonic stem cell lines derived from discarded embryos. Stem Cells 21:521–526

Munné S, Lee A, Rosenwaks Z et al (1993) Diagnosis of major chromosome aneuploidies in human preimplantation embryos. Hum Reprod 8:2185–2192

Munné S, Fragouli E, Colls P et al (2010) Improved detection of aneuploid blastocysts using a new 12 chromosome FISH test. Reprod Biomed Online 20:92–97

Ozen RS (2005a) Advantage of haploid genome analysis. Presented at the 6th international symposium on preimplantation genetics, London, 19–21 May 2005

Ozen RS, Rechitsky S, Verlinsky Y (2002). Optimisation of high-resolution single cell comparative genomic hybridization to fit into a PGD framework. Presented at the 4th international symposium on preimplantation genetics, Limassol, 10–13 April 2002

Ozen RS, Rechitsky S, Sharapova T et al (2003) Polar body and blastomere analysis for spinal muscular atrophy; SMN1 and SMN2 exon 7–8 deletion detection together with single-copy and multi-copy flanking STR markers. Presented at the 53th annual meeting of the American society of human genetics, Los Angeles, 4–8 November 2003

Ozen RS, Rechitsky S, Sharapova T et al (2005b) Polar body and blastomere analysis for fragile X syndrome with inside gene and flanking VNTR and SNP. Presented at the 6th international symposium on preimplantation genetics, London, 19–21 May 2005

Ozen RS, Karadayi H, Kervancioglu E et al (2009) Preimplantation genetic diagnosis with blastomere analysis for Crigler-Najjar syndrome; UGT1A1 analysis with aneuploidy testing. Presented at the 9th international conference on preimplantation genetics, Miami, 23–25 April 2009

Peura T, Bosman A, Chami O et al (2008) Karyotypically normal and abnormal human embryonic stem cell lines derived from PGD-analyzed embryos. Cloning Stem Cells 10:203–216

Pickering SJ, Minger SL, Patel M et al (2005) Generation of a human embryonic stem cell line encoding the cystic fibrosis mutation deltaF508, using preimplantation genetic diagnosis. Reprod Biomed Online 10:390–397

Rechitsky S, Kuliev A, Sharapova T et al (2004) PCR-based aneuploidy testing for chromosome 2, 3, 4, 5, 6, 7, 8, 9, 11, 12, 13, 15, 16, 17, 18, 19, 21, and 22 in cleaving embryos combined with preimplantation genetic diagnosis (PGD) for mendelian disorders. Presented at the 54th annual meeting of the American society of human genetics, Toronto, 26–30 October 2004

Reubinoff BE, Pera MF, Fong CY et al (2000) Embryonic stem cell lines from human blastocysts: somatic differentiation in vitro. Nature Biotechnol 18:399–404

Schlenker Y, Stevens J, Rawlins M et al (2009) Clinical success with vitrification following trophectoderm biopsy for comprehensive chromosomal screening. Presented at the 65th annual meeting of the American society for reproductive medicine, Atlanta, 17–21 October 2009

Sher G, Keskintepe L, Keskintepe M et al (2009) Genetic analysis of human embryos by metaphase comparative genomic hybridization (mCGH) improves efficiency of IVF by increasing embryo implantation rate and reducing multiple pregnancies and spontaneous miscarriages. Fertil Steril 92:1886–1894

Stephenson EL, Mason C, Braude PR (2009) Preimplantation genetic diagnosis as a source of human embryonic stem cells for disease research and drug discovery. BJOG 116:158–165

Strelchenko N, Verlinsky O, Kukharenko V et al (2004) Morula derived human embryonic stem cells. Reprod Biomed Online 9:623–629

Taei A, Gourabi H, Seifinejad A et al (2010) Derivation of new human embryonic stem cell lines from preimplantation genetic screening and diagnosis-analyzed embryos. In Vitro Cell Dev Biol Anim 46:395–402

Thomson JA, Itskovitz-Eldor J, Shapiro SS et al (1998) Embryonic stem cell lines derived from human blastocysts. Science 282:1145–11477

Tropel P, Tournois J, Come J et al (2010) High-efficiency derivation of human embryonic stem cell lines following pre-implantation genetic diagnosis. In Vitro Cell Dev Biol Anim 46:376–385

Tur-Kaspa I, Rechitsky S, Ozen RS et al (2004) Sperm DNA genotyping for preimplantation genetic diagnosis (PGD). Presented at the 60th American society of reproductive medicine meeting, Phiadelphia, 16–20 October 2004

Umay B, Karadayi H, Mesci L et al (2009) Preimplantation genetics diagnosis with blastomere analysis for alpha-mannosidosis; MAN2B1 analysis with HLA typing. Presented at the 9th international conference on preimplantation genetics, Miami, 23–25 April 2009

Vanneste E, Voet T, Melotte C et al (2009) What next for preimplantation genetic screening? High mitotic chromosome instability rate provides the biological basis for the low success rate. Hum Reprod, 24(11):2679–2682. Epub 2009 Jul 24.

Verlinsky Y, Rechitsky S, Sharapova T et al (2004) Preimplantation HLA testing. JAMA 291:2079–2085

Verlinsky Y, Strelchenko N, Kukharenko V et al (2005) Human embryonic stem cell lines with genetic disorders. Reprod Biomed Online 10:105–110

Voullaire L, Slater H, Williamson R et al (2000) Chromosome analysis of blastomeres from human embryos by using comparative genomic hybridization. Hum Genet 106:210–217

Wells D, Alfarawati S, Fragouli E (2008) Use of comprehensive chromosomal screening for embryo assessment microarrays and CGH. Mol Hum Reprod 14:703–710

Wells D, Fragouli E, Alfarawaty S et al (2009) Highly significant improvement in embryo implantation and increased live birth rate achieved after comprehensive chromosomal screening: implications for single embryo transfer. Presented at the 65th annual meeting of the American society for reproductive medicine, Atlanta, 17–21 October 2009

Chapter 14
New Treatment Modalities by Disease-Specific and Patient-Specific Induced Pluripotent Stem Cells

Sibel Yildirim

Abstract The broadly accepted and deeply rooted belief in developmental biology was that terminally differentiated cells had lost the potential to produce other cell types. In 2006, however, mouse somatic cells were reprogrammed as induced pluripotent stem (iPS) cells that resembled embryonic stem cells. This therapeutic promise is being challenged by thousand of researchers worldwide to understand the ability of these cells to reverse biological clocks. Utilizing both "forward" and "reverse" genetic approaches with the aid of iPS cells offers exciting prospects for dissecting molecular mechanisms of commitment and differentiation in a cell lineage. This discovery will help clarify our understanding of the rewired regulatory networks active in somatic and pluripotent cells.

14.1 Introduction

During development, there is a gradual loss of differentiative potency, proceeding from totipotency to pluripotency and multipotency, in committed cell lineages toward terminal differentiation (Hemberger et al. 2009). The theory was broadly accepted and deeply rooted for many years in developmental biology that once a cell has terminally differentiated and become lineage-committed, it loses the potential to produce other cell types. However, the field was surprised when Takashi and Yamanaka (2006) reprogrammed mouse somatic cells into "induced pluripotent stem (iPS) cells" by the viral expression of four transcription factors: OCT4, SOX2, KLF4, c-MYC. Thus, today many belief systems in biology are shifting to accept

S. Yildirim, D.D.S., Ph.D. (✉)
Faculty of Dentistry, Department of Pediatric Dentistry, Selcuk University, Kampus 42031, Konya, Turkey
e-mail: ysibel@gmail.com

K. Turksen (ed.), *Adult and Embryonic Stem Cells*, Stem Cell Biology and Regenerative Medicine, DOI 10.1007/978-1-61779-630-2_14,

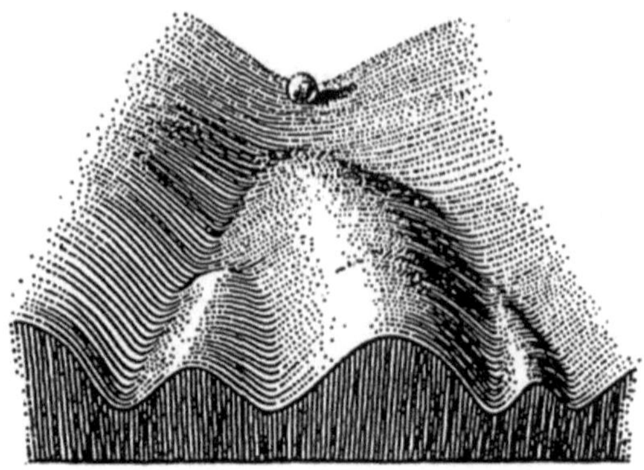

Fig. 14.1 In Waddington's landscape, the ball represents a totipotent fertilized egg. It will differentiate into various lineages while it rolls down the valley (Waddington 1957)

that a mature body cell can be reverted to an embryonic state without the help of eggs or embryos. In their changed identities, iPS cells are now ready to participate as exciting alternatives in the fields of disease pathogenesis, drug discovery, oncology, and cell transplantation. Since 2006, this therapeutic promise is being explored by hundred of researchers worldwide to understand the ability of these cells to reverse the biological clock.

During the late 1950s, Waddington (1959) introduced the term *epigenetics* to describe the unfolding of the development of the genetic program. To Waddington, epigenetics was not very different from embryology, but it was a theory of development proposing that the early embryo was undifferentiated but it was changed by epigenetics (Waddington 1959). His epigenetic landscape is a metaphor used to represent the way that developmental decisions are made. One common metaphor was of a ball placed on a landscape, where the shape of the landscape "attracts" the ball so it is more likely to follow certain channels and end up in certain places (Fig. 14.1). These lowest points represent the eventual fates of cells (i.e., the tissue types they form). According to this theory, cells in the embryo would evolve according to the same laws, but because of the existence of inducing signals cells in different regions would follow different pathways and end up at different attractors, which can be elegantly associated with different states of terminal differentiation. Once in its final valley, the ball cannot easily cross the mountain into neighboring valleys or return to the beginning (Waddington and Robertson 1966; Slack 2002).

However, the recent groundbreaking reversion of this assumingly and potentially irreversible developmental process by the derivation of mouse iPS cells from adult dermal fibroblasts (Takahashi and Yamanaka 2006) has surprised many cell biologists. Since the nuclear transfer (NT) experiments showed that the nucleus of most, if not all, adult cells retains nuclear plasticity and can be rebooted to an embryonic state (Byrne et al. 2007), Takahashi and Yamanaka (2006) showed that overexpression of defined transcription factors can also convert cells (or nuclei) to the pluripotent state. In this first report, they reprogrammed mouse fibroblasts through retroviral transduction with 24 transcription factors that are highly expressed in embryonic stem cells (ESCs). The combination of genes was gradually reduced to four transcription factors: octamer 3/4 (Oct4), SRY box-containing gene 2 (Sox2), Kruppel-like factor 4 (Klf4), and c-Myc (Takahashi and Yamanaka 2006). The cells were selected because of their ability to express the gene F-box protein 15 (Fbx15) and were similar to ESCs in morphology, growth properties, and ability to form teratomas in immunodeficient mice. They differed in terms of global gene expression profiles and certain DNA methylation patterns, and they failed to produce adult chimeric mice. With the improved endpoints for the reprogramming process, the resulting iPS cells were even more similar to ESCs and could contribute to adult chimeras (Okita et al. 2007). After a year, human fibroblasts were reprogrammed with the same transcription factor genes (Takahashi et al. 2007; Wernig et al. 2007; Yu et al. 2007; Park et al. 2008a). Subsequently, several groups independently replicated the reprogramming of human fetal, neonatal, and adult somatic cells into iPS cells (Takahashi et al. 2007; Aasen et al. 2008; Lowry et al. 2008; Park et al. 2008a, b). Sophisticated techniques for making and testing iPS cells have advanced rapidly over the past few years, and finally iPS cells were obtained from patients suffering from the neurodegenerative disease amyotrophic lateral sclerosis (Ralston and Rossant 2010; Dimos et al. 2008) as well as patients with other diseases, including juvenile onset type 1 diabetes mellitus, Parkinson's disease (PD) (Park et al. 2008c), and spinal muscular atrophy (SMA) (Ebert, Yu et al. 2009).

14.2 Modeling Human Diseases "in a Dish"

The main idea for therapeutic approaches by iPS cells at the beginning was the fact that patient-specific iPS cells provide important information for inherited human disorders because pluripotent stem cells are capable of differentiating into most, if not all, cell types. This idea is deeply reliant on the studies of directed differentiation of subtypes and genetically defined ESCs from animal models (Gearhart 1998). Moreover, human ESC biology has been pursuing the generation of mutant human ESC lines as disease models since Thomson et al. (Thomson et al. 1998) derived human ESC lines in 1998. With the known disease-associated genetic loci and explicit disease phenotype, genetically modified human ESCs are being explored for use in cell replacement therapy and for modeling human diseases (Saha and Jaenisch 2009).

However, the use of ESCs has several limitations, not only in regard to political, religious, ethical, and moral concerns about the need to destroy human embryos but also the inefficiency of the methods used to generate genetically modified human ESCs. For example, human ESC generation via preimplantation genetic diagnosis (PGD) embryos is available for only a limited number of diseases, and lack of proper techniques is still challenging to human ESC genetic modifications. Also, only a few monogenic diseases are detectable via PGD, and there is no consistency between the severity and clinical symptoms of those diseases from patient to patient due to variable penetrance (Colman and Dreesen 2009). Other alternatives, such as generating individual pluripotent stem cells by NT, cell fusion with ESCs, or treatment with extracts of pluripotent cells are highly restrictive for several reasons, and there are still only a few diseases that have been explored in these ways (Wakayama et al. 2001; Cowan et al. 2005; Taranger et al. 2005).

The other alternative—animal models for human diseases—have been used for decades. However, they also have limitations such as showing no or only an approximate resemblance to the human disease, differences in physiology and anatomy between animals and humans, no mirroring for cognitive or behavioral defects of neurological diseases, and the different genetic backgrounds of animals and humans in terms of the resulting phenotype of disease-associated mutations (Colman and Dreesen 2009; Saha and Jaenisch 2009).

To overcome these drawbacks, iPS cells offer disease- and patient-specific cells with knowledge of the clinical history of the donor, and they can be made with cells taken from persons of all ages, even elderly patients with chronic disease (Dimos et al. 2008). Although there are still many challenges regarding their identity, a couple reports are available for an overview of a disease phenotype in vitro (Ebert et al. 2009; Lee et al. 2009; Raya et al. 2009; Ye et al. 2009b). Because human ESC lines display variable outcomes in regard to differentiation into specific lineages (Osafune et al. 2008), multiple iPS cell lines generated from a single patient are favored because they have an identical genetic background.

14.3 Disease-Specific iPS Cells

Disease-specific iPS cells have been generated from individuals with such disorders as neurodegenerative disease, including ALS (Dimos et al. 2008), Parkinson's disease (Soldner et al. 2009), SMA (Ebert et al. 2009), familial dysautonomia (Lee et al. 2009), and inherited diseases, including adenosine deaminase deficiency-related severe combined immunodeficiency, Shwachman-Bodian-Diamond syndrome, Gaucher's disease type III, Duchenne and Becker muscular dystrophy, Huntington's disease, juvenile-onset type 1 diabetes mellitus, Down syndrome/trisomy 21, the carrier state of Lesch-Nyhan syndrome (Park et al. 2008c), and Fanconi anemia (Raya et al. 2009) (Table 14.1).

Saha and Jaenisch (2009) have reported existing human iPS cell lines from several diseases. According to their classification there are many disease-specific iPS cell lines available. Specific iPS cells are available for familial neurodegenerative diseases such as dysautonomia (Lee et al. 2009), SMA type 1 (Ebert et al. 2009), Huntington's disease (Park et al. 2008c), ALS (Dimos et al. 2008); from sporadic neurodegenerative diseases such as Parkinson's disease (Park et al. 2008a; Soldner et al. 2009); and from the neurodevelopmental disease group such as Rett syndrome (Hotta et al. 2009) and Down syndrome (Park et al. 2008a). Although nothing has been published on neurobehavioral/physiological disease-specific human iPS cells for hematologicaly/oncological or endocrinological diseases, several human iPS cell

Table 14.1 Disease-specific iPS cells

Disease model	Genetic disorder	Cell source	Reprogramming method	Detection of specificity of iPS line	Correction	Differentiation into disease-relevant cell types	Reference
Down Syndrome (DS)	Trisomy 21 anomaly	Fibroblast line	Retroviruses carrying four factors (*OCT4, SOX2, KLF4, MYC*)	47, XZ +21 kartotype	No attempt	Patient-specific cells	Park et al. (2008c)
Adenosine deaminase deficiency (ADA)	Mutations in ADA gene	Fibroblast line	Retroviruses carrying four factors (*OCT4, SOX2, KLF4, MYC*)	GGG to GAA transition mutation at exon 7	No attempt	Patient-specific cells	Park et al. (2008c)
Shwachman-Bodian-Diamond Syndrome		Bone marrow mesenchymal cells	Retroviruses carrying four factors (*OCT4, SOX2, KLF4, MYC*)	Point mutations at the IV2+2T>C intron 2 splice donor site and IVS3-1G>A mutation	No attempt	Patient-specific cells	Park et al. (2008c)
Gaucher disease type III	Mutations in acid β-glucosidase gene	Fibroblast line	Retroviruses carrying four factors (*OCT4, SOX2, KLF4, MYC*)	1226A>G point mutation	No attempt	Patient-specific cells	Park et al. (2008c)
Lesch-Nyhan syndrome	Heterozygous deficiency of the *HPRT* gene	Female carrier fibroblast line	Five doxycycline-inducible lentiviral vectors (*OCT4, SOX2, KLF4, cMYC, NANOG*)	Heterozygosity of HPRT1 gene	No attempt	Patient-specific cells	Park et al. (2008c)
Duchenne muscular dystrophy	Dystrophin gene	Dermal fibroblasts	Retroviruses carrying four factors (*OCT4, SOX2, KLF4, MYC*)	Several intragenic intervals, deletion of exons 45–52	No attempt		Park et al. (2008c)

(continued)

Table 14.1 (continued)

Disease model	Genetic disorder	Cell source	Reprogramming method	Detection of specificity of iPS line	Correction	Differentiation into disease-relevant cell types	Reference
Huntington disease	Huntington gene		Retroviruses carrying four factors (*OCT4, SOX2, KLF4, MYC*)	Presence of expanded (CAG) n polyglutamine triplet repeat sequences in the proximal portion of the related gene	No attempt	Patient-specific cells	Park et al. (2008c)
Parkinson disease	No defined genetic basis	Fibroblasts (Coriell Cell Repository)	DOX-inducible lentiviruses transducing either four (OCT4, SOX2, c-MYC, KLF4) or three (OCT4, SOX2, KLF4)	No genetic verification		Generation dopaminergic neurons	Soldner et al. (2009)
Amyotrophic lateral sclerosis	For a slowly progressing form of ALS L144F polymorphism of the superoxide dismutase (SOD1) gene	Skin fibroblasts	Moloney-based retroviruses carrying KLF4, SOX2, OCT4, and c-MYC	L144F (Leu144 → Phe) dominant allele of the *SOD1* gene		Generation of spinal motor neurons and glia	Dimos et al. (2008)
Spinal muscular atrophy (SMA)	Mutations in the survival motor neuron 1 gene (SMN1)	Skin fibroblasts	Lentiviruses encoding OCT4, SOX2, NANOG and LIN28	Significantly reduced SMN protein levels due to loss of the *SMN1* gene		Generation of motor neurons	Ebert, Yu et al. (2009)

Familial dysautonomia	Point mutation in the *IKBKAP8* gene	Skin fibroblasts	Lentiviruses encoding OCT4, SOX2, KLF4 and c-MYC	Mis-splicing of *IKBKAP*	Patient-specific cells	Lee et al. (2009)
Fanconi anemia		Skin fibroblasts	Lentiviruses encoding OCT4, SOX2, KLF4 and c-MYC		Patient-specific cells	Raya et al. (2009)
β-Thalassemia	*β-globin* gene mutation	Skin fibroblasts	Retroviruses carrying OCT4, SOX2, KLF4 and c-MYC	Homozygous codon 41/42 4-bp (CTTT) deletion		Ye et al. (2009a)
Myeloproliferative disorders	JAK2-V617F somatic mutation	Peripheral blood CD34 cells	Retroviruses carrying OCT4, SOX2, KLF4 and c-MYC	Homozygous codon 41/42 4-bp (CTTT) deletion	Differentiation to hematopoietic cells	Ye et al. (2009a)
Juvenile diabetes mellitus	Unknown	Skin fibroblasts	Retroviruses encoding OCT4, SOX2, KLF4		Differentiation to pancreatic beta cell	Park et al. (2008a, c), Maehr et al. (2009), Zhang et al. (2009a)
Leopard Syndrome	Mutation in the *PTPN11* gene	Skin fibroblasts	Moloney-based retroviruses carrying KLF4, SOX2, OCT4, and c-MYC	T468M *PTPN11* mutation	Differentiate into hematopoietic and cardiac lineages	Carvajal-Vergara et al. (2010)
Duchenne muscular dystrophy	Mutation in *dystrophin* gene	Skin fibroblasts	Retroviruses encoding OCT4, SOX2, KLF4, c-MYC	Deletion of exons 4–43	Complete correction of *dystrophin* deficiency	Kazuki et al. (2010)

lines have been generated, including those for β-thalassemia (Ye et al. 2009a), Fanconi anemia (Raya et al. 2009), myeloproliferative disorders (Ye et al. 2009b), and juvenile diabetes mellitus (Park et al. 2008c; Maehr et al. 2009; Zhang et al. 2009a) (Table 14.1).

14.3.1 Choosing Cell Sources

To generate patient-specific iPS cells, the first step is derivation of human iPS cells from somatic cells of a patient. iPS cells can be created with cells taken from patients of all ages with full medical records. There are more than 5000 known genetic diseases, whether simple or complex (Colman and Dreesen 2009). There are human tissues available with no ethical or surgical concerns, although they are limited. Those tissues include fat (Sun et al. 2009), blood (Ye et al. 2009b), biopsy specimens, skin, plugged hair (Aasen et al. 2008), and extracted teeth (Aasen et al. 2008; Sun et al. 2009; Ye et al. 2009b; Yan et al. 2010). Thus far in humans, skin fibroblasts and bone marrow mesenchymal cells (Takahashi et al. 2007; Yu et al. 2007; Huangfu et al. 2008), keratinocytes (Aasen et al. 2008; Maherali et al. 2008), peripheral blood cells (Loh et al. 2009; Ye et al. 2009b), melanocytes (Utikal et al. 2009), neural stem cells (Kim et al. 2009b), amniotic fluid-derived cells (Li et al. 2009), adipose stem cells from lipoaspirate (Sun et al. 2009), dental stem cells (Tamaoki et al. 2010; Yan et al. 2010), and mesenchymal stem cells from umbilical cord matrix and amniotic membrane (Cai et al. 2010) have been used for reprogramming. Frozen banked tissues and cell lines can also be used, although there is often very little clinical information on the specific donor (Colman and Dreesen 2009).

Because most disease phenotypes are observed only in differentiated cells, for many diseases only iPS cell generation can provide a source for pluripotency. Monogenic diseases are the most fruitful targets because the gene and often its product are known (Colman and Dreesen 2009). Extending this experimental paradigm to diseases with either unknown or more complex, multifactorial phenotypes or diseases, and to model disease with a long latency such as Alzheimer's or Parkinson's disease, would be challenging (Colman and Dreesen 2009). For diseases that exhibit a late onset in humans, the kinetics of the disease pathology can be stimulated in the cell culture dish by exposing the cells to in vitro experimental stress (e.g., serum starvation, oxygen reduction, heat shock) (Saha and Jaenisch 2009). On the other hand, Colman and Dreesen (2009) emphasized that "late onset" does not reflect subclinical developments that may occur much earlier and may be captured by the in vitro methodology.

It is obvious that the disease pattern directs us to alternatives for donor cell type. As the detection of mutations in diseases such as SMA is possible in all of the cell types of the patient, skin biopsies provide readily accessible donor cells. On the other hand, a heterozygous genotype of most of the hematopoietic disorders can be detected only in particular progenitors. Hence, those progenitors should be chosen as cell sources for reprogramming (Ebert et al. 2009; Saha and Jaenisch 2009; Ye et al. 2009b).

Recent reports about persistent epigenetic imprinting in iPS cells (Kim et al. 2010; Polo et al. 2010) provides an opportunity to research sporadic and multifactorial diseases. Especially for diseases that have a combination of genetic and environmental factors, persistent epigenetic memory would be advantageous. In those diseases, any epigenetic alterations would be studied via iPS cells carrying parental imprinting. The cells derived from patients suffering the same disease but living in different geographical regions could give us important clues for environmental factors, such as toxic metals and pesticides, lifestyles, and dietary habits, which may have an effect on the epigenome and reflect risk factors (Jaenisch and Bird 2003). Lastly, for non-cell-autonomous diseases, possible success with one cell type may affect the other pathological mediators (Colman and Dreesen 2009).

Although it is still unclear whether any of the iPS cell lines can be used for future cell therapy, it should be quite useful to establish clinical-grade iPS cell banks with a sufficient repertoire of HLA types. Nakatsuji (2010) estimated that a collection of unique iPS cell lines with homozygous alleles of the three HLA loci (A, B, DR) would enable full matching for 80–90% of the Japanese population with a perfect match of these three loci. In addition, Tamaoki et al. (2010) attempted to use dental pulp stem cells to generate iPS banking with a sufficient repertoire of HLA types. They also reported the possibility of identifying homozygous donors for human iPS cell lines for the construction of such HLA-type banking. The practical isolation and handling of dental pulp cells may make it easy to expand the size of the bank in multiple institutions and even establish a number of iPS cell lines homozygous for the three HLA loci (Tamaoki et al. 2010).

14.3.2 Reprogramming

The current methods used to generate iPS cells utilize retroviral/lentiviral, adenoviral, plasmid, protein delivery of transcription factors (Wang et al. 2010). The methods, which use nonintegrating vectors, excisable genetic elements, and small chemicals and/or proteins, leave no genetic footprint (Feng et al. 2009; Yu et al. 2009; Kim et al. 2009a). Those methods for iPS generation reveal the inherent robustness of this process. Yet, there are two main unresolved drawbacks: low efficiency and genomic integration. Residual transgene expression or later reactivation of exogenously applied transcription factors could cause tumor formation at an alarming rate (Okita et al. 2007; Miura et al. 2009). Incomplete transgene silencing also compromises the differentiation of iPS cells (Brambrink et al. 2008). To minimize the risk of chromosomal disruptions, reprogramming protocols are refined to eliminate genetic integration. Moreover, reprogramming efficiencies have been enhanced by supplementing defined factors with additional genes or small chemicals (for detailed information refer to the provided references) (Masip et al. 2010; Wang et al. 2010). Using fewer transcription factors omitting oncogenes (KLF4 and c-MYC) (Nakagawa et al. 2008; Utikal et al. 2009; Kim et al. 2009b) or replacing one or two transcription factors with small chemicals such as VPA (a histone deacetylase inhibitor)

(Huangfu et al. 2008), SB431542, and pd0325901 [inactivators of the respective receptors for transforming growth factor β (TGFβ) and MEK-ERK pathways] in combination with thiazovivin (Lin et al. 2009) and vitamin C (Esteban et al. 2010) not only offers safer clinical potential but also significantly enhances the efficiency of deriving iPS cells. More recently iPS cells from human fibroblasts were successfully derived utilizing adenoviral or sendai viral gene delivery systems without viral or transgene integration (Fusaki et al. 2009; Zhou and Freed 2009). Another virus-free method utilizing oriP/EBNA1 (Epstein-Barr nuclear antigen-1)-based episomal vectors and a piggyBac-based single vector reprogramming system have been used successfully for reprogramming human fibroblasts (Kaji et al. 2009; Yu et al. 2009). Moreover, reprogramming cells with proteins of four factors also offers efficient reprogramming (Zhou et al. 2009; Kim et al. 2009a).

In a recent report, Bhutani et al. (2010) provided new insights into the mechanisms regulating the path to pluripotency by demonstrating rapid, efficient induction of pluripotency-associated genes in human fibroblasts after fusion to mouse ESCs and the perturbation of this induction when a candidate factor (activation-induced cytidine deaminase, or AID) is disrupted. The latest reports also suggest NT is not only more effective at establishing the ground state of pluripotency but also safer (in terms of erasing epigenetic memory) than factor-based reprogramming (Kim et al. 2010).

More recently Warren et al. (2010) showed efficient reprogramming using synthetic modified mRNA, which does not modify the genome. Although the protocol is technically complex because the technology is RNA-based, it completely eliminates the risk of genomic integration and insertional mutagenesis inherent in all DNA-based methodologies.

Although reprogramming factors reset the cellular phenotype from the inside, it clearly also requires extrinsic signals provided by the ESC culture conditions. These conditions include growth factors, cytokines, and other signals provided by the cell culture medium, fetal bovine serum, and feeder cells. How extrinsic signals are integrated with intrinsically acting factors is not entirely clear but is an active area of investigation (Ralston and Rossant 2010).

14.3.3 Characterization of the iPS Cell Lines

The generation of patient/disease-specific iPS cells follows standard methods. In brief, target cells are infected with a reprogramming system carrying reprogramming factors. After several days, infected cells are trypsinized and replated onto feeder layer cells, and the medium is replaced with standard human ESC cell medium the next day and changed every day thereafter. When human ESC-like colonies appear, they are picked mechanically or selected by drug-inducing systems and are passaged. Pluripotency is then evaluated according to the similarity of putative iPS cells to ESCs. To confirm proper and complete reprogramming, the activity of the cell cycle profile, maintenance of the normal karyotype, alkaline phosphatase activity, and expression of several ESC-associated antigens (e.g., SSEA-3, SSEA-4, TRA1-81, Nanog), down-regulation or lack of immunreactivity against parental cell-specific factors (e.g., fibroblast-associated antigen TE-7), expression of pluripotency genes (*REX1/ZFP42*, *FOXD3*, *TERT*, *NANOG*, *CRIPTO/TDGF1*) should be checked. Moreover, the stem cell marker genes *SOX2* and *OCT4* should not be expressed in the patient donor cells, whereas the endogenous loci in the putative iPS cells should become activated to levels similar to those in ESCs (Maherali and Hochedlinger 2008). To achieve differentiated lineage-specific cells, embryoid bodies (EBs) are generally created as a first step by aggregating or placing clumps of iPS cells in suspension culture. The resulting EBs are plated onto plastic gelatin-coated dishes and allowed to attach for the outgrowth culture. Thereafter, iPS cell lines spontaneously differentiate into cell types representative of the three embryonic germ layers (Maherali and Hochedlinger 2008).

Although EB formation shows the in vitro differentiation capacity of iPS cells to have three germ layers, the differentiation capacity should also be assessed by teratoma formation assay in vivo. Immunocytochemistry analyses can be used to detect expression of smooth muscle α-actin (α-SMA), desmin, and vimentin for mesoderm, α-fetoprotein (AFP) for endoderm, glia fibrillary acidic protein (GFAP) and βIII-tubulin for ectoderm markers. To determine pluripotency in vivo, iPS cells should be injected into immunocompromised NOD-SCID mice. Histological analyses of the resulting teratomas should show cell types representative of the three germ layers, including, for example, pigmented cells for ectodermal differentiation; lung, respiratory, and gut-like epithelia for endodermal differentiation; and mesenchyme, adipose tissue, and cartilage for mesodermal differentiation (Carvajal-Vergara et al. 2010).

Although criteria and standards are important to allow cross-laboratory data comparisons, the proposed stringent criteria may not be fully required for applications in which reprogrammed human cell lines are used to model disease processes in vitro or to screen for novel drugs or drug toxicity (Ellis et al. 2009). Whereas germline competence after chimera formation offers the most stringent functional criteria for mouse iPS cells, teratoma formation in immunodeficient mice seems to be acceptable so far for human iPS cells (Daley et al. 2009). Moreover, because current teratoma assays are qualitative in nature, full quantitative assessment of differentiation capacity of generated cell lines could rapidly allow whether those lines retain the capacity for differentiation into three germ layers or have restricted differentiation capacity, even with the lines that fail at the functional level of forming a teratoma. One group of researchers agrees that the most desirable iPS cells to use for transplantation might be those that do not form teratomas in vivo but retain the capacity to differentiate to desired cell types in vitro. They proposed that this possibility can be determined only if non-teratoma-forming iPS cell lines are fully studied in vitro (Ellis et al. 2009).

It was unclear whether reprogramming female human cells reactivates the inactive X chromosome, as in the mouse. It has been shown that human iPS cells derived from several female fibroblasts carry an inactive X chromosome, in contrast to mouse iPS cells, which carry two active X chromosomes. Although those data indicate that reversal of X chromosome inactivation is not required for human cell programming, the implication is that X chromosome inactivation should be taken into consideration for the use of female iPS cells with devastating X-linked genetic diseases, such as fragile X syndrome (mutation in *FMR1*), α-thalassemia (*ATRX*), Rett syndrome (*MECP2*), Coffin-Lowry syndrome (*RSK2*), DMD, Lesch-Nyhan syndrome (*HPRT*), and Wiskott-Aldrich syndrome (*WASP*) (Tchieu et al. 2010).

14.3.4 Characterization of Genetic Mutation

The generation of iPS cells from patients with a variety of genetic diseases offers an opportunity to recapitulate both normal and pathological human tissue formation in vitro. It hence provides a tool for investigating disease pathology. If the cells taken from patients having classic mendelian inherited disorders, point mutations in known genes essential for the given function, molecular mutation analysis such as karyotyping, and fingerprinting analysis should be carried out (Park et al. 2008c).

To verify that the patient-specific iPS cell lines are genetically matched to the donor cells, DNA fingerprinting analysis of the iPS cell lines and the donor cells from which they were derived should be done. Additionally, direct sequencing and an allele-specific restriction fragment length polymorphism should be used to compare the genotype of the iPS cell line with that of the donated host cells. Furthermore, polymerase chain reaction (PCR) analysis of genomic DNA from iPS cell lines can reveal if they carry integrated copies of the transgenes that have been transduced (Dimos et al. 2008).

14.3.5 Differentiation of Obtained iPS Cells to Desired Cell Types

Beginning with the first cell lineage decisions, gene expression and mutual interactions between lineage-determining transcription factors with antagonizing functions show stochasticity (Hemberger et al. 2009). The orchestrating role of the gene regulatory network points out biological patterns for differentiation. To date, ESC studies have mostly focused on the derivation of subsets of tissue-specific cell populations. Thus, lineage-specific differentiation of murine and human ESCs has been shown as a powerful tool for studying early embryonic events and lineage restriction for generating an unlimited cellular supply for cell therapies and tissue engineering. To produce either progenitors or more mature cells, various exogenous factors were applied in a sequence and time course that is highly reminiscent of normal development.

Much of the hope invested in patient-specific stem cells is based on the assumption that it will be possible to differentiate them into disease-relevant cell types. To differentiate desired cell types in the mixed population of differentiating ESCs, there are several well-established stem cell differentiation protocols mimicking the proper timeline of normal human organogenesis. There are many well-established protocols used for studying hematopoietic, endothelial, osteoblastic, osteoclastic (Tsuneto et al. 2003; Kawaguchi et al. 2005; Grigoriadis et al. 2010), cardiac (Arbel et al. 2010), neural commitment (Ying and Smith 2003), pancreatic (Banerjee et al. 2011), hepatic (Gerbal-Chaloin et al. 2010), chondrogenic, and adipogenic (Wdziekonski et al. 2003) differentiation. Moreover, cell fate specification and maturation can be selectively altered via manipulation of endogenous developmental signaling pathways (Rathjen and Rathjen 2003; Meyer et al. 2009). The ability to differentiate iPS cells in vitro to specific lineages efficiently and reproducibly has been achieved using those described protocols with certain modifications (Dimos et al. 2008; Narazaki et al. 2008; Tateishi et al. 2008; Wernig et al. 2008; Ebert et al. 2009; Hu and Zhang 2009; Jin et al. 2009; Karumbayaram et al. 2009; Pfannkuche et al. 2009; Senju et al. 2009; Tanaka et al. 2009; Taura et al. 2009b; Zhang et al. 2009b; Carvajal-Vergara et al. 2010; Comyn et al. 2010; Dick et al. 2010; Gamm and Meyer 2010; Huang et al. 2010; Kaichi et al. 2010; Lamba et al. 2010; Lee et al. 2010; Martinez-Fernandez et al. 2010; Parameswaran et al. 2010; Swistowski et al. 2010; Teramura et al. 2010; Zhou et al. 2010).

There are many typical examples for directed differentiation of iPS cells to the cell types influenced by the disease (Tateishi et al. 2008; Meyer et al. 2009; Grigoriadis et al. 2010). In these studies, the researchers showed that it is possible to generate and fine-tune desired lineages from human somatic cells. In brief, after creating EBs from generated iPS cells, chemically defined differentiation media were used to promote the stepwise production of organ-specific cell types. Using a targeted, stepwise differentiation process that follows a normal developmental timeline, the researchers modeled cell and/or organ development with human iPS cells (Ueda et al. 2010).

Differentiation should be confirmed by showing the expression of transcription factors or surface markers. The function of those differentiated cells are the last point for evaluation (Dimos et al. 2008). Dimos et al. (2008) demonstrated that skin cells from patients with amyotrophic lateral sclerosis could be reprogrammed and subsequently differentiated into disease-free motor neurons. Also of note, Ebert et al. (2009) created iPS cells from the fibroblasts of a SMA patient and his unaffected mother. They were the first to demonstrate a preserved patient-specific disease phenotype in motor neurons generated from fibroblast iPS cells. Treatment of these cells in vitro with valproic acid and tobramycin led to up-regulation of survival motor neuron protein synthesis and revealed selective deficits when compared with normal motor neurons.

Today, many iPS cell lines that were directed into differentiated functional cell types are available. Examples are auditory retinal cells (Jin et al. 2009; Comyn et al. 2010; Lamba et al. 2010; Parameswaran et al. 2010), cardiomyocytes (Narazaki et al. 2008; Pfannkuche et al. 2009; Tanaka et al. 2009; Zhang et al. 2009b; Carvajal-Vergara et al. 2010; Kaichi et al. 2010; Martinez-Fernandez et al. 2010), insulin-secreting islet-like clusters (Tateishi et al. 2008), motor neurons (Dimos et al. 2008; Ebert et al. 2009; Hu and Zhang 2009; Karumbayaram et al. 2009), dopaminergic neurons (Wernig et al. 2008; Cai et al. 2010; Swistowski et al. 2010), auditory spinal ganglion neurons (Nishimura et al. 2009), smooth muscle cells (Taura et al. 2009b; Xie et al. 2009), vascular endothelial cells (Taura et al. 2009b), dendritic cells and macrophages (Senju et al. 2009, 2010), adipocytes (Tashiro et al. 2009; Taura et al. 2009a), osteoblasts (Tashiro et al. 2009), hematopoietic cells (Hanna et al. 2007; Eminli et al. 2009; Kaufman 2009; Lu et al. 2009; Okabe et al. 2009; Raya et al. 2009; Kaneko et al. 2010), and endothelial progenitor cells (Xu et al. 2009; Abaci et al. 2010; Alipio et al. 2010; Ho et al. 2010; Homma et al. 2010).

Hanna et al. (2007) showed that differentiated iPS cells can be used to rescue organ function in a humanized mouse model of sickle cell anemia. Also, several other groups demonstrated the therapeutic potential of iPS cells, both alone and in combination with genetic corrective therapy. The study included the generation of disease-free hematopoietic progenitors from keratinocytes obtained from patients with Fanconi anemia (Raya et al. 2009), correction of hemophilia in mice using iPS cell-derived endothelial progenitors (Xu et al. 2009), and multilineage functional repair of diseased heart tissue in immunocompetent mice using undifferentiated iPS cells (Nelson et al. 2009). Finally, functional dopamine neurons were generated from reprogrammed mouse fibroblasts, and transplantation of these neurons restored dopamine function when grafted in parkinsonian rats (Wernig et al. 2008).

To determine if any observations obtained from derived iPS cells are specific to a given cell line or to pluripotency, healthy wild-type controls should be used. Although established human ESC or iPS cell lines can be used for this purpose, an additional panel of lines derived from the same patient or unrelated patients suffering from the same disease would give valuable information. On the other hand, for single-gene diseases, genetically modified iPS cells could represent an ideal isogenic control (Colman and Dreesen 2009; Saha and Jaenisch 2009).

Utilizing both "forward" and "reverse" genetic approaches with the aid of iPS cells offers exciting prospects for dissecting molecular mechanisms of commitment and differentiation in a cell lineage. These approaches will help clarify our understanding of rewiring regulatory networks active in somatic and pluripotent cells (Hemberger et al. 2009; Huang 2009; Saha and Jaenisch 2009). Future work should focus on finding out when and how the lineage-specific genetic and epigenetic markers arise.

14.4 Similarities and Differences Between iPS and Embryonic Stem Cells

The characteristics of somatic cells are limited proliferation, methylation of pluripotency genes, tissue-specific cell morphology, inactivation of the X chromosome, an active G_1 cell cycle checkpoint, and expression of somatic cell-specific markers. On the other hand, pluripotent cells robustly show self-renewal, ESC morphology, reactivation of pluripotency genes by demethylation, X chromosome reactivation in female cells, telomerase activity, and loss of the G_1 checkpoint (Cox and Rizzino 2010).

Although a number of studies have clearly shown that iPS cells are quite similar to ESCs (Mikkelsen et al. 2008; Okita et al. 2008; Soldner et al. 2009), the degree of transcriptional similarity between ESCs and iPS cells is still not completely elucidated. It was recently demonstrated that gene expression signatures of virally programmed iPS cells are distinguished from those of ES cells (Lowry et al. 2008; Maherali et al. 2008; Chin et al. 2009; Soldner et al. 2009). However, human iPS cells generated without viral vectors or genomic insertions displayed retained potential transcriptional signatures (Marchetto et al. 2009). Wang et al. (2010) compared transcriptomes of fibroblasts, partially reprogrammed iPS cells, ESCs, and iPS cells using microarray data. They concluded that most of the reprogrammed iPS cell lines were similar to ESC lines. Especially, the transcriptomes of the iPS cells derived by episomal-based nonintegrating plasmids (Yu et al. 2009), unlike that of retroviral-derived iPS cells (Lowry et al. 2008; Maherali et al. 2008; Chin et al. 2009; Soldner et al. 2009), were much closer to that of ESCs (Wang et al. 2010). Moreover, the extent of overlapping implemented in the microarray platform was determined showing that ESCs and iPS cells express 2390 common genes, with only 684 and 249 genes expressed exclusively in iPS or ES cells, respectively. In general, iPS cells share more genes in common with the fibroblasts from which they were derived (Wang et al. 2010).

Given that the reprogramming process is expected to remove any epigenetic alterations associated with the original cells (e.g., NT cells and ESCs), which have reportedly faithfully erased any epigenetic marks present in donor cells (Okita et al. 2007; Maherali et al. 2008; Mikkelsen et al. 2008), the data showing that iPS cells share more common genes with their parental fibroblasts was quite surprising.

In fact, in July 2010, two simultaneously published papers proved that even rigorously selected early-passage iPS cells could retain epigenetic marks characteristic of the donor cell. Although the persistent imprinting manifests as differential gene expression and alters differentiation capacity, it may be utilized in potential therapeutic applications to enhance differentiation into desired cell lineages (Kim et al. 2010; Polo et al. 2010).

14.5 Challenges to Therapeutic Potential of Human iPS Cells

In a metazoan body, all cells possess the same set of genes. There are only exceptions of postmeiotic germ cell lines, mature lymphocytes, and cells in species that exhibit chromosome diminution (Kloc and Zagrodzinska 2001). Therefore, generating a pluripotent cell in vitro and directing its conversion to a specific differentiated cell fate (which means rewinding the internal clock of a mammalian cell to an embryonic state) and then forwarding these high-potential cells to diseased cells represents a rational ongoing approach in regenerative medicine. On the other hand, quality control and safety are the main concerns, and there are several technical challenges when using human iPS cells to treat several irreparable human diseases. Factor-free human iPS cells are necessary to minimize or eliminate genetic alterations in the derived iPS cell line created. Defining a disease-relevant phenotype needs in vitro and in vivo models. To generate markers for differentiation and gene corrections, gene-targeting strategies are necessary. Cell-type specific lineage reporters, lineage-tracking tools, and tools to disrupt, repair, or overexpress genes should be developed to model many human diseases (Saha and Jaenisch 2009). Yet, iPS cells would have profound implications for both basic research and clinical therapeutics by providing a patient-specific model system to study the pathogenesis of disease and test the effectiveness of pharmacological agents. It also would be an ample source of autologous cells that could be used for transplantation (Robbins et al. 2010).

Because cellular functions are influenced by microenvironmental stimuli, it is important to evaluate the results obtained from iPS studies so far conducted regarding the reprogramming methods, culture conditions, and differentiation protocols, all of which influence the outcome (Daley et al. 2009). Kim et al. (2010) revealed several important facts about reprogramming and resultant iPS cells: First, the tissue source influences the efficiency and fidelity of reprogramming. Second, there are substantial differences between iPS cell-derived and embryo-derived ESCs. Third, the differentiation propensity and methylation profile of iPS cells can be reset. Finally, and most strikingly, NT-derived ESCs are more faithfully reprogrammed than most iPS cells generated from adult somatic tissues (Kim et al. 2010).

14.6 Why Are Reprogramming Events Rare But Robust?

Although Waddington's epigenetic landscape metaphor is available for understanding the differentiation status of a cell, further explanations are needed to understand molecular nature and epigenetic barriers for determining cell fate. Hochedlinger and Plath (2009) modified Waddington's landscape metaphor for explaining developmental potential and epigenetic status of a cell. In this modified diagram a marble (zygote) stays uphill (dedifferentiation) and then moves across valleys (lineage conversion) along with the developmental pathways. Yet, further explanations are needed to understand the molecular nature and epigenetic status of cells because classic biological theories failed to explain many phenomenon such as epigenetic barriers, the ground state of pluripotency, and cell fate commitment, among others.

Colebrook (2002) claimed that we are in a postlinguistic era and that we need to develop theories and approaches that are not language-dependent. In biology, we are using language to describe extremely complex and interactive events. If we think that mathematics is the language of nature, we should be able to represent and understand everything around us through numbers. When we turn those numbers of any system into graphs, patterns emerge. Therefore, there are patterns everywhere in nature (Aronofsky 1998). In philosophy, systems theory, science, and art, emergence is the way complex systems and patterns arise out of a multiplicity of relatively simple interactions. Emergence is central to the theories

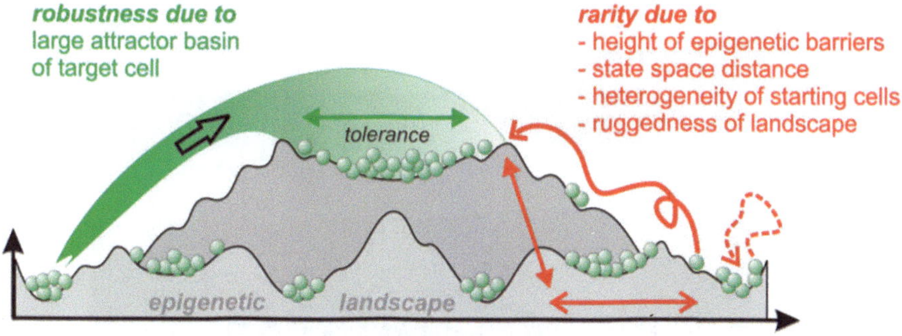

Fig. 14.2 Coexistence of rarity and robustness during reprogramming the pluripotent state ("jumping back"). Note the "subattractors" ("washboard potentials") as a manifestation of the ruggedness of the epigenetic landscape, which imposes intermediate states that slow down the reprogramming events (Huang 2009)

Fig. 14.3 The ball is moving in the bowl until it finds the lowest point (the attractor)

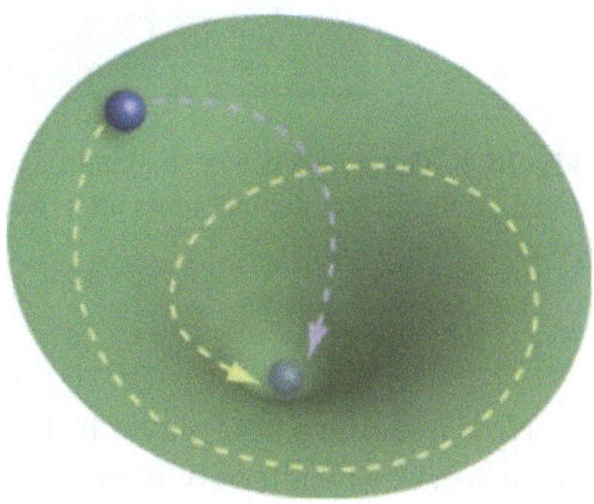

of integrative levels and of complex systems; therefore, it can be defined as "the arising of novel and coherent structures, patterns, and properties during the process of self-organization in complex systems" (Corning 2002). The complexity of cell fate determination then transports us to an emerging pattern from divergent genetic and epigenetic signals.

Fortunately, Huang (2009) tried to set up a pedagogical framework to describe sources of cellular states with the help of an integrated dynamic system. By doing so, he invited experimental biologists to walk through a new path called dynamic systems, which have been using to describe the behavior of complex dynamic systems. Accordingly, Huang (2009) described the natural and expected "ground state" character of pluripotency to explain the rarity and robustness of reprogramming events (Fig. 14.2).

To understand the basics of the dynamic system theory, we may take the ball from Waddington's landscape and throw it into any bowl. The ball moves around the bowl until it eventually comes to rest at the lowest point of the bowl. In dynamic systems, that lowest point is called the *attractor* (because the ball was "attracted" to that point) (Fig. 14.3).

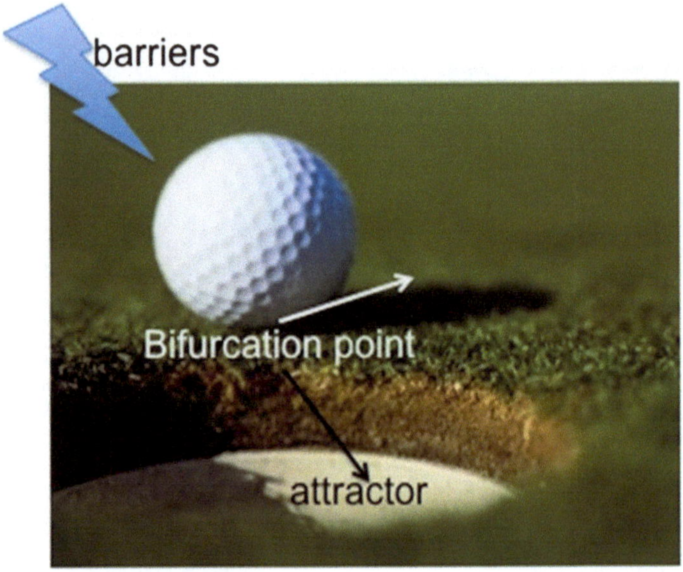

Fig. 14.4 The properties of the landscape and the ability of the player influence the result. A talented golfer directs the ball to its attractor, the targeted hole. Once the ball exceeds the bifurcation point, it drops in the hole

Now imagine a golf player trying to get the ball in a hole. The hole is an attractor, but it works as such only if the ball gets close enough to the rim. The properties of landscape, the ability of player, and many other local attractors and barriers (e.g., small holes, ridges) influence the result. Each ridge creates a point of decision, wherein the ball can fall to one side or the other. Moreover, which path the ball takes is highly dependent on the initial conditions. A minor change can switch the path from one attractor (targeted) to a nearby one (missed). Once this "bifurcation" point has been passed, it may take a large perturbation (e.g., a hidden fountain at the bottom of the hole) to change the attractor again (Fig. 14.4). Otherwise. we can say the system has become locked into a particular attractor.

The entire hole is what the dynamic system calls the basin of attraction of that system (Kauffman 1993). In cell biology, the basins are separated by some unstable states, which constitute the epigenetic barriers. Once an attractor is reached, the associated expression pattern is maintained (Huang 2009).

State space is simply an imaginary map of all the possibilities open to the system; for example, with a coin toss it is just two points: heads or tails. In cell biology, it is gene regulatory circuits. With a bistable gene regulatory circuit, if for instance gene 1 (unconditionally) inhibits gene 2 or vice versa, there is only two possibilities: $X_1 \geq X_2$ or $X_1 \leq X_2$. This particular gene then generates two distinct attractor states: S_A has the expression pattern $X_1 \geq X_2$, and S_B has the complementary pattern $X_1 \leq X_2$. Because the two attractors can coexist under the same environmental conditions, the

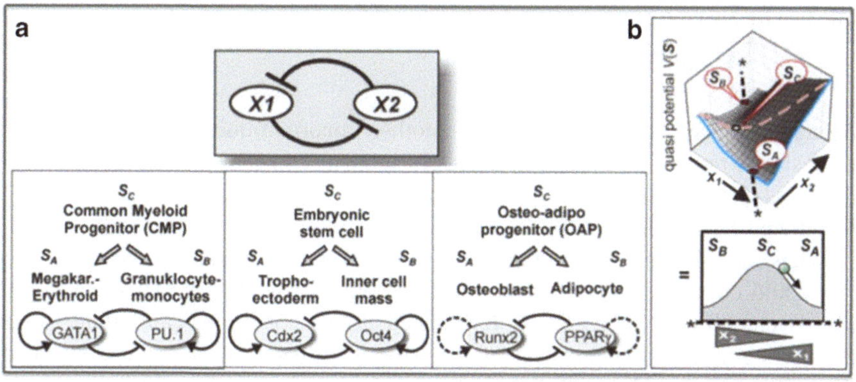

Fig. 14.5 Dynamics of a two-gene regulatory circuit. (**a**) Circuit architecture of two mutually inhibitory genes and examples of gene regulatory circuits using the same binary decisions for cell differentiation in multipotent cells. (**b**) Quasi-potential landscape. States S_A and S_B (*red circles*) are stable steady states (*attractors*). S_C (*empty circle*) is an unstable steady state. *Dashed line* represents the separatrix, dividing the state space into the basins of attraction. *Bottom*: Simplified representations obtained from cross sections along the *dashed line* (*- - - - -*) (Huang 2009)

circuit is said to be bistable. Attractor states are robust, "self-stabilizing," distinct states (Fig. 14.5) (Huang 2009).

If we go back to Huang's epigenetic landscape in Fig. 14.3, we are able to see that in the dynamic perspective individual cells in a clonal population show fluctuations in their expression levels within the attractor basin. Also, the cells that at a given time happen to be near the rim of the basin are most responsive to differentiating signals that kick them out of the stem cell attractor or destabilize the latter (Huang 2009). Overall, the efforts for inducing differentiation pathways in both somatic and embryonic cells indicate that once the epigenetic barriers are exceeded by any reprogramming events the cell fate changes (Hochedlinger and Plath 2009). If stem cells contain a heterogeneous mixture of microstates, each primed for a distinct fate, transitioning into each other in a dynamic equilibrium (within the attractor basin) is highly possible when no fate-committing external cue is present (Huang 2009). Since Ying et al. (2008) stated ESCs cells have an innate program for self-replication that does not require extrinsic instruction, the ground state character of pluripotency is accepted as the natural default state. With the help of Huang's (2009) complex high-dimensional dynamic system, we can understand why reprogramming pluripotency is robust although rare: "Since the pluripotent state is an attractor state with a rather large basin of attraction, it is robust—a ground state." The rarity of reprogramming events can be explained as due to the ruggedness of the attractor landscape: "Only a small fraction of the cells in the population, namely those whose fluctuating microstate maps into a gene expression pattern that fulfills some particular priming requirement, may actually be responsive to the nature of the reprogramming signals" (Fig. 14.3) (Huang 2009).

14.7 Is Reprogramming Necessary for Regenerative Therapies?

Along with the epigenetic discussion, another important question is whether it is necessary to reprogram cells back to the pluripotent stem cell state. For regenerative therapies, pluripotency may not be a prerequisite for the generation of certain differentiated cell types.

It has been shown that overexpression or deletion of individual transcription factors could change the cell fate in somatic cells (Hochedlinger and Plath 2009). Moreover, culturing stem cells under defined culture conditions can initiate differentiation programs (Kocaefe et al. 2010). Also, it has been shown that there is direct conversion of fibroblasts to functional neurons with no prior pluripotent stage (Vierbuchen et al. 2010). Masip et al. (2010) thus attempted to describe "induced transdifferentiated (iT) cells" as a novel tool for modifying the adult cell fate. Interconversion between adult cells from ontogenically different lineages by an induced transdifferentiation process is based on overexpression of a single or cocktail of transcription factors. As there is no attempt to reach it through an ESC-like state, iT cells may provide an alternative for regenerative medicine. On the other hand, like iPS cells, they require safe methods with transient and/or nonintegrative tools for generation. They also need in vivo assays to determine the suitability of their transplantation and applicability in regenerative medicine (Masip et al. 2010).

Thus far, pluripotent stem cells seem to be the only sources with the unique potential to provide comprehensive model systems for the earliest stages of human ontogenesis. Generating iPS cells from almost all cell types in the human body would aid in the investigation of the molecular events governing cellular specification from human pluripotent stem cells. Generated iPC cells must undergo a targeted, stepwise differentiation process that follows a normal developmental timeline (Meyer et al. 2009).

Differentiation and lineage commitment warrant a ground state model. If ontogenically closed cells could be gathered in a common place, mimicking differentiation pathways would follow the pathways through cellular specification aimed at identifying the earliest lineage precursors. A tooth and alveolar bone duo may serve as a model to search for close differentiation pathways because they are closely specialized tissues in an anatomical localization. Such tissues are alveolar bone of jaws, the periodontal ligament, and cementum and dentin of the tooth root. These tissues are secreting by closely related cells: osteoblasts, periodontal ligament fibroblasts, cementoblasts, and odontoblasts respectively. They are all mesenchymal in origin except that odontoblasts have a neuroectodermal origin (Koussoulakou et al. 2009). Although those tissues display many functional and physiological differences, there are no specific markers available to identify their specificity. On the other hand, mesenchymal stem cell properties of the periodontal ligament and dental pulps cells are almost (if not completely) identical (Huang et al. 2009). However, the attempt to reprogram them to the iPS state showed varied phenotypes that can be moderated through passages (Fig. 14.6) (Yildirim unpublished observation).

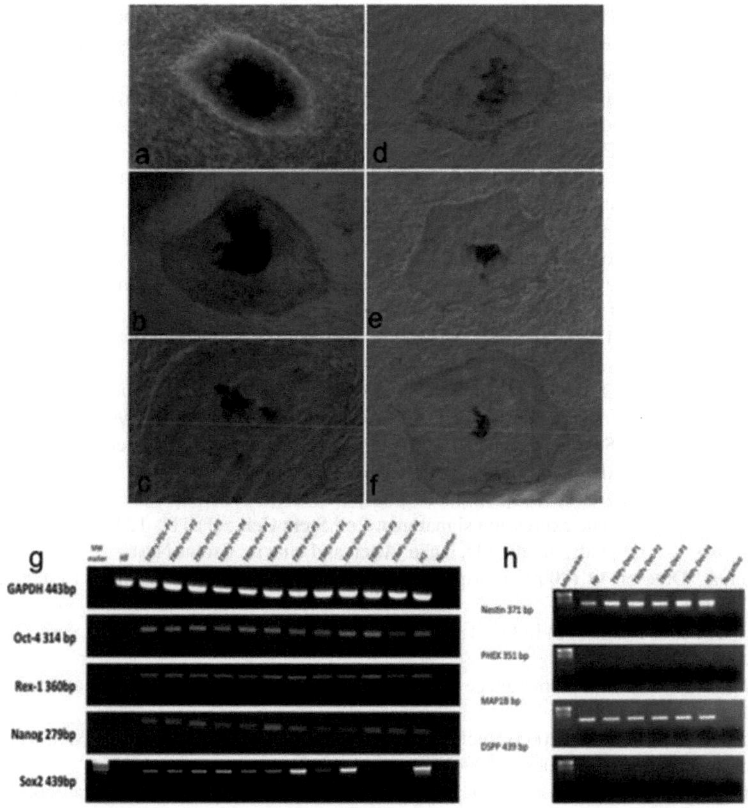

Fig. 14.6 Colonies generated from closely related dental cells. (**a**, **d**) Deciduous dental pulp stem cells. (**b**, **e**) Permanent dental pulp stem cells. (**c**, **f**) Tooth-matched permanent periodontal ligament stem cells showed different morphologies during early passages (**a–c** passage 1; **d–f** passage 5). Phenotypic uniformity was established through further passaging. All colonies expressed pluripotency genes throughout the passages (**g**), although they also expressed tissue-specific genes (**h**), indicating retention of parental imprinting

It would be interesting to determine if those morphological discrepancies reflect an epigenetic status of different but closely related cellular origins from which they arose.

References

Aasen T, Raya A et al (2008) Efficient and rapid generation of induced pluripotent stem cells from human keratinocytes. Nat Biotechnol 26(11):1276–1284

Abaci HE, Truitt R et al (2010) Adaptation to oxygen deprivation in cultures of human pluripotent stem cells, endothelial progenitor cells, and umbilical vein endothelial cells. Am J Physiol Cell Physiol 298(6):C1527–C1537

Alipio Z, Adcock DM et al (2010) Sustained factor VIII production in hemophiliac mice 1 year after engraftment with induced pluripotent stem cell-derived factor VIII producing endothelial cells. Blood Coagul Fibrinolysis 21(5):502–504

Arbel G, Caspi O et al (2010) Methods for human embryonic stem cells derived cardiomyocytes cultivation, genetic manipulation, and transplantation. Methods Mol Biol 660:85–95

Aronofsky D (1998) Pi. USA: 84 minutes

Banerjee I, Sharma N et al (2011) Impact of co-culture on pancreatic differentiation of embryonic stem cells. J Tissue Eng Regen Med 5(4):313–323

Bhutani N, Brady JJ et al (2010) Reprogramming towards pluripotency requires AID-dependent DNA demethylation. Nature 463(7284):1042–1047

Brambrink T, Foreman R et al (2008) Sequential expression of pluripotency markers during direct reprogramming of mouse somatic cells. Cell Stem Cell 2(2):151–159

Byrne JA, Pedersen DA et al (2007) Producing primate embryonic stem cells by somatic cell nuclear transfer. Nature 450(7169):497–502

Cai J, Yang M et al (2010) Dopaminergic neurons derived from human induced pluripotent stem cells survive and integrate into 6-OHDA-lesioned rats. Stem Cells Dev 19(7):1017–1023

Carvajal-Vergara X, Sevilla A et al (2010) Patient-specific induced pluripotent stem-cell-derived models of LEOPARD syndrome. Nature 465(7299):808–812

Chin MH, Mason MJ et al (2009) Induced pluripotent stem cells and embryonic stem cells are distinguished by gene expression signatures. Cell Stem Cell 5(1):111–123

Colebrook C (2002) Understanding Deleuze. Allen and Unwin, Crows Nest

Colman A, Dreesen O (2009) Pluripotent stem cells and disease modeling. Cell Stem Cell 5(3):244–247

Comyn O, Lee E et al (2010) Induced pluripotent stem cell therapies for retinal disease. Curr Opin Neurol 23(1):4–9

Corning PA (2002) The re-emergence of “emergence”: a venerable concept in search of a theory. Complexity 7(6):18–30

Cowan CA, Atienza J et al (2005) Nuclear reprogramming of somatic cells after fusion with human embryonic stem cells. Science 309(5739):1369–1373

Cox JL, Rizzino A (2010) Induced pluripotent stem cells: what lies beyond the paradigm shift. Exp Biol Med (Maywood) 235(2):148–158

Daley GQ, Lensch MW et al (2009) Broader implications of defining standards for the pluripotency of iPSCs. Cell Stem Cell 4(3):200–201, author reply 202

Dick E, Rajamohan D et al (2010) Evaluating the utility of cardiomyocytes from human pluripotent stem cells for drug screening. Biochem Soc Trans 38(4):1037–1045

Dimos JT, Rodolfa KT et al (2008) Induced pluripotent stem cells generated from patients with ALS can be differentiated into motor neurons. Science 321(5893):1218–1221

Ebert AD, Yu J et al (2009) Induced pluripotent stem cells from a spinal muscular atrophy patient. Nature 457(7227):277–280

Ellis J, Bruneau BG et al (2009) Alternative induced pluripotent stem cell characterization criteria for in vitro applications. Cell Stem Cell 4(3):198–199, author reply 202

Eminli S, Foudi A et al (2009) Differentiation stage determines potential of hematopoietic cells for reprogramming into induced pluripotent stem cells. Nat Genet 41(9):968–976

Esteban MA, Wang T et al (2010) Vitamin C enhances the generation of mouse and human induced pluripotent stem cells. Cell Stem Cell 6(1):71–79

Feng B, Ng JH et al (2009) Molecules that promote or enhance reprogramming of somatic cells to induced pluripotent stem cells. Cell Stem Cell 4(4):301–312

Fusaki N, Ban H et al (2009) Efficient induction of transgene-free human pluripotent stem cells using a vector based on Sendai virus, an RNA virus that does not integrate into the host genome. Proc Jpn Acad Ser B Phys Biol Sci 85(8):348–362

Gamm DM, Meyer JS (2010) Directed differentiation of human induced pluripotent stem cells: a retina perspective. Regen Med 5(3):315–317

Gearhart J (1998) New potential for human embryonic stem cells. Science 282(5391):1061–1062

Gerbal-Chaloin S, Duret C et al (2010) Isolation and culture of adult human liver progenitor cells: in vitro differentiation to hepatocyte-like cells. Methods Mol Biol 640:247–260

Grigoriadis AE, Kennedy M et al (2010) Directed differentiation of hematopoietic precursors and functional osteoclasts from human ES and iPS cells. Blood 115(14):2769–2776

Hanna J, Wernig M et al (2007) Treatment of sickle cell anemia mouse model with iPS cells generated from autologous skin. Science 318(5858):1920–1923

Hemberger M, Dean W et al (2009) Epigenetic dynamics of stem cells and cell lineage commitment: digging Waddington's canal. Nat Rev Mol Cell Biol 10(8):526–537

Ho PJ, Yen ML et al (2010) Endogenous KLF4 expression in human fetal endothelial cells allows for reprogramming to pluripotency with just OCT3/4 and SOX2: brief report. Arterioscler Thromb Vasc Biol 30:1880–1881

Hochedlinger K, Plath K (2009) Epigenetic reprogramming and induced pluripotency. Development 136(4):509–523

Homma K, Sone M et al (2010) Sirt1 plays an important role in mediating greater functionality of human ES/iPS-derived vascular endothelial cells. Atherosclerosis 212:36–39

Hotta A, Cheung AY et al (2009) Isolation of human iPS cells using EOS lentiviral vectors to select for pluripotency. Nat Methods 6(5):370–376

Hu BY, Zhang SC (2009) Differentiation of spinal motor neurons from pluripotent human stem cells. Nat Protoc 4(9):1295–1304

Huang S (2009) Reprogramming cell fates: reconciling rarity with robustness. Bioessays 31(5):546–560

Huang GT, Gronthos S et al (2009) Mesenchymal stem cells derived from dental tissues vs. those from other sources: their biology and role in regenerative medicine. J Dent Res 88(9):792–806

Huang HP, Yu CY et al (2010) Factors from human embryonic stem cell-derived fibroblast-like cells promote topology-dependent hepatic differentiation in primate embryonic and induced pluripotent stem cells. J Biol Chem 285:33510–33519

Huangfu D, Osafune K et al (2008) Induction of pluripotent stem cells from primary human fibroblasts with only Oct4 and Sox2. Nat Biotechnol 26(11):1269–1275

Jaenisch R, Bird A (2003) Epigenetic regulation of gene expression: how the genome integrates intrinsic and environmental signals. Nat Genet 33(Suppl):245–254

Jin ZB, Okamoto S et al (2009) Induced pluripotent stem cells for retinal degenerative diseases: a new perspective on the challenges. J Genet 88(4):417–424

Kaichi S, Hasegawa K et al (2010) Cell line-dependent differentiation of induced pluripotent stem cells into cardiomyocytes in mice. Cardiovasc Res 88:314–323

Kaji K, Norrby K et al (2009) Virus-free induction of pluripotency and subsequent excision of reprogramming factors. Nature 458(7239):771–775

Kaneko S, Otsu M et al (2010) Reprogramming adult hematopoietic cells. Curr Opin Hematol 17(4):271–275

Karumbayaram S, Novitch BG et al (2009) Directed differentiation of human-induced pluripotent stem cells generates active motor neurons. Stem Cells 27(4):806–811

Kauffman SA (1993) Self-organization and adaptation in complex system. Oxford University Press, New York/Oxford

Kaufman DS (2009) Toward clinical therapies using hematopoietic cells derived from human pluripotent stem cells. Blood 114(17):3513–3523

Kawaguchi J, Mee PJ et al (2005) Osteogenic and chondrogenic differentiation of embryonic stem cells in response to specific growth factors. Bone 36(5):758–769

Kazuki Y, Hiratsuka M et al (2010) Complete genetic correction of ips cells from Duchenne muscular dystrophy. Mol Ther 18(2):386–393

Kim D, Kim CH et al (2009a) Generation of human induced pluripotent stem cells by direct delivery of reprogramming proteins. Cell Stem Cell 4(6):472–476

Kim JB, Greber B et al (2009b) Direct reprogramming of human neural stem cells by OCT4. Nature 461(7264):649–653

Kim K, Doi A et al (2010) Epigenetic memory in induced pluripotent stem cells. Nature 467(7313):285–290
Kloc M, Zagrodzinska B (2001) Chromatin elimination: an oddity or a common mechanism in differentiation and development? Differentiation 68(2–3):84–91
Kocaefe C, Balci D et al (2010) Reprogramming of human umbilical cord stromal mesenchymal stem cells for myogenic differentiation and muscle repair. Stem Cell Rev 6(4):512–522
Koussoulakou DS, Margaritis LH et al (2009) A curriculum vitae of teeth: evolution, generation, regeneration. Int J Biol Sci 5(3):226–243
Lamba DA, McUsic A et al (2010) Generation, purification and transplantation of photoreceptors derived from human induced pluripotent stem cells. PLoS One 5(1):e8763
Lee G, Papapetrou EP et al (2009) Modelling pathogenesis and treatment of familial dysautonomia using patient-specific iPSCs. Nature 461(7262):402–406
Lee G, Chambers SM et al (2010) Derivation of neural crest cells from human pluripotent stem cells. Nat Protoc 5(4):688–701
Li C, Zhou J et al (2009) Pluripotency can be rapidly and efficiently induced in human amniotic fluid-derived cells. Human Mol Genet 18(22):4340–4349
Lin T, Ambasudhan R et al (2009) A chemical platform for improved induction of human iPSCs. Nat Methods 6(11):805–808
Loh YH, Agarwal S et al (2009) Generation of induced pluripotent stem cells from human blood. Blood 113(22):5476–5479
Lowry WE, Richter L et al (2008) Generation of human induced pluripotent stem cells from dermal fibroblasts. Proc Natl Acad Sci USA 105(8):2883–2888
Lu M, Kardel MD et al (2009) Enhanced generation of hematopoietic cells from human hepatocarcinoma cell-stimulated human embryonic and induced pluripotent stem cells. Exp Hematol 37(8):924–936
Maehr R, Chen S et al (2009) Generation of pluripotent stem cells from patients with type 1 diabetes. Proc Natl Acad Sci USA 106(37):15768–15773
Maherali N, Hochedlinger K (2008) Guidelines and techniques for the generation of induced pluripotent stem cells. Cell Stem Cell 3(6):595–605
Maherali N, Ahfeldt T et al (2008) A high-efficiency system for the generation and study of human induced pluripotent stem cells. Cell Stem Cell 3(3):340–345
Marchetto MC, Yeo GW et al (2009) Transcriptional signature and memory retention of human-induced pluripotent stem cells. PLoS One 4(9):e7076
Martinez-Fernandez A, Nelson TJ et al (2010) c-MYC independent nuclear reprogramming favors cardiogenic potential of induced pluripotent stem cells. J Cardiovasc Transl Res 3(1):13–23
Masip M, Veiga A et al (2010) Reprogramming with defined factors: from induced pluripotency to induced transdifferentiation. Mol Hum Reprod 16(11):856–868
Meyer JS, Shearer RL et al (2009) Modeling early retinal development with human embryonic and induced pluripotent stem cells. Proc Natl Acad Sci USA 106(39):16698–16703
Mikkelsen TS, Hanna J et al (2008) Dissecting direct reprogramming through integrative genomic analysis. Nature 454(7200):49–55
Miura K, Okada Y et al (2009) Variation in the safety of induced pluripotent stem cell lines. Nat Biotechnol 27(8):743–745
Nakagawa M, Koyanagi M et al (2008) Generation of induced pluripotent stem cells without Myc from mouse and human fibroblasts. Nat Biotechnol 26(1):101–106
Nakatsuji N (2010) Banking human pluripotent stem cell lines for clinical application? J Dent Res 89(8):757–758
Narazaki G, Uosaki H et al (2008) Directed and systematic differentiation of cardiovascular cells from mouse induced pluripotent stem cells. Circulation 118(5):498–506
Nelson TJ, Martinez-Fernandez A et al (2009) Repair of acute myocardial infarction by human stemness factors induced pluripotent stem cells. Circulation 120(5):408–416
Nishimura K, Nakagawa T et al (2009) Transplantation of mouse induced pluripotent stem cells into the cochlea. Neuroreport 20(14):1250–1254

Okabe M, Otsu M et al (2009) Definitive proof for direct reprogramming of hematopoietic cells to pluripotency. Blood 114(9):1764–1767

Okita K, Ichisaka T et al (2007) Generation of germline-competent induced pluripotent stem cells. Nature 448(7151):313–317

Okita K, Nakagawa M et al (2008) Generation of mouse induced pluripotent stem cells without viral vectors. Science 322(5903):949–953

Osafune K, Caron L et al (2008) Marked differences in differentiation propensity among human embryonic stem cell lines. Nat Biotechnol 26(3):313–315

Parameswaran S, Balasubramanian S et al (2010) Induced pluripotent stem cells generate both retinal ganglion cells and photoreceptors: therapeutic implications in degenerative changes in glaucoma and age-related macular degeneration. Stem Cells 28(4):695–703

Park IH, Arora N et al (2008a) Disease-specific induced pluripotent stem cells. Cell 134(5):877–886

Park IH, Lerou PH et al (2008b) Generation of human-induced pluripotent stem cells. Nat Protoc 3(7):1180–1186

Park IH, Zhao R et al (2008c) Reprogramming of human somatic cells to pluripotency with defined factors. Nature 451(7175):141–146

Pfannkuche K, Liang H et al (2009) Cardiac myocytes derived from murine reprogrammed fibroblasts: intact hormonal regulation, cardiac ion channel expression and development of contractility. Cell Physiol Biochem 24(1–2):73–86

Polo JM, Liu S et al (2010) Cell type of origin influences the molecular and functional properties of mouse induced pluripotent stem cells. Nat Biotechnol 28(8):848–855

Ralston A, Rossant J (2010) The genetics of induced pluripotency. Reproduction 139(1):35–44

Rathjen J, Rathjen PD (2003) Lineage specific differentiation of mouse ES cells: formation and differentiation of early primitive ectoderm-like (EPL) cells. Methods Enzymol 365:3–25

Raya A, Rodriguez-Piza I et al (2009) Disease-corrected haematopoietic progenitors from Fanconi anaemia induced pluripotent stem cells. Nature 460(7251):53–59

Robbins RD, Prasain N et al (2010) Inducible pluripotent stem cells: not quite ready for prime time? Curr Opin Organ Transplant 15(1):61–67

Saha K, Jaenisch R (2009) Technical challenges in using human induced pluripotent stem cells to model disease. Cell Stem Cell 5(6):584–595

Senju S, Haruta M et al (2009) Characterization of dendritic cells and macrophages generated by directed differentiation from mouse induced pluripotent stem cells. Stem Cells 27(5):1021–1031

Senju S, Hirata S et al (2010) Pluripotent stem cells as source of dendritic cells for immune therapy. Int J Hematol 91(3):392–400

Slack JM (2002) Conrad Hal Waddington: the last Renaissance biologist? Nat Rev Genet 3(11):889–895

Soldner F, Hockemeyer D et al (2009) Parkinson's disease patient-derived induced pluripotent stem cells free of viral reprogramming factors. Cell 136(5):964–977

Sun N, Panetta NJ et al (2009) Feeder-free derivation of induced pluripotent stem cells from adult human adipose stem cells. Proc Natl Acad Sci USA 106(37):15720–15725

Swistowski A, Peng J et al (2010) Efficient generation of functional dopaminergic neurons from human induced pluripotent stem cells under defined conditions. Stem Cells 28(10):1893–1904

Takahashi K, Yamanaka S (2006) Induction of pluripotent stem cells from mouse embryonic and adult fibroblast cultures by defined factors. Cell 126(4):663–676

Takahashi K, Tanabe K et al (2007) Induction of pluripotent stem cells from adult human fibroblasts by defined factors. Cell 131(5):861–872

Tamaoki N, Takahashi K et al (2010) Dental pulp cells for induced pluripotent stem cell banking. J Dent Res 89(8):773–778

Tanaka T, Tohyama S et al (2009) In vitro pharmacologic testing using human induced pluripotent stem cell-derived cardiomyocytes. Biochem Biophys Res Commun 385(4):497–502

Taranger CK, Noer A et al (2005) Induction of dedifferentiation, genomewide transcriptional programming, and epigenetic reprogramming by extracts of carcinoma and embryonic stem cells. Mol Biol Cell 16(12):5719–5735

Tashiro K, Inamura M et al (2009) Efficient adipocyte and osteoblast differentiation from mouse induced pluripotent stem cells by adenoviral transduction. Stem Cells 27(8):1802–1811
Tateishi K, He J et al (2008) Generation of insulin-secreting islet-like clusters from human skin fibroblasts. J Biol Chem 283(46):31601–31607
Taura D, Noguchi M et al (2009a) Adipogenic differentiation of human induced pluripotent stem cells: comparison with that of human embryonic stem cells. FEBS Lett 583(6):1029–1033
Taura D, Sone M et al (2009b) Induction and isolation of vascular cells from human induced pluripotent stem cells: brief report. Arterioscler Thromb Vasc Biol 29(7):1100–1103
Tchieu J, Kuoy E et al (2010) Female human iPSCs retain an inactive X chromosome. Cell Stem Cell 7(3):329–342
Teramura T, Onodera Y et al (2010) Induction of mesenchymal progenitor cells with chondrogenic property from mouse-induced pluripotent stem cells. Cell Reprogram 12(3):249–261
Thomson JA, Itskovitz-Eldor J et al (1998) Embryonic stem cell lines derived from human blastocysts. Science 282(5391):1145–1147
Tsuneto M, Yamane T et al (2003) In vitro differentiation of mouse ES cells into hematopoietic, endothelial, and osteoblastic cell lineages: the possibility of in vitro organogenesis. Methods Enzymol 365:98–114
Ueda T, Yamada T et al (2010) Generation of functional gut-like organ from mouse induced pluripotent stem cells. Biochem Biophys Res Commun 391(1):38–42
Utikal J, Maherali N et al (2009) Sox2 is dispensable for the reprogramming of melanocytes and melanoma cells into induced pluripotent stem cells. J Cell Sci 122(Pt 19):3502–3510
Vierbuchen T, Ostermeier A et al (2010) Direct conversion of fibroblasts to functional neurons by defined factors. Nature 463(7284):1035–1041
Waddington C (1957) The strategy of genes. Geo Allen and Unwin, London
Waddington CH (1959) Canalization of development and genetic assimilation of acquired characters. Nature 183(4676):1654–1655
Waddington CH, Robertson E (1966) Selection for developmental canalisation. Genet Res 7(3):303–312
Wakayama T, Tabar V et al (2001) Differentiation of embryonic stem cell lines generated from adult somatic cells by nuclear transfer. Science 292(5517):740–743
Wang Y, Mah N et al (2010) A transcriptional roadmap to the induction of pluripotency in somatic cells. Stem Cell Rev 6(2):282–296
Warren L, Manos PD et al (2010) Highly efficient reprogramming to pluripotency and directed differentiation of human cells with synthetic modified mRNA. Cell Stem Cell 7:618–630
Wdziekonski B, Villageois P et al (2003) Development of adipocytes from differentiated ES cells. Methods Enzymol 365:268–277
Wernig M, Meissner A et al (2007) In vitro reprogramming of fibroblasts into a pluripotent ES-cell-like state. Nature 448(7151):318–324
Wernig M, Zhao JP et al (2008) Neurons derived from reprogrammed fibroblasts functionally integrate into the fetal brain and improve symptoms of rats with Parkinson's disease. Proc Natl Acad Sci USA 105(15):5856–5861
Xie CQ, Huang H et al (2009) A comparison of murine smooth muscle cells generated from embryonic versus induced pluripotent stem cells. Stem Cells Dev 18(5):741–748
Xu D, Alipio Z et al (2009) Phenotypic correction of murine hemophilia A using an iPS cell-based therapy. Proc Natl Acad Sci USA 106(3):808–813
Yan X, Qin H et al (2010) iPS cells reprogrammed from human mesenchymal-like stem/progenitor cells of dental tissue origin. Stem Cells Dev 19(4):469–480
Ye L, Chang JC et al (2009a) Induced pluripotent stem cells offer new approach to therapy in thalassemia and sickle cell anemia and option in prenatal diagnosis in genetic diseases. Proc Natl Acad Sci USA 106(24):9826–9830
Ye Z, Zhan H et al (2009b) Human-induced pluripotent stem cells from blood cells of healthy donors and patients with acquired blood disorders. Blood 114(27):5473–5480
Ying QL, Smith AG (2003) Defined conditions for neural commitment and differentiation. Methods Enzymol 365:327–341

Ying QL, Wray J et al (2008) The ground state of embryonic stem cell self-renewal. Nature 453(7194):519–523

Yu J, Vodyanik MA et al (2007) Induced pluripotent stem cell lines derived from human somatic cells. Science 318(5858):1917–1920

Yu J, Hu K et al (2009) Human induced pluripotent stem cells free of vector and transgene sequences. Science 324(5928):797–801

Zhang D, Jiang W et al (2009a) Highly efficient differentiation of human ES cells and iPS cells into mature pancreatic insulin-producing cells. Cell Res 19(4):429–438

Zhang J, Wilson GF et al (2009b) Functional cardiomyocytes derived from human induced pluripotent stem cells. Circ Res 104(4):e30–e41

Zhou W, Freed CR (2009) Adenoviral gene delivery can reprogram human fibroblasts to induced pluripotent stem cells. Stem Cells 27(11):2667–2674

Zhou H, Wu S et al (2009) Generation of induced pluripotent stem cells using recombinant proteins. Cell Stem Cell 4(5):381–384

Zhou J, Su P et al (2010) High-efficiency induction of neural conversion in hESCs and hiPSCs with a single chemical inhibitor of TGF-beta superfamily receptors. Stem Cells 28(10):1741–1750

Chapter 15
Cancer Stem Cells: Current Concepts and Therapeutic Implications

A. Ugur Ural

Abstract Cancer stem cells are the cells in a tumor that possess the capacity for self-renewal and to produce the heterogeneous lineages of cancer cells that compromise the tumor, act like normal stem cells, and render tumors resistant to chemotherapy. Like normal stem cells, cancer stem cells are marked by their ability to unlimited self-renewal, to differentiate, and to regenerate a phenocopy of the original tissue. In addition to playing a role in primary tumor formation, it has been shown that cancer stem cells are also key players in the metastatic process, with their unlimited capacity for self-renewal, the requirement for a specific microenvironment in which to grow, use of the CXCL12/CXCR4 axis for homing and quiescence. They also enhance resistance to apoptosis and have an increased capacity for drug resistance. Developing targeted therapies that are selectively toxic to cancer stem cells, and their microenvironment while sparing normal stem cells may lead to more effective treatment options for eradicating this crucial population of cells.

15.1 Introduction

Most cancers are difficult to eradicate completely, and they recur easily and metastasize. Some cancers are resistant to chemotherapy, and some are highly aggressive. Are we targeting the right cells by conventional cancer therapy? These facts raised

A.U. Ural, M.D. (✉)
Gulhane Medical Faculty, Department of Hematology,
Medical and Cancer Research Center, Etlik- Ankara 06010, Turkey
e-mail: aural@gata.edu.tr; auural@gmail.com

K. Turksen (ed.), *Adult and Embryonic Stem Cells*, Stem Cell Biology and Regenerative Medicine, DOI 10.1007/978-1-61779-630-2_15,

the hypothesis that tumors arise from a rare population of cells with stem cell properties that have been termed cancer stem cells (CSCs). According to this hypothesis, first proposed in 1983 by Mackillop, only a small fraction of cells in certain tumors are tumorigenic; that is, only the CSCs can produce all of the cells necessary to repopulate the tumor (Yang and Chang 2008).

The CSCs have normal stem cell characteristics that render them resistant to classic cancer chemotherapy, and they are responsible for tumor initiation, propagation, and metastasis. The CSC hypothesis assumes that CSCs exhibit the following characteristics: (1) self-renewal; (2) multidirectional differentiation; (3) tumor-initiating capacity; (4) resistance to apoptosis. These properties of CSCs have been tested with limiting dilution and serial transplantation experiments.

Lapidot and colleagues (Lapidot et al. 1994) offered the first evidence of CSCs from patients with acute myeloid leukemia (AML) in 1994. They isolated a CD34+CD38– cell fraction from these patients and showed that transfer of the CD34+CD38– cells at a density of 5×10^3 into nonobese diabetic, severe combined immunodeficient (NOD/SCID) mice resulted in leukemic proliferation that could be serially transplanted into secondary recipients, whereas injection of more differentiated cells (5×10^3) expressing CD34+CD38+ failed to do so. Al-Hajj et al. also reported that CD24–/lowCD44+ cell fraction from metastatic pleural effusion had significantly higher tumorigenic potential when injected into the mammary fat pad of female NOD/SCID mice than CD24+CD44+/– cells (Al-Hajj et al. 2003). Since then, CSCs have been widely studied and isolated in many tumors including multiple myeloma (Matsui et al. 2004), brain cancer (Galli et al. 2004), colon cancer (O'Brien et al. 2007), head and neck cancer (Prince et al. 2007), prostate cancer (Collins et al. 2005), and melanoma (Fang et al. 2005). These studies in murine models provided evidence that CSCs can arise from stem cells that acquire transforming mutations or from transformed downstream progenitor or precursor cells, which reacquire stem cell characteristics (e.g., self-renewal) (Deshpande and Buske 2007). The transformation of stem cells, their progenitors, or cells differentiated to CSCs may be due to the accumulation of genetic modifications (mutations in oncogenes, suppressor genes, mismatched repair genes) and epigenetic alterations (abnormal methylation, histone modification). The fact that leukemia stem cells and glioblastoma stem cells have markers similar to those of normal stem cells also support this hypothesis (Gil et al. 2008). The cells that acquire the ability for self-renewal accumulate genetic changes over long periods of time, escape from the control of their microenvironment, and give rise to cancerous growth (Shipitsin and Polyak 2008). Currently available chemotherapeutic agents and radiation therapy target the bulk tumor mass, but quiescent CSCs show resistant to such therapy because of the high expression of antiapoptotic proteins.

Elucidating the biological properties of CSCs may provide new insights into the factors that drive tumor initiation and propagation and may help to develop novel therapeutic approaches to overcome drug resistance, to improve therapeutic efficacy, and to develop novel cancer treatments which target CSCs while sparing normal stem cells.

15.2 Transformation of Normal Cells to Cancer Stem Cells

If CSCs arise from normal stem cells then cancer cells could require the stem cell regulatory pathways for self-renewal. On the other hand, if these cells arise from more differentiated cells, genetic modifications and epigenetic alterations would be required to drive differentiation and for the cells to regain self-renewal properties.

As is well known, differentiated cells have a limited life span, and thus it is unlikely that all the necessary mutations could occur during their short life. In contrast, the self-renewal capacity of normal stem cells means these cells may be the only cells that are around long enough to accumulate the necessary mutations (Allan et al. 2007). Hence, stem or progenitor cells must be the initial targets for malignant transformation as the CSCs must be able to self-renew, and it would be more difficult for a mature cell to regain the ability to self-renew through genetic mutations.

15.3 Signaling Pathways in Cancer Stem Cells

Stem cells use multiple signaling pathways to control normal stem cell self-renewal or differentiation. Constitutive activation or dysregulation of these pathways may lead to neoplastic proliferation, with the development of CSCs. Numerous signaling pathways have been implicated in this process, including *Bmi-1*, Wnt/β-catenin, Sonic Hedgehog (SHH), and Notch (Kakarala and Wicha 2007; Hambardzumyan et al. 2008; Chumsri et al. 2007; Yang and Chang 2008). These pathways are tightly regulated in determining normal stem cell fate whether a stem cell self-renews or differentiates.

Bmi-1 has been shown to be required for the self-renewal of both neural stem cells and hematopoietic stem cells (HSCs) (Lessard and Sauvageau 2003; Cui et al. 2007). Overexpression of *Bmi-1* results in down-regulation of p16 and p19, which are important antiproliferative proteins (Yang and Chang 2008). *Bmi-1* also plays an essential role in the self-renewal of CSCs; for example, *Bmi-1–/–* deficient mice display failure of hematopoiesis (Lessard and Sauvageau 2003). Recent experiments showed that CSCs lacking *Bmi-1* were enable to engraft and proliferate, and they displayed differentiation and apoptosis. Reconstitution of the *Bmi-1* gene was found to completely abrogate these defects (Lessard and Sauvageau 2003).

In the Wnt pathway, stimulation of cells with Wnt ligands leads to stabilization of β-catenin, a protein with dual function as a cytoskeletal component and a latent transcription factor that subsequently translocates to the nucleus and activates specific gene transcription. Overexpression of β-catenin in HSCs induces their expansion, whereas expression of the β-catenin antagonist axin impairs HSC proliferation. Wnt/β-catenin signaling is involved in regulating HoxB4 and Notch1, two critical regulators of HSC self-renewal activity (Reya et al. 2003). The self-renewal role of Wnt/β-catenin signaling was also conserved in self-renewing of CSCs, which show increased nuclear β-catenin activity (O'Brien et al. 2007). Furthermore, it has been implicated in chronic lymphocytic leukemia and the progression chronic myeloid leukemia blast crisis (Chumsri et al. 2007).

SHH signaling regulates multiple aspects of central nervous system development, controlling both cell proliferation and cell differentiation. Damage to components of the SHH pathway can lead to birth defects and some types of cancer. SHH has also been shown to regulate the self-renewal of neural stem cells and HSCs and to be up-regulated and associated with tumorigenesis and a poor clinical outcome (Ahtar et al. 2006; Hooper and Scott 2005).

Notch signaling appears particularly capable of affecting both tumorigenesis and self-renewal, and it is necessary for maintaining a pool of undifferentiated stem cells (Hambardzumyan et al. 2008; Morrison et al. 2000). Deletion of Notch1, one of the transmembrane heterodimeric receptors, results in a decrease in neural stem cells and a reduction in their proliferation (Hambardzumyan et al. 2008). In T-cell acute lymphoblastic leukemia, Notch1 is found to be constitutively activated in patients with the t(7;9) abnormality (Weng et al. 2004).

15.4 Isolation and Markers of Cancer Stem Cells

The most widely accepted assay to validate a candidate CSC population is tumor initiation and serial transplantation in NOD/SCID mice, where the tumor that grows in the mice recapitulates the heterogeneity of the primary patient tumor. To identify CSCs, patient-derived tumor cells are stained with antibodies to various cell surface markers, and techniques such as fluorescence-activated cell sorting are used to separate labeled versus unlabeled populations of cells. In situ distribution of proposed markers for CSCs— including CD44 or CD24 for breast, colon, or pancreatic carcinomas; CD133 for glioblastomas or colon, prostate, and kidney carcinomas; CD20 for melanoma; CD166 for colon carcinomas—failed to reveal defined localization patterns that reliably identify only CSCs in tumors (Zhou and Zhang 2008; Hill and Perris 2007).

Lapidot and colleagues isolated and identified CD34+CD38– leukemia stem cells (LSCs) from patients with AML and demonstrated that these cells initiated leukemia in NOD/SCID mice compared with the CD34+CD38+ and CD34– fractions (Lapidot et al. 1994). An engrafted leukemia could be serially transplanted into secondary recipients, providing functional evidence for self-renewal.

CD133+ cells were also demonstrated in human brain cancers that possess differentiative and self-renewal capacities and can initiate tumor growth in vivo, whereas CD133– cells cannot. In vivo, transfer of as few as 100 CD133+ cells could regenerate a serially transplantable phenocopy of the original tumor in the brains of NOD/SCID mice (Singh et al. 2004). CSCs in the brain express not only CD133 but also nestin, which is known to be a marker for a poor prognosis in patients with a glioma and is activated by Notch signaling (Shih and Holland 2006; Strojnik et al. 2007).

Al-Hajj and colleagues demonstrated marked difference in tumorigenicity between CD44+CD24– expressing cells and nonexpressing cells in breast cancer (Al-Hajj et al. 2003). The number of putative CSCs identified as CD44+CD24– cells did not achieve a statistically significant association between the frequency of these cells and clinical outcome, although the number of the cells was associated with distant metastasis (Abraham et al. 2005). In addition, the markers presently used to sort putative CSCs (including CD24, CD44, and CD133) are not specific for CSCs because they are expressed on normal cells (Hill and Perris 2007). CD133 is a marker expressed by many types of normal stem cell, including neural and HCSs. CD44 is a cell surface receptor for hyaluronic acid and is involved in cell migration, adhesion, and metastasis (Croker and Allan 2008).

15.5 Cancer Stem Cells, Niches, and Metastasis

In light of the significant role of the normal stem cell niche in controlling the stem cell fate (e.g., self-renewal, quiescence, homing, engraftment, proliferation), it has been proposed that a CSC niche exists and that interactions with this tumor niche may have a similar role in specifying a self-renewing population of stem cells. There is mounting evidence that CSCs receive important signals from the microenvironment that support self-renewal and may exploit the normal homeostatic mechanisms (Lane et al. 2009). Factors derived from the tumor environment serve to regulate cancer cells. Specialized microenvironments of bone marrow endothelial cells, known as vascular niches, appeared to be required for the homing and engraftment of both HSCs and leukemic cells (Sipkins et al. 2005). Xenograft transplantation assays have been able to support the role of niche signaling in CSC engraftment, chemotherapy resistance, and cell cycle regulation (Ishikawa et al. 2007).

There are several possible models for CSC–niche interactions. First, the CSC may not require a distinct niche for expansion and may instead be capable of surviving in the normal stem cell niche. For example, leukemic SCs may impair the function of the normal HSC niche by direct invasion or secretion of substances such as stem cell factor. Leukemic SCs can infiltrate these niches and hijack the normal homeostatic processes, leading to enhanced self-renewal and proliferation signaling, enforced quiescence, and resistance to conventional chemotherapy including secretion of antagonists (Lane et al. 2009). Second, a distinct CSC niche may be necessary for activation. CSCs may also exhibit dysregulated homing and engraftment, leading to alternative niche formation. As with normal stem cells, the niche may be important for maintaining asymmetrical division of CSCs and for tethering CSCs close to signals that maintain stem-like properties (Hermann et al. 2008).

To form a metastasis, cells should require features similar to those of the cells initiating the primary tumor. Therefore, CSCs represent the only cells capable of spreading from the primary tumor and of giving rise to metastases (Hermann et al. 2008). The chemokine CXCL12 is believed to play a critical role in stem cell migration in cooperation with its receptor CXCR4 (Ma et al. 2007). CXCL12 is an ideal candidate for helping metastasis because its major biological effects are related to the ability of this chemokine to induce motility, chemotactic responses, adhesion, secretion of matrix metalloproteinases (MMPs), and secretion of angiopoietic factors such as vascular endothelial growth factor (VEGF) in cells that express CXCR4

(Ratajczak et al. 2006). In cancer development, fibroblast expression of CXCL12 and tumor cell expression of CXCR4 is often increased in hypoxic areas of the tumor, subsequently triggering tumor growth and motility. A fundamental role of the CXCR4/CXCL12 axis in metastatic spread through the body according to a CXCL12 gradient has also been suggested for a variety of cancers (Smith et al. 2004; Su et al. 2005; Saur et al. 2005). This axis may also help to explain the organ-specific nature of metastatic growth because CXCR4-expressing cancer cells may home to organs that express high levels of CXCL12. For example, CXCR4-expressing breast cancer cells preferentially metastasize to CXCL12-expressing organs such as lymph nodes, liver, and bone (Chambers et al. 2002; Pantel and Brakenhoff 2004; Muller et al. 2001). This may be one of several potential explanations for the increased incidence of widespread metastases from breast cancer (Spillane and Henderson 2007).

15.6 Therapeutic Implications of Cancer Stem Cells

In current clinical practice, standard chemotherapeutic agents are still used with the intent to kill the bulk tumor mass. Unfortunately, the drugs target not only tumor cells but also normal cells. Additionally, most of these approaches fail to eradicate CSCs, resulting in disease relapse.

For example, CD34+CD38– LSCs were found to be less sensitive to daunorubicin than the more committed CD34–CD38+ cells (Costello et al. 2000). Myeloma CSCs were relatively more resistant to standard chemotherapy (Matsui et al. 2004). The resistance of CSCs to chemotherapy may be due to increased expression of proteins from the BCL-2 antiapoptotic family, which leads to an increase in the expression of membrane proteins responsible for drug resistance (Al-Hajj et al. 2003). CSCs can also resist apoptosis by a number of mechanisms, including transforming growth factor β signaling and activation of the SHH pathway (Thayer et al. 2003). Increased expression of transporting proteins, such as MDR1 and ABC transporters, is also an important factor in chemotherapy resistance (Jordan et al. 2006). The resistance of CSCs to radiotherapy may be due to activated and increased expression of the checkpoint proteins in response to DNA damage by radiotherapy (Bao et al. 2006). Also, cells resistant to radiotherapy show expression of prominin-1 (CD133+), which also appears on the surface of neuronal and brain SCs (Gil et al. 2008). It has been proposed that CSCs also express these proteins at higher levels than the bulk population of tumor cells and may be more resistant to chemotherapy, permitting reproliferation of tumors after chemoradiotherapy (Spillane and Henderson 2007). Because of these mechanisms, CSCs tend to not only take up less drug, but they also diminish drug effects through DNA repair systems or antiimmune mechanisms. They therefore resist most chemotherapeutic attempts (Zhou and Zhang 2008).

Although CSC-specific treatments will be effective in curing cancer, the signal transduction pathways and phenotypes are similar in normal stem cells and CSCs, making it difficult to target the culprits without toxicity to normal stem cells. Fortunately, some studies have suggested that it may be possible to target CSCs using new agents. It is reasonable that inhibition of ABC transporters could at least partially reverse the drug resistance phenotype of CSCs. In particular, the combination of ABC transporter inhibitors and chemotherapy may help eliminate CSCs (Lou and Dean 2007). Pharmacological inhibition of checkpoint proteins (e.g., Chk1 and Chk2), results in decreased resistance to ionizing radiation of CD133+ CSCs (Bao et al. 2006). In another study, monoclonal antibody targeting of CD44, a CSC surface molecule, eradicated human leukemic SCs without affecting HSCs in a xenograft model in NOD/SCID mice. The authors hypothesized that this result was due to interference with transport to the stem cell microenvironment and alteration of the CSC fate toward differentiation (Jin et al. 2006).

Interestingly, chronic myeloid leukemia (CML) stem cells are resistant to kinase inhibitors because they do not require BCR/ABL signaling for survival in the quiescent state. Indeed, these quiescent CML progenitors are resistant to multiple agents used alone or in combination with imatinib, including cytarabine, etoposide, and arsenic, each of which induces apoptosis in cycling leukemia cells (Jorgensen et al. 2005; Holtz et al. 2005). If quiescent CML stem cells could be induced to enter the cell cycle, it might increase their sensitivity to BCR/ABL kinase inhibitors (Krause and Van Etten 2007).

Other studies have shown that treatment of prostate and breast cancer cell lines with a siRNA against CD44 can decrease cancer cell adhesion to bone marrow stromal cells (Draffin et al. 2004). Combination of idarubicin and proteasome inhibitor, leading to inhibition of NF-κB activity, has been shown to mediate selective apoptosis of leukemic CSs while sparing normal stem cells (Yang and Chang 2008). Breast cancer stem cells have also been preferentially inhibited by NF-κB specific inhibitors, including parthenolide, pyrrolidinedithiocarbamate and diethyldithiocarbamate (Zhou et al. 2007). Several Notch inhibitors are in clinical development, including MK-0752, a γ-secretase inhibitor in Phase I trials in patients with T-cell acute lymphoblastic leukemia and other leukemias (Deangelo et al. 2006). Recently, breast cancer cells treated with a CXCR4 antagonist (AMD3100) show significantly inhibited metastatic ability, and intracranial glioblastoma xenografts treated with AMD3100 show increased tumor cell apoptosis (Liang et al. 2004; Rubin et al. 2003).

In conclusion, at the molecular level, alterations in signaling pathways responsible for self-renewal of stem cells are crucial in the transformation of any cell into CSCs. Progress can be made only by discovering the mechanisms of control of signaling pathways. Ultimately, new prognostic and predictive markers, as well as more specific therapeutic strategies targeted to CSCs or signaling pathways, may be developed to force tumors into permanent remission.

References

Abraham BK, Fritz P, KcClellan M et al (2005) Prevalence of CD44+/CD24−/low cell in breast cancer may not be associated with clinical outcome but may favor distant metastasis. Clin Cancer Res 11:1154–1159

Ahtar M, Tang X, Lee JL et al (2006) Hedgehog signaling in skin development and cancer. Exp Dermatol 15:667–677

Al-Hajj M, Wicha MS, Bertino-Hernandez A et al (2003) Prospective identification of tumorigenic breast cancer cells. Proc Natl Acad Sci USA 100:3983–3988

Allan AL, Vantyghern SA, Tuck AB et al (2007) Tumor dormancy and cancer stem cells: implications for the biology and treatment of breast cancer metastasis. Breast Dis 26:87–98

Bao S, Wu Q, McLendon RE et al (2006) Glioma stem cells promote radioresistance by preferential activation of the DNA damage response. Nature 444:756–760

Chambers AF, Groom AC, MacDonald IC (2002) Dissemination and the growth of cancer cells in metastatic sites. Nat Rev Cancer 2:563–572

Chumsri S, Matsui W, Burger AM (2007) Therapeutic implications of leukemic stem cell pathways. Clin Cancer Res 13:6549–6554

Collins AT, Barry PA, Hyde C et al (2005) Prospective identification of tumorigenic prostate cancer stem cells. Cancer Res 65:10946–10951

Costello RT, Mallet F, Gaugler B et al (2000) Human acute myeloid leukemia CD34+/CD38– progenitor cells have decreased sensitivity to chemotherapy and Fas-induced apoptosis, reduced immunogenicity, and impaired dendritic cell transformation capacities. Cancer Res 60:4403–4411

Croker AK, Allan AL (2008) Cancer stem cells: implications for the progression and treatment of metastatic disease. J Cell Mol Med 12:374–390

Cui H, Hu B, Li T et al (2007) Bmi-1 is essential for the tumorigenicity and neuroblastoma cells. Am J Pathol 170:1370–1378

Deangelo DJ, Stone RM, Silverman LB et al (2006) A phase I clinical trial of the notch inhibitor MK-0752 in patients with T-cell acute lymphoblastic leukemia/lymphoma and other leukemias. J Clin Oncol 24:6585

Deshpande AJ, Buske C (2007) Lymphoid progenitors as candidate cancer stem cells in AML. Cell Cycle 6:543–545

Draffin JE, McFarlane S, Hill A et al (2004) CD44 potentiates the adherence of metastatic prostate and breast cancer cells to bone marrow endothelial cells. Cancer Res 64:5702–5711

Fang D, Nguyen TK, Leishear K et al (2005) A tumorigenic subpopulation with stem cell properties in melanomas. Cancer Res 65:9328–9337

Galli R, Binda E, Orfanelli U et al (2004) Isolation and characterization of tumorigenic, stem-like neural precursors from human glioblastoma. Cancer Res 64:7011–7021

Gil J, Stembalska A, Pesz KA et al (2008) Cancer stem cells: the theory and perspectives in cancer therapy. J Appl Genet 49:193–199

Hambardzumyan D, Becher OJ, Holland EC (2008) Cancer stem cells and survival pathways. Cell Cycle 10:1371–1378

Hermann PC, Huber SL, Heeschen C (2008) Metastatic cancer stem cells. Cell Cycle 7:188–193

Hill RP, Perris R (2007) "Desteming" cancer stem cells. J Natl Cancer Inst 99:1435–1440

Holtz MS, Forman SJ, Bhatia R (2005) Nonproliferating CML CD34+ progenitors are resistant to apoptosis induced by a wide range of proapoptotic stimuli. Leukemia 19:1034–1041

Hooper JE, Scott MP (2005) Communicating with Hedgehogs. Nat Rev Mol Cell Biol 6:306–317

Ishikawa F, Yoshida S, Saito Y et al (2007) Chemotherapy-resistant human AML stem cells home to and engraft with the bone marrow endosteal region. Nat Biotechnol 25:1315–1321

Jin L, Hope KJ, Zhai Q et al (2006) Targeting of CD44 eradicates human acute myeloid leukemic stem cells. Nat Med 12:1167–1174

Jordan CT, Guzman ML, Noble M (2006) Cancer stem cells. N Engl J Med 355:1253–1261

Jorgensen HG, Allan EK, Graham SM et al (2005) Lonafarnib reduces the resistance of primitive quiescent CML cells to imatinib mesylate in vitro. Leukemia 19:1184–1191

Kakarala M, Wicha MS (2007) Cancer stem cells: implications for cancer treatment and prevention. Cancer J 13:271–275

Krause DS, Van Etten RA (2007) Right on target: eradicating leukemic stem cells. Trends Mol Med 13:470–481

Lane SW, Scadden DT, Gilliland DG (2009) The leukemic stem cell niche: current concepts and therapeutic opportunities. Blood 114:1150–1157

Lapidot T, Sirard C, Vormoor J et al (1994) A cell initiating human acute leukemia after transplantation into SCID mice. Nature 367:645–648

Lessard J, Sauvageau G (2003) Bmi-1 determines the proliferative capacity of normal and leukemic stem cells. Nature 423:255–260
Liang Z, Wu T, Lou H et al (2004) Inhibition of breast cancer metastasis by selective synthetic polypeptide against CXCR4. Cancer Res 64:4302–4308
Lou H, Dean M (2007) Targeted therapy for cancer stem cells: the patched pathway and ABC transporters. Oncogene 26:1357–1360
Ma S, Lee TK, Zheng BJ et al (2007) CD133(+) HCC cancer stem cells confer chemoresistance by preferential expression of the Akt/PKB survival pathway. Oncogene 27:1749–1758
Matsui W, Huff CA, Wang Q et al (2004) Characterization of clonogenic multiple myeloma cells. Blood 103:2332–2336
Morrison SJ, Perez SE, Qiao Z et al (2000) Transient Notch activation initiates an irreversible switch from neurogenesis to gliogenesis by neural crest stem cell. Cell 101:499–510
Muller A, Homey B, Soto H et al (2001) Involvement of the chemokine receptors in breast cancer metastasis. Nature 410:50–56
O'Brien CA, Pollett A, Gallinger S et al (2007) A human colon cancer cell capable of initiating tumour growth in immunodeficient mice. Nature 445:106–110
Pantel K, Brakenhoff RH (2004) Dissecting the metastatic cascade. Nat Rev Cancer 4:448–456
Prince ME, Sivanandan R, Kaczorowski A et al (2007) Identification of a subpopulation of cells with cancer stem cell properties in head and neck squamous cell carcinoma. Proc Natl Acad Sci USA 104:973–978
Ratajczak MZ, Zuba-Surma E, Kucia M (2006) The pleitropic effects of the SDF-1-CXCR4 axis in organogenesis, regeneration and tumorigenesis. Leukemia 20:1915–1924
Reya T, Duncan AW, Ailles E et al (2003) A role for Wnt signaling in self-renewal of hematopoietic stem cells. Nature 423:409–414
Rubin JB, Kung AL, Klein RS et al (2003) A small-molecule antagonist of CXCR4 inhibits intracranial growth of primary brain tumors. Proc Natl Acad Sci USA 100:13513–13518
Saur D, Seidler B, Schneider G et al (2005) CXCR4 expression increases liver and lung metastasis in a mouse model of pancreatic cancer. Gastroenterology 129:1237–1250
Shih AH, Holland EC (2006) Notch signaling enhances nestin expression in gliomas. Neoplasia 8:1072–1082
Shipitsin M, Polyak K (2008) The cancer stem cell hypothesis: in search of definitions, markers, and relevance. Lab Invest 88:459–463
Singh SK, Hawkins C, Clarke I et al (2004) Identification of human brain tumor initiating cells. Nature 432:396–401
Sipkins DA, Wei X, Wu J et al (2005) In vivo imaging of specialized bone marrow endothelial microdomains for tumour engraftment. Nature 435:969–973
Smith MC, Luker KE, Garbow JR et al (2004) CXCR4 regulates growth of both primary and metastatic cancer. Cancer Res 64:8604–8612
Spillane JB, Henderson MA (2007) Cancer stem cells: a review. ANZ J Surg 77:464–468
Strojnik T, Rosland GV, Sakariassen PO et al (2007) Neural stem cell markers nestin, and musashi proteins, in the progression of human glioma: correlation of nestin with prognosis of patient survival. Surg Neurol 68:133–143
Su L, Zhang J, Hu X et al (2005) Differential expression of CXCR4 is associated with the metastatic potential of human non-small cell lung cancer cells. Clin Cancer Res 11:8273–8280
Thayer SP, Pasca di Magliano M, Heiser PW et al (2003) Hedgehog is an early and late mediator of pancreatic cancer tumorigenesis. Nature 425:851–856
Weng AP, Ferrando AA, Lee W et al (2004) Activating mutations of Notch1 in human T cell acute lymphoblastic leukemia. Science 306:269–271
Yang YM, Chang JW (2008) Current status and issues in cancer stem cell study. Cancer Invest 26:741–755
Zhou J, Zhang Y (2008) Cancer stem cells. Cell Cycle 7:1360–1370
Zhou J, Zhang H, Gu P et al (2007) NF-KB pathway inhibitors preferentially inhibit breast cancer stem-like cells. Breast Cancer Res Treat 115:269–277

Chapter 16
Problems to Be Solved in Molecular Oncology

Ayfer Haydaroğlu

Abstract Cancer ranks as the second leading cause of death following heart disease. However, considering the increase in its incidence, cancer is expected to take over first place in causes of death within the next 10–15 years. Despite major advances in medicine, we still have not achieved the desired results in the treatment of cancer. The biggest reason is the fundamental problems in oncology with many unknowns to be solved. They need more research. Oncologists know well the unresolved issues in the clinical world, but they are not successful enough to direct basic scientists toward these issues, and basic scientists often stay away from the problems encountered in the clinic. Collaboration between oncologists and basic scientists is essential to cope with these problems. The main causes of failure in oncology include resistance to the basic methods of treatment such as radiotherapy and chemotherapy, deactivation of treatment mechanisms by DNA repair mechanisms, oxygenation problems (hypoxia, reoxygenation), metabolic biotransformation and/or inactivation of drugs, decreased apoptosis sensitivity of cancer cells (escape from apoptosis), molecular and genetic pathways involved in drug resistance, genetic predispositions and vulnerabilities, angiogenesis (neovascularization), the blood–brain barrier, intolerance to treatments, side effects, failure to prevent metastasis, and invasion. The aim of this chapter is to present the basic problems encountered in oncology and contribute to the research studies of the basic scientists on related issues.

A. Haydaroğlu (✉)
Radiation Oncology Department, Ege University Hospital, Bornova, Izmir 35100, Turkey
e-mail: ayfer.haydaroglu@ege.edu.tr; haydaroglua@gmail.com

K. Turksen (ed.), *Adult and Embryonic Stem Cells*, Stem Cell Biology and Regenerative Medicine, DOI 10.1007/978-1-61779-630-2_16,

16.1 Introduction

One of the most important causes of failure in oncology is resistance to treatment. Molecular and genetic mechanisms are the underlying factors of the resistance. The basic elements of cancer treatment—radiotherapy, chemotherapy, hormonotherapy, immunotherapy—have distinctive biological, molecular, and genetic problems. In this review, we discuss the molecular problems, which are considered among the sources of failure, particularly of adjuvant radiotherapy and chemotherapy treatment modalities for cancer and that are still waiting to be solved.

16.2 Molecular Problems in Radiation Therapy That Need to Be Solved

We can understand the resistance problems encountered in cancer radiotherapy by explicating the effects of radiation treatment at the cellular level and the molecular and genetic mechanisms of resistance.

16.2.1 Radiation-Induced Cell Death

Physicochemical changes in living individuals caused by incidents caused by ionizing radiation occur in less than a second, whereas biological consequences such as genetic mutations, cancer formation, and cell death need hours, days, months, or even years to emerge. Depending on the radiation, biological target molecules are influenced and damaged in direct or indirect ways. With the *direct* effect, radiation directly interacts with biological target molecules, and energy is transferred directly. With the indirect effect, radiation energy transfer to biological molecules is realized through free radicals. Radiobiologically, *indirect* effects of radiation are more important than the direct effects because radiation damage is mainly due to these effects.

As a result of absorption, free radicals and induced molecules are formed in living cells. The cells largely consist of water, and radiation energy is likely absorbed by water molecules. Hydroxyl and hydrogen radicals are formed following exposure to radiation. These free radicals cause radiobiological damage by interacting with biologically important molecules (e.g., DNA, enzymes). Damage caused by ionizing radiation in the cell is not selective. All cellular structures exposed to radiation are affected, but DNA damage is the main process leading to cell death (Jonathan et al. 1999).

Radiobiologically, there are two types of cellular death: mitotic and apoptotic. Mitotic death is the loss of cell viability due to accumulated genetic damage in the cells exposed to radiation. The main cause of cell death is thought to be the DNA damage. Mitotic cell death depends on the status and rate of cellular proliferation.

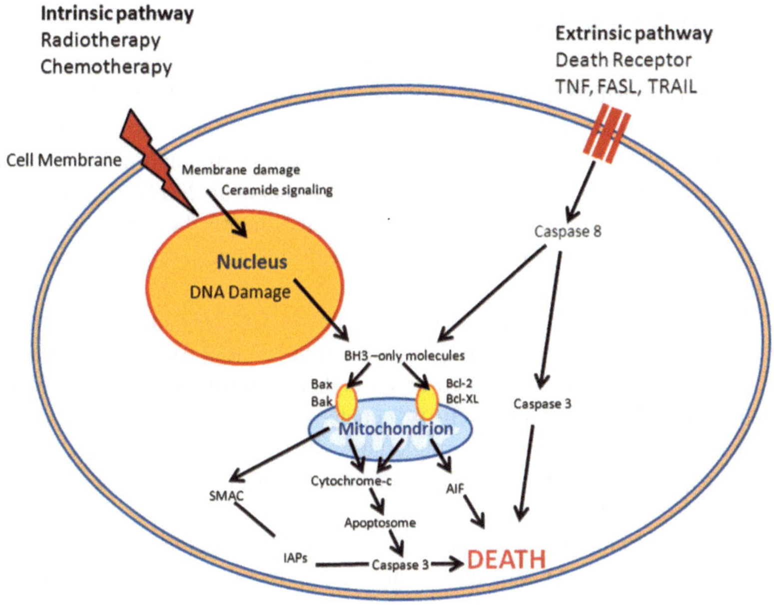

Fig. 16.1 Mechanisms of cell death

It may occur at either an early or a late phase. A few more divisions may be observed in the cell. Apoptotic death (interphase death) is observed by irradiation of the cells that are not divided or less divided and can be explained by apoptosis. The exact mechanism of radiation-induced cell death is variable. Radiation-induced apoptosis is definitively displayed in most of the systems (Illidge 1998). However, it is different for common solid tumors of epithelial origin. As tumors grow, they become hypoxic and are subject to mutation. In this environment, apoptosis-resistant cells have a selective advantage; and when exposed to radiation, they become dominant in the tumor. In this type of tumor, "mitotic" death is considered more important (Hatfield et al. 2005). Mitotic death occurs after irradiation, following mitosis, as a consequence of chromosome aberrations. Apoptosis is programmed cell death. It is cell death that occurs during normal cell growth realized throughout embryonic development. With apoptosis, the cytoplasm thickens, the nuclear membrane and the cell membrane undergo convolution, and apoptotic bodies are formed. They may be digested by neighboring cells (Fig. 16.1). Following radiotherapy, apoptosis becomes an important issue in lymphomas, varies in carcinomas, and is not an issue in sarcomas. Rapid response of a tumor to small doses of radiation is indicative of apoptotic cell death (Hall and Giaccio 2006).

The type of radiation-induced death of a tumor becomes important when determining the radiation dose and the number of fractions (Brenner and Hall 1999; Bentzen et al. 2008). Therefore, studies of radiation-induced cell death should be further developed in more detail.

16.2.2 Radiation and Hypoxia

Cells are more sensitive to X-rays in the presence of oxygen. In the phase of radiation in which free radicals are produced, oxygen exposure leads to the formation of peroxide radicals, which have a highly destructive potential and increase the indirect effects of radiation (Hall and Giaccio 2006; Joiner et al. 2009). Hypoxia reduces the effectiveness of radiation treatment and consequently becomes an important question. As the tumor grows, the blood circulation becomes insufficient, some cells become hypoxic, and necrosis occurs.

Different tumors possess cells that are oxygenated at different rates. Particularly tumor cells surrounding the central necrotic area are less oxygenated and so are more resistant than well-oxygenated tumor cells and normal tissue cells. When the tumor is small, there are more well-oxygenated cells; but as the tumor grows, the hypoxic area increases as well. A single large dose kills the oxygenated cells, and the hypoxic portion increases. The hypoxic part then decreases again owing to reoxygenation (Wilson 2007).

Chronic hypoxia is the result of a limited oxygen diffusion distance in a tissue actively using oxygen. As a result of chronic hypoxia over a long period of time, cells die and undergo necrosis. Acute hypoxia occurs as a result of temporary closure of blood vessels. If the vessels are reopened, the cells are exposed to intermittent hypoxia.

A mixture of oxygenated and hypoxic cells are found in tumors. Oxygenated cells are more radiosensitive, and most of them are killed by X-rays. Hypoxic cells are often left alone. Due to reoxygenation, the tumor reverts to a previous form. If the time period between the two fractions is sufficient for reoxygenation, the presence of hypoxic cells does not heavily affect the result. The degree and the rate of reoxygenation are variable and unpredictable (Wilson 2007).

Tumor hypoxia and reoxygenation are important issues to be examined in terms of treatment efficacy.

16.2.3 Radiation and the Immune Response

It has been shown in some studies that radiation has potential beneficial effects on the cells in terms of increasing immunogenity (Hall and Giaccio 2006). Radiotherapy-induced cell death has some important consequences. However, contradictory and controversial results have been achieved in the studies conducted. In some studies, it is reported that radiation-induced cell death may trigger an immune response (Steinman et al. 2000), and in others it was suggested that apoptosis has an inhibitory effect in terms of dendritic cell activation (Voll et al. 1997). Sauter et al. took another step and showed that dendritic cells may phagocytose both necrotic and apoptotic cells in different malignant cell lines, whereas only necrotic cells may activate dendritic cells. All of these studies support the tolerigenic nature of apoptotic death and the inhibition of autoimmunity (Sauter et al. 2000). Whether tumor death is due to apoptosis or necrosis, the immune stimulation caused by cell death is not sufficient to induce a complete immune response. Many questions related to radiation-induced cell death are waiting to be answered.

Concerning radiation therapy, further detailed studies are needed regarding the effects of radiation on the immune response.

16.2.4 DNA Repair Mechanisms

Irradiation or chemotherapy may lead to double- and single-chain fractures of DNA, insertions and deletions, abasic areas, and DNA–protein cross-link formations (Martin 2008). DNA damage may be due to endogenous factors. Throughout DNA replication and DNA recombination processes, endogenous agents such as free radicals that are produced as a by-product of cellular metabolism may cause DNA damage as well (Sancar and Lindsey-Boltz 2004). Minor injuries are mostly repaired by DNA repair systems (Ece et al. 2009). High levels of DNA damage cause "cell death" by inducing apoptosis. Thus, the organism may protect itself. A moderate level of DNA damage may cause mutations. Cells respond to DNA damage via various metabolic pathways. Severe DNA damage causes cell death by activating the apoptosis pathway. Cells may repair DNA damage by "DNA repair mechanisms." Klein and Glazer (2010) if DNA damage cannot be repaired during replication, it may cause mutation and consequently genomic instability, cancer, and/or aging.

More than 100 genes play an important role in the DNA repair system, and the proteins encoded by these genes are involved in the repair mechanisms. Approximately 104 injuries occur daily in human cell DNA, causing noncoded or miscoded genes. Accumulation of point mutations in mitochondrial DNA is thought to increase in parallel with aging, and therefore particularly mitochondrial DNA damage is associated with aging. DNA repair mechanisms comprise the systems maintaining genomic stability with different types of repair.

16.2.4.1 Direct Repair Mechanisms: Reversal of Damage

- *Photoreactivation*: When the cells with ultraviolet (UV)-induced mutations are exposed to visible light containing the blue spectrum (300–500 nm), they eliminate the damage by reversing it and recover. This phenomenon is called photoreactivation (Sancar 2003). During the course of the evolutionary process, this system has been preserved. Through light (300–500 nm) and two chromophores, the "DNA Phototolyase" enzyme is activated, and conversion occurs. Eukaryotic organisms have this enzyme.
- *O^6-Methylguanin Repair*: O^6-Methylguanin (mG) is formed in the presence of alkylating agents and is considered highly mutagenic (Ece et al. 2009).
- *Basic Single-Strand Break Ligation*: Agents such as X-rays or peroxides may cause basic fractures in the DNA chain. These fractures are immediately repaired

by DNA ligase enzyme. DNA ligase forms a phosphodiester bond between the 5′ phosphate group and the 3′ OH group in a energy-required reaction (Powell and Bindra 2009; Kulaksiz and Sancar 2007).

16.2.4.2 Cut–Remove Repairs (Excision)

- *Base Excision Repair*: Base excision repair (BER) is concerned with natural hydrolysis of DNA bases or repair of inappropriate bases occurred due chemical agents (Powell and Bindra 2009).
- *Nucleotide Excision Repair*: DNA lesions causing extensive deterioration in the helix structure of the DNA are repaired by the nucleotide excision repair (NER) system. Pyrimidine dimers originated from UV light, base changes such as benzo-pyrene-guanine originating from smoking, and base changes originating from chemotherapeutic drugs are excised as single bases by BER, whereas damaged bases are excised as oligonucleotide pieces using NER. As a result of defects in proteins involved in the mechanism of NER, some rare diseases such as xeroderma pigmentosum are defined (Powell and Bindra 2009; Kulaksiz and Sancar 2007).
- *Mismatch Excision Repair*: DNA polymerase has the capability of reading errors during replication (proofreading). Mismatch excision repair (MER) is the repair mechanism concerned with mispairs that remained after proofreading (Martin et al. 2010; Gomez 2010).

16.2.4.3 Recombination Repair

Recombination repair is the mechanism activated following replication, when DNA is not repaired by other repair mechanisms.

16.2.4.4 SOS Repair

The SOS repair is the mechanism taking part in cases of high DNA damage rates and when other repair mechanisms result in failure. During DNA synthesis, the system allows DNA polymerase replication rather than bypassing the lesion. However, the accuracy of replication is somewhat sacrified. Therefore, it is considered an error-prone system (Janion 2001; Schlacher et al. 2006).

16.2.4.5 Double-Strand Breaks Repair

DNA double-strand breaks, occur as a result of ionizing radiation, oxidative damage, or naturally. They are repaired in two ways (Powell and Bindra 2009): homologous and nonhomologous recombination. DNA damage repair is desirable in normal tissues when there is radiotherapy- or chemotherapy-induced DNA damage. It reduces toxicity. DNA damage repair is not wanted in tumor tissue, however. If it could be designed that way, it would increase the effectiveness of cancer treatment. Numerous DNA repair enzymes have helped to improve DNA repair improve, leading to reduced effectiveness of cancer treatment. If DNA repair mechanisms could be deactivated in tumor tissue, we could provide more effective treatment. More research studies should be conducted on DNA repair mechanisms in cancer tissue.

16.3 Molecular Problems in Chemotherapy That Need to Be Solved

One of the most important problems currently encountered in cancer treatment is primary or acquired drug resistance. Therefore, sequential or simultaneous combined use of cytotoxic agents having different mechanisms are used to help break the emerging resistance mechanisms. In recent years, an important focus has been clarify the drug response/resistance relation using genomic analysis methods performed with tissues obtained from cancer cell lines and tumors.

16.3.1 Multidrug Resistance

For the first time, three decades ago, during preliminary studies conducted with cancer cell lines, an efflux system was found to be responsible for drug resistance. It was called the permeability (P-) glycoprotein and was observed to be responsible for the structural and functional resistance of many drugs. P-glycoprotein is a plasma membrane protein belonging to the ATP binding cassette (ABC) transporter family. P-glycoprotein reduces the intracellular therapeutic level of the drug. Other carrier proteins, such as MRP1 and BRCP, also belong to this family. Overexpression of these proteins in some tissues was found to be associated with primary drug resistance. It is paradoxical that expression of these proteins increase in parallel with cellular damage and the stress response. In this case, an increase in these transporter proteins and development of resistance are observed as the best response to chemotherapy during the early period (Patwardhan et al. 2010).

16.3.2 DNA Repair Mechanisms

Endogenous DNA repair mechanisms include base excision, nucleotide excision, mismatch repair gene activation, and direct repair of DNA damage. ERCC-1 (excision repair cross-complementing) protein, belonging to the Nucleotide Excision

Repair (NER) family, is an important and commonly examined molecule related to chemotherapy resistance (Lee et al. 2008). In many studies dealing with in vivo/ in vitro resistance, ERCC-1-mediated DNA repair mechanism is reported to be responsible especially for resistance to cisplatin (Vilmar and Sorenson 2009).

Among the DNA repair enzymes, O_6-guanine DNA methyltransferase (MGMT) is frequently and extensively investigated in cancer research studies. The MGMT promoter region is silenced by hypermethylation in many tumors. This has a negative influence on DNA repair and leads to an increase in sensitivity particularly to alkylating agents (Hegi et al. 2005).

The prognostic and predictive roles of DNA repair enzymes in cancer treatment should be further investigated.

16.3.3 p53 Gene Status of Tumor and Drug Resistance

In all cellular stress conditions such as DNA damage, activation of oncogenes, loss of cellular contact, and hypoxia, the *p53* gene is activated and endeavors to protect genetic information. However, if the damage is beyond the control capacity of P53 protein, apoptosis is induced through proapoptotic proteins. A *p53* mutation is the most common mutation encountered in cancer biology, and in the presence of this mutation resistance to apoptosis is seen in many tumors. Accordingly, *p53* mutation leads to resistance to many chemotherapeutic agents. Drug resistance could not be restored in many clinical trials in which the gene become functional. Consequently, further studies on the subject are on the way.

16.3.4 Decrease in Apoptosis Sensitivity of Cancer Cells and Its Molecular Basis

Apoptosis is a type of cell death that occurs regularly during an organism's normal life cycle. Accelerated or slowed down apoptosis is pathological. *Necrosis* is a pathological cell death seen in nonphysiological conditions. Excessive tissue damage and inflammatory response develop. *Apoptosis* is programmed cell death seen under physiological conditions. There is no tissue damage, local cell loss develops, but there is no inflammatory response. With the apoptosis signal, chromatin condensation begins in the cell. The cytoplasm appears dense, and cell dimensions shrink. After a period of time, the cell is divided into smaller pieces called apoptotic bodies. Apoptotic bodies reveal new signal transmitting structures on the surface; and with the excitation of this signal, the cell is phagocytosed and removed by the nearby cells (which are usually histiocytes) (Lipponen et al. 1994). The apoptosis process starts with factors such as toxins and radiation from outside of the cell (extrinsic) or with exciting stimuli such as hormones and growth factors within the cell (intrinsic).

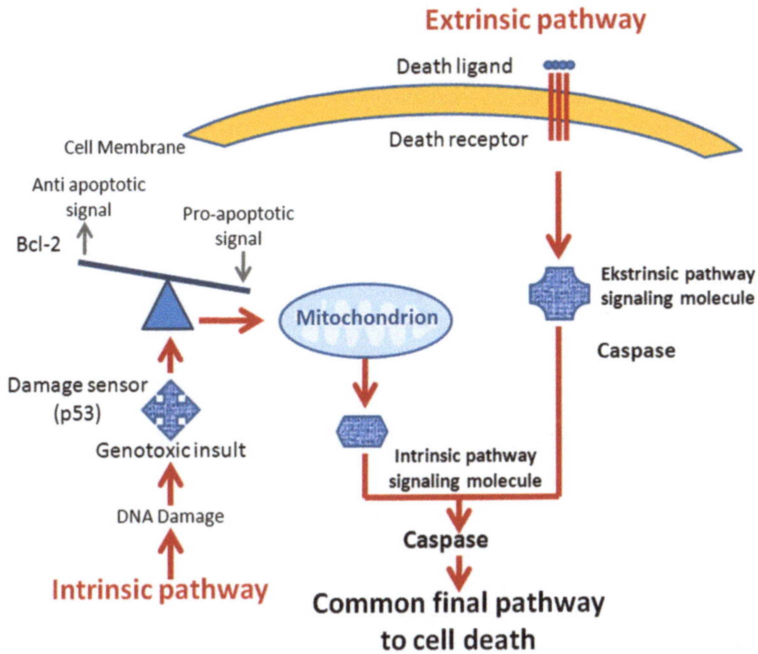

Fig. 16.2 Balance between proapototic and antiapoptotic signals in programmed cell death

The genes involved in the mechanisms of apoptosis and the proteins they encode are directly related to the development of cancer. The Bcl-2 family has an important role in apoptosis. The proapoptotic members promote apoptosis, and antiapoptotic members inhibit apoptosis;, the two function in balance. This balance between pro- and antiapoptotic members determines the choice between life and death (Fig. 16.2). It has been observed that overexpression of antiapoptotic members suppresses apoptosis, whereas overexpression of proapoptotic members kills the cells. Proapoptosis causes the release of cytochrome C from mitochondria into the cytoplasm, whereas antiapoptosis suppresses the release of cytochrome C. As a result of mitochondrial activation, cytochrome c migrates through cytoplasm and activates the caspase system. Caspases are a family of cystein proteases that play essential roles in apoptosis and proteolysis. They are inactive (zymogene) in the cell and proteolytically activate each other. Thus, they operate in a cascade. A balance between proapoptotic and antiapoptotic actions ensures integrity of the mitochondrial membrane (Plati et al. 2011).

Many studies have shown that overexpression of Bcl-2/Bcl-XL proteins is correlated with resistance to chemotherapy. Recently, drug sensitivity is being investigated by microarray and proteomic analyses on clinical materials. By determining the differences in cancer cells in terms of apoptotic molecules, it would be possible to differentiate cancer cells from normal cells, and we could develop drugs that are able to trigger apoptosis.

16.3.5 *Changing Drug Targets in Drug Resistance*

In drug-resistant cancer cells, target molecules of chemotherapy undergo genetic mutation. "Epigenetic regulation" is one of the mechanisms of this mutation. Transcriptional inhibition occurs by DNA hypermethylation without any changes in the base sequence. It is common in tumor suppressor genes. In many types of cancer, apoptosis and drug resistance would be triggered, for example, by methylation of the caspase 8 promoter region. Mutations or decreased expression of topoisomerases I and II is known to lead to intracellular targets of drugs to undergo mutations and develop drug resistance against various drugs.

16.3.6 *Metabolic Biotransformation and/or Inactivation of Drugs*

Metabolic biotransformation is the determinant of how long and at which therapeutic level active molecules of chemotherapeutic agents should contact cancer cells. CYP2C8 and CYP3A4 are the most commonly used enzymes in the metabolism of chemotherapeutic agents, and a decrease in the expression of CYP3A4 is thought to be associated with resistance to docetaxel in breast cancer (Noguchi 2006).

16.4 Main Indicators of Cancer That Need to Be Solved

With rapid advances during the last quarter century, cancer is displayed as the disease with dynamic changes in the genome. Hanahan and Weinberg suggested that a broad range of cancer cell genotypes is an indicator of six fundamental changes in cell physiology (Hanahan and Weinberg 2000). These six indicators are also the six problems to be resolved in oncology. The indicators include self-sufficiency in growth signals, insensitivity to growth-inhibitory signals, evasion from programmed apoptosis, limitless growth potential, sustained angiogenesis, and tissue invasion and metastasis.

16.4.1 *Self-sufficiency in Growth Signals*

Normal cells require mitogenic growth signals before they move from a quiescent state to an active, proliferating state. These signals are transmitted into the cell by transmembrane receptors that bind the signaling molecules. In the absence of such stimulating signals, no proliferation can occur. In cancer, normal growth signals in the cell somehow accomplish the signaling by mimicking (Hanahan and Weinberg 2000). Relating to this issue, the problem to be solved is to define clearly the details of the growth signaling phase.

16.4.2 Insensitivity to Growth-Inhibitory Signals

Multiple antiproliferative signals maintain homeostasis and cellular quiescence. These growth-inhibitory signals are received by transmembrane cell surface receptors, associated with intracellular signaling circuits. Anti-proliferation signals can inhibit proliferation by two distinct mechanisms: cells may be enforced from active proliferative state to quiescent state (G0) or to abandon their proliferative potential constantly by bringing to enter post-mitotic states (Hanahan and Weinberg 2000). The reasons of insensitivity of cancer tissue to the signals that inhibit growth should be further investigated.

16.4.3 Evasion of Programmed Apoptosis

The capacity of tumor cells to expand in number is determined by the rate of cell proliferation, and also by the rate of cell attrition. Apoptosis, which is the programmed cell death, is the main source of this cell attrition. Acquired resistance to apoptosis is one of the most striking features of all types of cancer. Apoptotic program is actually present as latent form in all cell types of the body. Cell membranes are interrupted, cytoplasmic and nuclear skeleton are broken, chromosomes are degraded. Resistance to apoptosis can be achieved by cancer cells through various strategies. The loss of a proapoptotic regulator occurs most commonly through mutation of p53 tumor suppressor gene. As a result, the functional inactivation of p53 protein occurs in more than 50% of human cancers, resulting in the removal of key component of DNA damage which can induce the apoptotic effector cascade (Hanahan and Weinberg 2000). Methods should be developed to eliminate the problem of evasion from apoptosis in cancer.

16.4.4 Limitless Replicative Potential

Telomeric DNA is lost from the ends of each chromosome during replication during each cell cycle. Progressive erosion of telomeres at the end of successive replication cycles causes them to lose the ability to protect the ends of the chromosomal DNA. In malignant cells, telomere maintenance is in evident. Around 85–90% of these cells achieve it by up-regulating expression of the telomerase enzyme. The remaining cells activate a mechanism called ALT. In most of the human cells, both mechanisms seem to be strongly suppressed.

16.4.5 Sustained Angiogenesis

The oxygen and nutrients supplied by vascular structure are vital for cell survival, and no cells of any type in a tissue can reside farther than 100 μm from a capillary blood vessel (Folkman 1971). In normal tissues, angiogenesis is controlled by proangiogenic and antiangiogenic mechanisms. Tumor angiogenesis is associated with specific growth factors, endothelial cell receptor activation and endothelial cell proliferation capacity, and extracellular matrix components. Many factors are involved in angiogenesis. They are released from tumor cells and other cells (e.g., monocytes, fibroblasts) and by degradation of the collagen matrix (Klagsbrun 1996).

Tumor angiogenesis offers a therapeutic target commonly shared by all tumor types. A catalog of angiogenic regulatory molecules expressed by various tumor types is expected to be available during the next decade (Hanahan and Weinberg 2000).

16.4.6 Tissue Invasion and Metastasis

Primary tissue masses produce pioneer cells that move out and invade adjacent tissues first and then distant tissues. This distant localization of tumor cells—otherwise known as metastasis—cause 90% of the deaths due to cancer. Invasion and metastasis are highly complex processes, and their genetic and biochemical determinants are not well understood. Understanding these mechanisms will contribute to the fight against cancer (Hanahan and Weinberg 2000).

16.5 Cancer Stem Cells

Cancer stem cells (CSCs) are a small number of tumor-forming cells found in tumor tissue. They possess the ability to self-renew. CSCs and normal stem cells (NSCs) have similar original signal transduction systems. These systems play a functional role in self-renewal and differentiation. The main difference between them is the program change in CSCs regarding the regulation process of signal transduction. Signal transduction systems regulating the self-renewal system of NSCs lead to tumor formation when there is a systemic problem. For example, pentaerythritol tetranitrate (PTEN) causes proliferation of stem cells. On the other hand, the loss of the tumor suppressor gene in PTEN initiates myeloproliferative disease (Reya et al. 2001).

The CSC hypothesis states that CSCs arise from mutations of specific stem cells or progenitor cells (Fig. 16.3). According to the hypothesis, not all of the cells are capable of proliferating and ensure continuity of tumor. A small subset of cells known as CSCs may be replicated and renew itself. In recent studies, CSCs have been reported in blood, breast, brain, spleen, head and neck, colon, skin, and ovarian cancers. Tumors caused by mutations of stem cells are heterogeneous, and more metastatic. On the other hand, tumors caused by mutations of the progenitors are more homogeneous and less metastatic (Woodward and Bristow 2009).

Data derived from recent studies suggest that solid tumors are heterogeneous in terms of proliferation and differentiation. With metastatic spread, malignant clusters of heterogeneity similar to that of the primary tumor are observed. This feature can be explained by the CSC hypothesis.

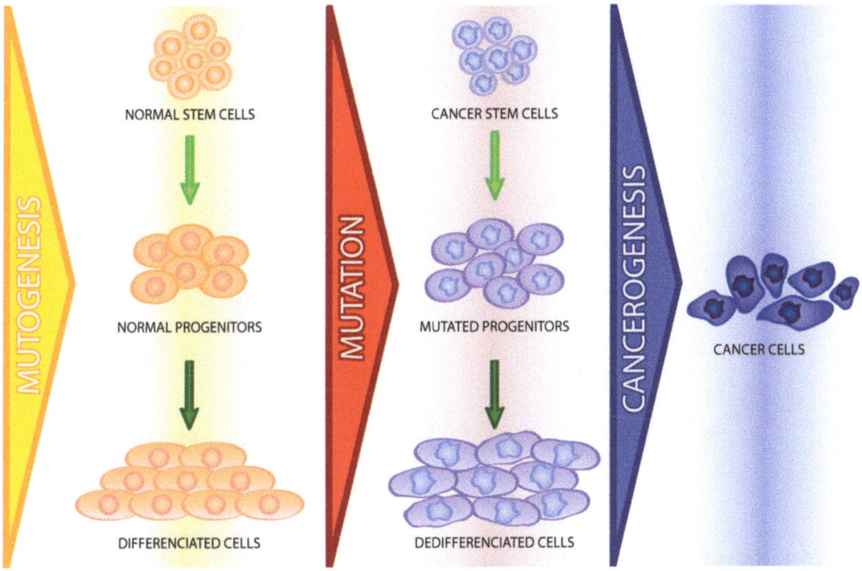

Fig. 16.3 Cancer stem cell development

The microenvironment allows the homeostasis of stem cells. There is a continuous interaction between stem cells and other elements of the microenvironment, meeting the needs of differentiated cells. the microenvironment protects stem cells from differentiation and apoptotic stimuli and provides a stem cell reserve lifelong. It also protects from excessive proliferation of stem cells. It regulates the rest–activity cycles of stem cells. CSCs escaping cancer therapy may sleep in microenvironment niches for many years. These cells may be awakened with a stimulating signal, reactivated, replicated, and differentiated to metastatic foci similar to the primary tumor. CSCs and niches form the "CSC compartment" (Hinohara and Gotoh 2010).

The CSC compartment plays an important role in the onset of disease, resistance to treatment, and development of metastatic disease. One of the most important aspects of treatment is targeting the CSC compartment without causing any toxic effects to normal stem cells. Cancer therapy that reduces the number of stem cells, without radically destroying the stem cell niche, may cause reproliferation of stem cell niche compartment with addition of new cancer stem cells (Morrison et al. 2008). Another problem encountered in the treatment of cancer is the difficulty of destroying the CSC compartment.

There are a number of features common to normal SCs and CSCs. Both cells have the ability of self-renew and overproliferate. Tumor cells occurs as uncontrolled cell proliferation, whereas normal cells appear in a tightly controlled process regarding embryogenesis, organogenesis, protection, and repair of adult tissues. Both cell types have long-lasting life cycles with active antiapoptotic pathways and telomerase activity. These features lead to mutations in stem cells that make them prone to genomic instability (Hiyama and Hiyama 2007).

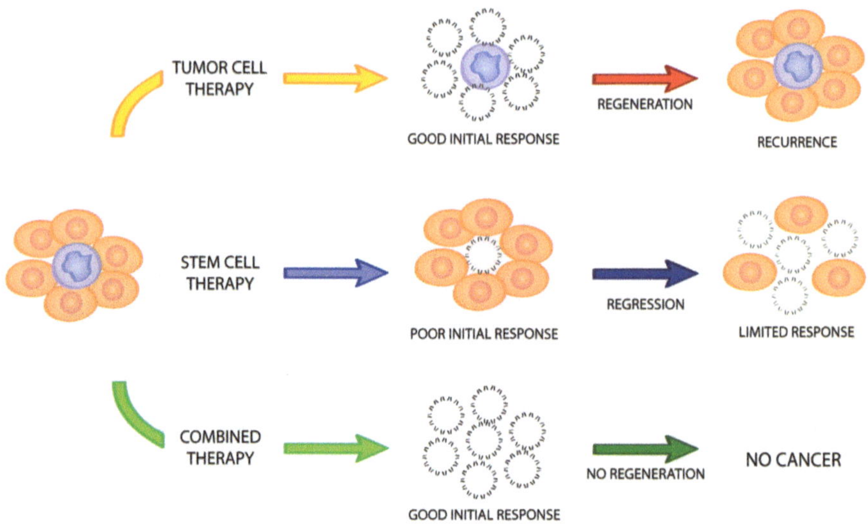

Fig. 16.4 Role of cancer stem cells in cancer therapy

16.6 Cancer Stem Cells and Cancer Therapy

The reason for only temporary improvements after conventional cancer treatments, such as cytotoxic chemotherapy and radiotherapy, is the development of resistance in CSCs and the inefficacy of the current treatment methods. One recent study showed that CSCs are more resistant to treatment with chemotherapy and radiation. The initial response to these treatment models are generally satisfactory, but CSCs that could not be destroyed with the treatment may proliferate and develop into cancer (Fig. 16.4). Although an inadequate response is initially achieved with the treatment of stem cells, the tumor nevertheless declines and eventually disappears (Diehn and Cho 2009). Any method targeting CSCs will definitely contribute to the radical destruction of cancer.

16.7 Conclusion

Molecular and genetic bases of resistance that may emerge in response to cancer treatments should be emphasized. Research studies on DNA repair mechanisms and oxygenation problems should be continued. Decreased apoptosis sensitivity of cancer cells should be managed. Genetic tests investigating genetic predisposition and sensitivity should keep pace with progress. Neovascularization should be avoided, the blood–brain bottleneck should be overcome, intolerance to treatments and side effects of treatments should be reduced, and metastasis and invasion control mechanisms should be developed. Cancer immunotherapy research studies should be done. Studies focusing on target molecules to destroy CSCs should be accelerated.

As we increase our knowledge about the molecular biology of cancer and develop a collaboration between basic scientists and clinicians, it will be easier to control the resistance mechanisms and solve the problems with cancer diagnosis and treatment.

References

Bentzen SM, Agrawal RK, Aird EG, Barrett JM, Barrett-Lee PJ, Bliss JM et al (2008) The UK standardisation of breast radiotherapy (START) trial B of radiotherapy hypofractionation for treatment of early breast cancer: a randomised trial. Lancet 371:1098–1107

Brenner DJ, Hall EJ (1999) Fractionation and protraction for radiotherapy of prostate carcinoma. Int J Radiat Oncol Biol Phys 43:1095–1101

Diehn M, Cho RW, Clarke MF (2009) Therapeutic implications of the cancer stem cell hypothesis. Semin Radiat Oncol 19(2):78–86

Folkman J (1971) Tumor angiogenesis: therapeutic implications. N Engl J Med 285:1182–1186

Hall EJ, Giaccio AJ (2006) Radiobiology for the radiologist, 6th edn. Lippincott Wilkins & Williams, Philadelphia, USA, ISBN: 0-7817-4151-3

Hanahan D, Weinberg RA (2000) The hallmarks of cancer. Cell 100:57–70

Hatfield P, Merrick A, Harringtony K, Vilez R, Batemanx A, Selby P, Melcher A (2005) Radiation-induced cell death and dendritic cells: potential for cancer ımmunotherapy? Clin Oncol 17:1–11

Hegi ME, Diserens AC, Gorlia T, Hamou MF, de Tribolet N, Weller M, Kros JM, Hainfellner JA, Mason M, Mariani L, Bromber JE, Hau P, Mirimanoff RO, Cairncroos JG, Janzer RC, Stupp R (2005) MGMT gene silencing and benefit from temozolomide in glioblastoma. N Engl J Med 352(10):997–1003

Hinohara K, Gotoh N (2010) Inflammatory signaling pathways in self-renewing breast cancer stem cells. Curr Opin Pharmacol 10:650–654

Hiyama E, Hiyama K (2007) Telomere and telomerase in stem cells. Br J Cancer 96:1020–1024

Illidge TM (1998) Radiation-induced apoptosis. Clin Oncol (R Coll Radiol) 10:3–13

Janion C (2001) Some aspects of the SOS response system: a critical survey. Acta Biochim Pol 48(3):599–610

Joiner M, Kogel A, Arnold H (2009) Basic clinical radiobiology, 4th edn. Oxford University Press, London, ISBN: 0340929669

Jonathan EC, Bernhard EJ, McKenna WG (1999) How does radiation kill cells? Curr Opin Chem Biol 3:77–83

Klagsbrun M, D' Amore P (1996) Vascular endothelial growth factor and its receptors. Cytokine Growth Factor Rev 7:259–270

Klein TJ, Glazer PM (2010) The tumor microenvironment and DNA repair. Semin Radiat Oncol 20(4):282–287

Kulaksiz G, Sancar A (2007) Nükleotid Eksizyon Onarımı ve Kanser Türk Biyokimya. Dergisi 32(3):104–111

Lee HW, Han JH, Kim JH, Lee MH, Jeong SH, Kang SY, Choi JH, Oh YT, Park KJ, Hwang SC, Sheen SS, Lim HY (2008) Expression of excision repair cross-complementation group 1 protein predicts poor outcome in patients with small cell lung cancer. Lung Cancer 59(1):95–104

Lipponen P, Aaltomaa S, Kosma VM, Syrjänen K (1994) Apoptosis in breast cancer as related to histopathological characteristics and prognosis. Eur J Cancer 30A(14):2068–2073

Martin LJ (2008) DNA Damage and repair: relevance to mechanisms of neurodegeneration. J Neuropathol Exp Neurol 67(5):377–387

Martin SA, Lord CS, Ashworth A (2010) Therapeutic targeting of the DNA mismatch repair pathway. Clin Cancer Res 16(21):5107–5113
Morrison BJ et al (2008) Breast cancer stem cells: implications for therapy of breast cancer. Breast Cancer Res 10:210
Noguchi S (2006) Predictive factors for response to docetaxel in human breast cancers. Cancer Sci 97(9):813–820
Onur E, Berrin Tuğrul, Ferda Bozyiğit (2009) DNA Hasarı ve Onarım Mekanizmaları. Türk Klinik Biyokimya Derg 7(2):61–70
Patwardhan G, Gupta V, Patwardhan G, Gupta V, Huang J, Gu X, Liu YY (2010) Direct assessment of P-glycoprotein efflux to determine tumor response to chemotherapy. Biochem Pharmacol 80(1):72–79
Plati J, Bucur O, Khosravi Far R (2011) Apoptotic cell signaling in cancer progression and therapy. Integr Biol. doi:10.1039/C0IB00144A
Powell SN, Bindra RS (2009) Targeting the DNA damage response for cancer therapy. DNA Repair 8(9):1153–1165
Reya T, Morrison SJ, Clarke MF, Weismann IL (2001) Stem cells, cancer, and cancer stem cells. Nature 414:105–111
Rivera E, Gomez H (2010) Chemotherapy resistance in metastatic breast cancer: the evolving role of ixabepilone. Breast Cancer Res 12(Suppl 2):S2
Sancar A (2003) Structure and function of DNA photolyase and cryptochrome blue-light photoreceptors. Chem Rev 103(6):2203–2237
Sancar A, Lindsey-Boltz LA, nsal-Kamaz K, Linn S (2004) Molecular mechanisms of mammalian DNA repair and the DNA damage checkpoints. Annu Rev Biochem 73:39–85
Sauter B, Albert ML, Francisco L, Larsson M, Somersan S, Bhardwaj N (2000) Consequences of cell death: exposure to necrotic tumor cells, but not primary tissue cells or apoptotic cells, induces the maturation of immunostimulatory dendritic cells. J Exp Med 191:423–434
Schlacher K, Pham P, Cox MM, Goodman MF (2006) Roles of DNA polymerase V and RecA protein in SOS damage-induced mutation. Chem Rev 106(2):406–419
Steinman RM, Turley S, Mellman I, Inaba K (2000) The induction of tolerance by dendritic cells that have captured apoptotic cells. J Exp Med 191:411–416
Vilmar A, Sorenson JB (2009) Excision repair cross-complementation group 1 (ERCC1) in platinum-based treatment of non-small cell lung cancer with special emphasis on carboplatin: a review of current literature. Lung Cancer 64(2):131–139
Voll RE, Herrmann M, Roth EA, Stach C, Kalden JR, Girkontaite I (1997) Immunosuppressive effects of apoptotic cells. Nature 390:350–351
Wilson GD (2007) Cell Kinetics. Clin Oncol 19:370–384
Woodward WA, Bristow RG (2009) Radiosensitivity of cancer-initiating cells and normal stem cells. Semin Radiat Oncol 19:87–95

About the Editor

Dr. Kursad Turksen is the Editor in Chief of the journal *Stem Cell Reviews and Reports* and the book series entitled *Stem Cells and Regenerative Medicine* published by Springer. He is also the editor of numerous protocol volumes published in Springer's Methods in Molecular Biology series.

Dr. Turksen is a Senior Scientist at the Ottawa Hospital Research Institute. His research interests are primarily in the areas of epidermal stem cells and the role of claudins in the epidermal lineage.

K. Turksen (ed.), *Adult and Embryonic Stem Cells*, Stem Cell Biology and Regenerative Medicine, DOI 10.1007/978-1-61779-630-2,

Index

MIX
Papier aus verantwortungsvollen Quellen
Paper from responsible sources
FSC® C105338

If you have any concerns about our products,
you can contact us on
ProductSafety@springernature.com

In case Publisher is established outside the EU,
the EU authorized representative is:
Springer Nature Customer Service Center GmbH
Europaplatz 3, 69115 Heidelberg, Germany

Printed by Libri Plureos GmbH
in Hamburg, Germany